T0136100

Cosmopolitics II

CARY WOLFE, SERIES EDITOR

ISABELLE STENGERS

TRANSLATED BY ROBERT BONONNO

Cosmopolitics II

posthumanities 10

UNIVERSITY OF MINNESOTA PRESS

MINNEAPOLIS · LONDON

The University of Minnesota Press gratefully acknowledges the generous assistance provided for the publication of this book by the Hamilton P. Traub University Press Fund.

Published in French as *Cosmopolitiques II*; copyright Éditions La Découverte, 2003. The contents of this book originally appeared as part of a seven-volume edition of *Cosmopolitiques*, also published by Éditions La Découverte in 1997. *Cosmopolitiques I* includes volumes I, II, and III of the original edition, and *Cosmopolitiques II* includes volumes IV, V, VI, and VII of the original edition.

Published by the University of Minnesota Press
111 Third Avenue South, Suite 290
Minneapolis, MN 55401-2520
http://www.upress.umn.edu

Library of Congress Cataloging-in-Publication Data

Stengers, Isabelle.
 [Cosmopolitiques. English]
 Cosmopolitics / Isabelle Stengers ; translated by Robert Bononno.
    v. cm. — (Posthumanities ; 10-)
 Includes bibliographical references and index.
 Contents: 1. Quantum Mechanics, In the Name of the Arrow of Time, Life and Artifice, The Curse of Tolerance
 ISBN 978-0-8166-5688-2 (v. 2 : hc : alk. paper) — ISBN 978-0-8166-5689-9 (v. 2 : pb : alk. paper)
 1. Science—History. 2. Science—Philosophy. 3. Science—Social aspects. I. Title.
 Q125.S742613 2010
 501—dc22

                            2010010387

Printed in the United States of America on acid-free paper

The University of Minnesota is an equal-opportunity educator and employer.

18 17 16 15 14 13 12 11   10 9 8 7 6 5 4 3 2 1

CONTENTS

PREFACE

How can we examine the discordant landscape of knowledge derived from modern science? Is there any consistency to be found among contradictory or mutually exclusive visions, ambitions, and methods? Is the hope of a "new alliance" that was expressed more than twenty years ago destined to remain a hollow dream?

I would like to respond to these questions by arguing for an "ecology of practices." I have constructed my argument in seven steps or parts, covering two separate volumes[1]—this is the second. Each section is self-contained and can be read on its own, but I would hope that readers view individual sections as an invitation to read the others, for the collection forms a unified whole. Step by step, I attempt to bring into existence seven problematic landscapes, seven endeavors to create the possibility of consistency where there is currently only confrontation. Whether the topic is the nature of physics and physical law, the debate over self-organization and emergence, or the challenges posed by ethnopsychiatry to the division between modern and archaic knowledge, in each case I tried to address the practices from which such knowledge evolves, based on the constraints imposed by the uncertainties they introduce and their corresponding obligations. No unifying body of knowledge will ever demonstrate that the neutrino of physics can coexist with the multiple worlds mobilized by ethnopsychiatry. Nonetheless, such coexistence has a meaning, and it has nothing to do with tolerance or disenchanted skepticism. Such beings can be collectively affirmed in a "cosmopolitical" space in which col-lide the hopes and doubts and fears and dreams they engender and which cause them to exist. That is why, through the explora-tion of knowledge, what I would like to convey to the reader is also a form of ethical experimentation.

BOOK IV

Quantum Mechanics

THE END OF THE DREAM

Atoms Exist!

In the early twentieth century a page was turned in the history of physics, which was now dedicated to going "beyond phenomena." The event had been researched, prepared, staged, and was intended to play the part of a "cause" in the history of physics as such, and not only as part of the content of physical understanding. And yet, it was a genuine event, not a "purely human" operation, analogous to a convention or a decision that, because produced by humans, could be called into question by them. And the majority of physicists who had gambled against atoms, against the legitimacy of hypotheses that invoked a discrete, unobservable reality "beyond phenomena," acquiesced. Even if they were the losers in this game, they couldn't use this as a pretext to retaliate. It was as if "reality" had accepted the offer, had agreed to the rendezvous proposed by those who intended to impose the existence of atoms on everyone.

When, at the turn of the century, the young Jean Perrin decided to study colloids—emulsions that the newest generation of microscopes enabled scientists to describe in terms of suspended grains—he was fully aware of what he was doing. He was trying to create a new kind of experimental device, one that allowed the physicist to demand that a discrete reality hypothetically situated "beyond" an observable phenomenon assume its place as the only reality capable of explaining the measurements directed at it. Until then, one spoke of atomic *models,* which meant that in spite of the fecundity of the atomic hypothesis, it was unable to silence those who considered it superfluous, dangerously irrational, or simply heuristic. Yet, "the only reality capable" reflects an ambition to provide atoms with a mode of existence that, following Bruno Latour, I have associated with "experimental factishes"—beings that we are compelled to describe as having been constructed by us and simultaneously endowed with an autonomous existence.[1]

For Perrin, a column of emulsion (where the density of grains of gum is a function of height) is a "miniature ponderable atmosphere; or, rather, it is an atmosphere of colossal molecules, which are actually visible. The rarefaction of this atmosphere varies with enormous rapidity, but it may nevertheless be perceived. In a world with such an atmosphere, Alpine heights might be represented by a few microns, in which case individual atmospheric molecules would be as high as the hills."[2] The challenge is obvious: if the column of emulsion indeed obeys the gas law (where the density varies as a function of atmospheric pressure), it will serve as an explicitly kinetic model of gas behavior, as defined by this law, a model that explains behavior in terms of molecular movement. Additionally, the device can be used to manipulate the "atmosphere" because Perrin can vary the viscosity of the liquid, the temperature, and the size of the grains of gum. This should allow him to assign a value to "Avogadro's number," the "N" that appears in every model in which "microscopic" actors and events are introduced. In other words, Perrin's manipulable atmosphere can create the difference between the physics of phenomena, limited to measurements directed at observable macroscopic behavior, and the physics that interrogates, through those phenomena, the invisible multitude of microscopic actors. And in doing so, it will have silenced everyone who, at the time, believed they could reduce physics to manipulative recipes, which, although convenient, were incapable of accessing reality.[3]

In other words, the column of emulsion whose grains Perrin is laboriously learning to count, and which are themselves laboriously prepared so they are of equal size, not only modifies the scale of atmospheric phenomena, it also modifies the scale of contemporary controversies. Among them, those that divide physicists over the vocation of physics and those that pit "ideologues" against one another concerning the truth-value of the sciences. Molecules become colossal in size and great ideas are associated with a specific challenge, the possibility of quantifying Avogadro's number. The column of emulsion transforms the relationships between ideas and things by inventing the means to make the postulated invisible visible. Its purpose is to obligate all physicists and, beyond them, anyone who might be interested in the question, to recognize that atoms, ions, molecules, and radioactive nuclei are all endowed with an autonomous existence and have the means to resist the claim that they are a product of human, and only human, interpretation. In other words, Perrin's demonstration implies and conjoins what are customarily distinguished as the language of physics and the language of the epistemology of physics. Perrin succeeded in coaxing atoms to bear witness to their numbers. In so doing, they not only quantify Avogadro's number, they affirm the legitimacy of seeking, beyond the phenomena we can measure, a reality that no one can reduce to a simple interpretation of those measurements, a reality that cannot be convinced of being relative to our measurements, but one that must be recognized as having the power to explain them.

Yet, here, measurement has little to do with the measurement that gave such prestige to mechanics, centered around the Galilean "equal" sign, and distributing the relative identities of cause and effect (see *Cosmopolitics*, Book II). The measurement

that allows atoms to bear witness is not relative to balancing
something against something else but to *detection*. The microscope enabled Perrin to observe the grains of emulsion and count
them at different heights. Other devices, such as the Wilson
cloud chamber, using cascade effects, would soon amplify the
events that were being identified. Today, high-energy physics is
primarily a physics of detection devices, which sort and amplify
the highly rare events postulated by theory. In one sense, we
might be tempted to say that the change of scale brought about
by the column of emulsion transformed Perrin into Maxwell's
demon, capable of observing a reality that is invisible to us (see
Cosmopolitics, Book III). But this is a risky analogy, for the demon
did not merely observe, it brought to bear on microscopic reality
the requirement that cause be conserved in effect. Its manipu-
lations corresponded to a knowledge of the gaseous milieu as a
"dynamic system." For Perrin, the column of emulsion was not a
dynamic system. It constrained him to a practice that was somewhat analogous to that of the naturalist. He had to prepare his
approach, the device that allowed him to count the grains with
maximum precision. He then had to *wait*—wait for the grains to
become still, and for the emulsion to reach equilibrium. Today,
at the bottom of mineshafts and beneath mountains, physicists
have installed metal plates sandwiched with numerous detectors. They too are waiting, this time for one of the protons in
the metal plates to demonstrate, through its disintegration, the
validity of the theories that assign it a finite lifetime. Although
superaccelerators accelerate the mad circular race of elementary
particles, they merely succeed in creating the conditions liable
to produce the desired particles through collisions. Once those
conditions are established, they too must wait, without knowing
when or where the events their detectors lie in wait for will be
produced.

To Galilean measurement's "full cause and total effect" corresponded the construction of an object capable of defining its

own rationale, or the definition of a state capable of communi-
cating instantaneous description and evolution over time. Such
measurement could lead to a physics of active rationalization,
similar to Pierre Duhem's, where the construction of a measurable object deliberately took precedence over realist ambition,
as well as to the dream of a physical reality "rational in itself,"
measurement of which would be limited to revealing its principles. The object of Hamiltonian dynamics is, in one sense, the
most integrally constructed of physical objects: all the variables
that define it, far from corresponding to any direct observation,
are themselves defined as a function of the system. And yet it
provides us with the most accomplished figure of an "autonomous" object we have, one that is ceaselessly self-determining,
excluding any judgment that might be associated with a particular point of view. For it imposes a systematic relationship of
reciprocal translation on all representations that, by conserving
the Hamiltonian, confirm causal measurement and are therefore defined as being "equivalent." To the autonomous object
corresponds a physicist-judge, who knows how to interrogate
the object, for she knows the principles to which that object corresponds. She is an ideal figure of justice, devoid of all arbitrariness, for the categories of law that arm the judge are, at the same
time, the principles governing the actions of the accused.

As for measurement by detector, it is partial by definition.
It is not the actor as such who becomes observable. What the
device makes observable is theoretically associated with the
actor but generally does not enable us to identify that actor as
an object, that is to say, does not enable us to judge it. That is
why Perrin can forget the numerous differences between his
grains of emulsion suspended in a liquid and the nearly freely
moving molecules that, following kinetic theory, constitute a gas
such as our atmosphere. For Perrin, neither the grains nor the
molecules need to be defined as wholly intelligible objects. He
withdraws only the information capable of confirming, and then

exploiting, the validity of the atmospheric model represented by the colloidal solution. More generally, the detector is addressed to events insofar as they are producers of information about entities to which those events "happen," which implies that the device is dependent on an event it can neither judge nor control, only await or provoke. Correlatively, the device is not "causal" but "logical." It does not introduce a cause that is exhausted in some measurable effect it produces but a chain of implications and consequences enabling us to follow traces of the event back to the actor that experienced or provoked it.

We must now distinguish two concepts of measurement and the reality to which measurement is addressed. This distinction is the price paid by physics's new identity, its vocation to encounter a reality "beyond phenomena."

As we have seen, causal measurement is intended to construct the object's identity, to authorize an objective judgment, but it also creates the question of the relevance of that judgment. Clausius's "rational" thermodynamics showed that it was not impossible to require that phenomena be presented in a way that gave meaning to causal measurement, although satisfying that requirement did not legitimize its relevance for phenomena. Measurement, in this case, assigned full responsibility for the presentation to the one who imposed it. The phenomenon was subject to a requirement through which it could be judged, but it did not provide confirmation that the judge was impartial. This leads to the following question: Wouldn't mechanics be merely a particular case, and thermodynamics a lucid generalization? In effect, mechanics would be the only case where the relevance of judgment could be confirmed by the object under investigation, the only case where the "remainder," the deviation between the mechanical system and the ideal being that would completely satisfy the requirements of measurement, could be reduced to "secondary" factors such as friction. In the general case, judgment, if it claimed to have the authorization of its object, would

resemble a misuse of power. We would then have to acknowledge that, in general, the judge questions the accused in a language the latter does not understand, and is in no way concerned
with finding out how the accused might describe what the judge
would define according to the categories of the "crime." The
requirement is satisfied by the phenomenon, of course, but we
can also say that this satisfaction has been extorted. The judge is
no longer authorized by reality.

That is why Duhem argued that greater attention should be
given to the conditions of relevance of the different theoretical constructs. Thus, the case of "false equilibrium" (think of
an apparently stable gas mixture that a mere spark can suddenly
ignite) challenges the relevance of the simple identification
between a chemically inert state and a state of equilibrium. Can
we use as a guide the analogy between "false equilibrium" and the
stone that, because of mechanical friction, remains motionless
on a slope until a flick of the finger causes it to tumble downhill?
And how can we modify the edifice of thermodynamic equilibrium to integrate the chemical analogues of "friction"? How can
we generalize the concept of chemical potential so as to enable
it to define false equilibrium and the "chemical friction" that
corresponds to it?[4] These are the kinds of stark questions that
Duhem claims are relevant, questions that obligate the physicist-judge to proceed with tact and skill, to negotiate the "cases"
before her according to her own rules, but not blindly, without
confusing those rules with a language that reality might speak.

The question of the relevance of the judgment that constructs an object can even go so far as to challenge the fundamental notion of trajectory found in dynamics. In *The Aim and
Structure of Physical Theory*, Duhem affirmed the limits of this concept even when the equations that appear to guarantee its legitimacy can be constructed. At the time he was nearly the only one
to draw radical conclusions based on the identification of the
first "pathological trajectories," trajectories that specialists in

dynamics today refer to as fractal or chaotic, and that are now
associated with the current "renewal of dynamics."[5]

For Duhem, trajectory had to be abandoned whenever it
failed to satisfy the requirements its use assumed—in this case,
whenever its mathematical representations ceased to be robust.
Trajectory is an instrument of mathematical deduction and
"a mathematical deduction is of no use to the physicist so long
as it is limited to asserting that a given *rigorously* true proposition has for its consequence the *rigorous* accuracy of some such
other proposition. To be useful to the physicist, it must still be
proved that the second proposition remains *approximately* exact
when the first is only *approximately* true. And even that does
not suffice . . . it is necessary to define the probable error that
can be granted the data when we wish to know the result within
a definite degree of approximation."[6] What the mathematician
Jacques Hadamard defined in 1898 is a situation that makes
deduction unusable. Take a complicated surface, such as the
forehead of a bull whose horns and ears proceed outward toward
infinity. We can draw a particular geodesic curve on that surface:
"after turning a certain number of times around the right horn,
this geodesic will go to infinity on the right horn, or on the left
horn, or on the right or left ear . . . No matter how we increase
the precision used to determine the practical data, reduce the
size of the spot where the initial position of the material point
is found, or tighten the bundle that contains the velocity's initial direction, the geodesic that remains at a finite distance by
turning continuously around the right horn can never be freed
of its unfaithful companions, which, after having turned around
the same horn with it, will deviate infinitely. The only effect
afforded by this greater precision in establishing the initial data
will be to require that the geodesics describe a greater number
of turns around the right horn before producing their infinite
branch. But this infinite branch can never be eliminated."[7]

I have cited Duhem at length to show that, in a book published in 1906, he quite clearly described the dynamic behavior now known as chaotic, characterized by a blend of qualitatively distinct trajectories regardless of the scale. Moreover, Duhem foresaw the fecundity of the problem. Although Hadamard's mathematical description signaled for him an irreconcilable divorce between the mathematical entity and physical representation, which had to be robust, which demanded a "mathematical representation of approximation," this divorce was not synonymous with the abandonment of mathematical construction. It created the challenge of having to construct a new mathematical representation that was physically pertinent: "But let us not be deceived about it; this 'mathematics of approximation' is not a simpler and cruder form of mathematics. On the contrary, it is a more thorough and more refined form of mathematics, requiring the solution of problems at times enormously difficult, sometimes even transcending the methods at the disposal of algebra today."[8]

Therefore, Duhem's perspective has nothing to do with that of a tranquil "instrumentalist," who claims to limit herself to the information provided by her measuring instruments. While "nothing is true in itself other than the results of experiment,"[9] mathematical physics cannot be purely and simply skeptical, for it has to be relevant. It is a Pascalian situation explicitly transposed to physics: "We have an impotence to prove invincible by any dogmatism, and we have an idea of truth invincible by any skepticism."[10] The solution to the dilemma is found in an ethics of obligation. To the extent that the physicist resists the "needs of the imagination" and fully accepts the twofold obligation imposed by mathematical consistency and experimental relevance, she may hope to create a theory capable of systematically translating phenomenal variety. It is then, and then only, that the order in which theory categorizes its results might lead us to conclude "that it is or tends to be a *natural classification.*"[11] Like any real physicist, Duhem can thus claim that "the belief in an order transcending physics is the sole justification of physical theory."[12] The difference is that, for Duhem, physics can be true to its purpose only to the extent that it avoids any confusion between what it constructs and what transcends it.

It is the ability to identify and judge, inseparable from the language of rational mechanics and thermodynamics, that has conferred its value as a critical concern to the question of relevance and its corresponding obligations. It is the judge, rather than the hunter or naturalist lying in wait, who must ceaselessly bear in mind that the crime defined in terms of legal categories must not be confused with the act that has led the accused before the court. That is why Jean Perrin had no problem with "reality." And at the time he succeeded in calculating Avogadro's number, he even took great pleasure in literally wallowing in the numbers that designate the microscopic reality whose autonomous existence he was now able to confirm: "In short, each molecule of the air we breathe is moving with the velocity of a rifle bullet; travels in a straight line between two impacts for a distance of nearly one ten-thousandth of a millimeter; is deflected from its course 5,000,000,000 times per second, and would be able, if stopped to raise a particle of dust just visible under the microscope by its own height. Three thousand million of them placed side by side in a straight line would be required to make up one millimeter. Twenty thousand million must be gathered together to make up one thousand-millionth of a milligram."[13] These figures all result from substituting Avogadro's number, N, by its value in the kinetic model of gases, and they satisfy the imagination of the physicist beyond all expectations. A Pascalian emotion, like Duhem's, gripped Perrin: "The conclusion we have just reached by considering a continuously diminishing center can also be arrived at by imagining a continually enlarging sphere, that successively embraces planets, solar systems, stars, and nebulae. Thus we find ourselves face to face with the now familiar conception developed by Pascal when he showed that man lies 'suspended between two infinities.'"[14] Two infinities that a particular species of "Man," the physicist who goes beyond phenomena, can now hope to approach through detection and numbers.

When Perrin's figures made their appearance, there disappeared from view the "transcendent" order Duhem hoped the edifice of physical theory might allude to. "Reality" seemed to invade the physicist's laboratory, dictating figures and more figures, which fed the most extravagant appetite for discovery. But it is at this point that we must pay close attention. The appetite that was fed was no longer that of a judge. The "reality" to which measurement by detection is addressed is assumed to be autonomous, but is no longer defined by the convergence between description and reason. The physicist's imagination, which reality authorizes, contrary to Duhem, is addressed to beings that allow themselves to be described but do not give the describer the power to identify them, or subject them to a law. Whereas causal measurement creates both an object and the judge of that object, providing the "equal" sign with the power to subject a phenomenon to a judgment, measurement by detection recognizes in what it addresses an "activity" that does not dictate how it should be described. Detection does not demand the power of judgment, only the ability to characterize unequivocally.

However, we cannot say that the requirement concerning reality has diminished—it has been transformed. To be valid, detection requires that the information produced has an unambiguous meaning, and that it can be associated with an actor. But the main point is that this requirement has become "nomadic." Unlike causal measurement, it does not singularize physics. Pasteur succeeded in transforming epidemic disease and fermentation into detectors of the presence of microorganisms. In order for disease to be recognized as able to "detect," he had to be able to demonstrate that different diseases are caused by different microorganisms and that each kind of microorganism results in a single disease. However, he did not have to prove that the microorganism constituted the "rationale" of the disease, the way in which the disease had to be understood.[15] It only had to be a "necessary condition." In other words, the "experimental factish" constructed by detection answers the questions addressed to it, but its responses, although veridical in the sense that they resist the accusation of being mere artifacts, are still not "truthful" if by this we mean the avowal a being can be led to make concerning its own truth.

Of course, measurement by detection hardly implies that the physicist is a pure observer, which is sometimes the position of the naturalist. Like causal measurement, it can provide the scientist with the power to characterize its objects in terms of well-defined variables and construct functions. Activity can be modulated, the detector can filter according to various criteria. But, in all cases, detection assumes that "reality" takes the initiative, that it is endowed with the ability to act. It has become "autonomous" in a way that differs completely from the dynamic object. The dynamic object is simultaneously and identically *autonomous and submissive*, "obeying its own laws," whereas the detected actor can, depending on circumstances, be considered "spontaneous," comparable to "natural" radioactive disintegration, whose rate of decay nothing can accelerate or decrease, or subject to laws to which detection does not provide access.

In one sense, we could say that detection implies a more "realist" approach than causal measurement, because it subjects itself to a reality it does not claim to reduce to a rational identity. Perrin's atoms, like Pasteur's microbes or the atomic nuclei that disintegrated over Hiroshima, exist "truly" in the sense that things "happen" to them and that they cause things "to happen," in the sense that they affect their environment, to which the detector belongs, and can be affected by it. But here "reality" has lost the power to dictate how it must be described. We can, of course, by extension, still speak of an object, but we must then overlook the legal connotations associated with the term and borrow the vocabulary of inquiry, distributing responsibilities without judging the culpable. Here, "object" celebrates the construction of a coherent representation of the actor responsible for everything detected about it. Between the object and the "reality" it designates, there exists an irreducible distance, separating detection and identification, one that new methods of detection may be able to reveal but not reduce, clearing the way for the discovery of new responsibilities.

The object of dynamics guaranteed the exhaustive nature of dynamic description because the cause, through its equivalence with effect, guaranteed the vanity of any attempt to describe "differently." The same is not true of the representation resulting from detection. For this relates the causal order to its original register, where cause is understood in several ways and supports nothing other than logical reasoning. The microorganism has the power to cause disease, true, but this power may vary in the laboratory, where living beings have become either reproductive media or instruments for detecting the virulence of the being under investigation.[16] But in spite of the accumulation of propositions concerning this power, they themselves are not sufficient to identify the cause, the microorganism, and the effect, the disease. On the contrary, they tend to multiply the problems presented by these two terms. This is the key difference between cause in the sense of dynamics and cause in the logical, or syllogistic, sense. Only the first has in itself the power to define its effect, for its identity is nothing other than this power; the second is relative to detection and its interpretation, and other questions, other detection techniques can transform it without ever achieving anything other than logical closure. Closure that is still liable to neglect unidentified ingredients or conditions, or to dissimulate the fact that the "cause" presented to us is liable to belong to quite different narratives under other conditions. Pasteur's microorganisms are not arbitrary, but their identity as a "cause of disease" is eminently variable. Whereas a dynamic "cause" imposes its "ontology," that is, its definition of what a cause is, a logical "cause," according to Latour's expression, is "ontologically variable": not only what it causes but the way it causes are subjects for history.

The reality quantified by Jean Perrin vanquished Duhem's "transcendent" order. But it too could have communicated with a form of transcendence, that of a world where responsibility cannot be used to identify the responsible party, that is, to penetrate its reasons. Which it did not do. And it is possibly because it did not do this that the "physical realism" Duhem had tried to disrupt was finally discovered to be in a *state of crisis*. The problem posed by Duhem may well have been part of an outmoded past, outmoded by the intentional and successful ambition to get "beyond phenomena"; nonetheless, it effected a rather strange return in a different form, the much more paradoxical and fascinating form given to it by quantum mechanics.

One of the aspects of anamnesis I attempted to develop throughout Book III was the break with the idea that the revolutions of twentieth-century physics had disturbed the tranquil course of "classical" physics, that is, a physics ruled by a form of ahistorical progress. Correlatively, here anamnesis is no longer necessary. The "crisis" of quantum physics and the "problem of the interpretation of quantum formalism" belong to contemporary culture. And I will approach them as such, as subjects of discussion rather than as more or less uplifting forms of pedagogy. I will begin by introducing a rather startling contrast: in 1913, physicists, whether enthusiastic or resigned, acknowledged that the world was indeed a confusion of atoms. Yet, less than fifteen years later, when those atoms were part of theory, theory led its interpreters to assert that the "real is veiled" and we can only access an "empirical" reality, which is to say, a *reality-for-us*.[17]

Some may feel that the fact that specialists of quantum mechanics speak of a "reality-for-us" should be a cause for rejoicing. For, doesn't it show that the physics they represent has finally been recognized as being relative, like all other areas of knowledge, to the questions, requirements, and obligations that allow it to exist? Haven't physicists finally abandoned the dream of knowledge that is able to access a reality that disqualifies all other forms of knowledge? In fact, as we shall see, this interpretation could have been legitimate. It would have been if what I will here call the "lesson of Niels Bohr" had been heard and accepted. And in that case, this entire exploration of physics would have been pointless. Physicists would today introduce themselves nonpolemically, as vectors of lucidity, if not as vectors of peace among fields of knowledge. Which is not exactly the case. The very expression "the real is veiled" is hardly one of peace. For here, the veil is not something a practice that brings actors into existence would create in the very act in which it "unveils." It is not there to signal, for example, the difference between the identification of responsibility and the identification of the one responsible. Very specifically, the veil is what specialists of quantum mechanics alone have to confront; it is the new, dramatic message of physics. What is celebrated is the disappointment of physics's leading ambitions—the impossibility of satisfying the physicist's vocation, access to a unique and intelligible vision of reality. Therefore, the veil consecrates the vocation whose impossibility it proclaims.

And thus, the "reality-for-us," which is relative only to "communicable human experience," designates neither an "us" nor an "experience," nor a "communication," which by that very fact might become *interesting* in their multiplicity, but an act of mourning that confirms the vocation of the physicist by announcing the impossibility of its fulfillment. And this proud vocation is still very much alive whenever a physicist feels justified in making human consciousness an inseparable correlate of what "we" nonphysicists "naively" refer to as an "independent" reality. The exploration of the questions of quantum mechanics, therefore, becomes part of my "cosmopolitical" project on the following basis: how can we transform the "mourning" that some physicists feel they can assume on behalf of all humans into a "history of physicists"?

2

Abandon the Dream?

Some people find it surprising that the problem of quantum mechanics remains a vital one even today. After all, quantum formalism is roughly sixty or seventy years old and, since then, quantum field theory, high-energy physics, and the unification of fundamental interactions have transformed physical representation. Is the survival of a long-standing debate symptomatic of the laziness of commentators, who limit themselves to things they can "understand"? Or is it the result of the sense of competitive urgency felt by the new generation of physicists, who no longer bother to take the time to elaborate or explain the nature of contemporary physics to outsiders? My position is that there is indeed a symptom, but this symptom may reflect an entirely different problem. Physicists representative of the older (and already obsolete) new generation, such as Richard Feynman, directed streams of irony at those who believed there was still something to "understand" in physics. Naturally, there are a great number of things to learn about the practices of contemporary physicists, especially concerning the difficult, strained relationships between experimental physicists, phenomenologists, and theoreticians that characterize high-energy physics. But the very complication of the field of practice, the obscure nature of the respective requirements and obligations of its various actors, makes it opaque from a point of view other than that—always situated—of its actors. In other words, there is nothing contingent about the lack of discussion concerning what is at play; it has nothing to do with the laziness of commentators or the competitive urgency of physicists. It is as if the only aspect of the physicist's vocation—finding a single intelligible representation of the physical world—that has remained truly invariant and explicit is the ambition to get "beyond," and in this case not only "beyond" the diversity of phenomena but beyond the ordinary conditions of experimentation. But this "beyond" comes at a price. For, whenever physicists today maintain the values of physics, it is above all to encourage us to accept the massive financial investments their equipment requires.

It is not impossible to tie this practical opacity to the current strange proliferation of statements from physicists who assume the role of thinkers and chroniclers of future revolutions. As if anticipating increasingly scandalous future revolutions held the place formerly occupied by the production of intelligibility. As if the rush toward such dizzying perspectives might enable them to transform the problems created by physical theory into the promise of new triumphs. In today's speculative physics, the effects of a possible revolution, far from being unexpected, are often already taken into account and exploited. So, for example, in *The Emperor's New Mind,* Roger Penrose, discussing the problems of interpreting quantum mechanics, introduces the "germ of an idea," which he uses to announce a "revolution" that is without doubt going to be more important than that of relativity or quantum mechanics.[1] "Revolution" is no longer an event, it is a component of the physicist's management of future perspectives. In Penrose's case, the problems raised by quantum mechanics will be resolved when gravitation has assumed its legitimate place.

This way of anticipating the future to settle the problems

of the past is, to say the least, surprising. As a theoretical-experimental science, quantum mechanics is recognized for its impressive success, that is, for the precision with which the data actually produced by experimental means have confirmed its predictions. The fact that these devices have not assigned gravitation a crucial role implies that no one has yet invented a way for gravitation to manifest the consequences of its existence in this experimental context. The idea that it would play a role—a role both secret and crucial because failing to account for it would be mysteriously responsible for the singular formal structure of the theory and the problems of interpretation this structure entails—amounts to claiming for this physical-mathematical language a kind of authentically transcendent truth. And that truth would be such a truthful rendition of reality that it would have the ability to manifest, through a quirk of its formal structure, that a dimension of that reality is lacking.

It is not impossible to maintain that an already dated quantum mechanics preserves its attraction for the simple reason that the physics that followed is marked both by the active rejection of any idea that is not a simple "reflection of the state of the question as defined by physicists" and by wild speculation uncoupled from practice, adopting as its only guideline a reference to the revolution that makes conceivable for tomorrow what is inconceivable today. Nor is it impossible that this dual singularity, the practical opacity of so-called high-energy physics and the grandiose perspectives it has been associated with, might be analyzed in terms of an "ecological isolate." It may possibly be related to the disappearance of any necessity, indeed, any possibility, of establishing relationships, even conflicting ones, with other practices that address the "same" reality.[2]

In any event, the quantum mechanics of old remains a crucial reference. It marks the moment when physicists once again debated what they were able to require of reality and the obligations imposed by that reality. And it is important to understand

why that moment resulted, in one way or another, in the physics of the "new generation" silencing those questions, indulging in ironic skepticism and unbridled speculation. "No one,"
wrote historian Stephen Brush, "has yet formulated a consistent
worldview that incorporates the Copenhagen Interpretation of
Quantum Mechanics while excluding what most scientists would
call pseudo-sciences—astrology, parapsychology, creationism,
and thousands of other cults and doctrines."[3]

What is this "Copenhagen interpretation" that Brush alludes
to? For some, this interpretation is that of Niels Bohr, who spent
his whole life thinking about the implications of the physics
he had helped redefine. For others, it is the "orthodox" interpretation of quantum mechanics, one that managed to displace
all the others. There is an alternative view, however, although
not shared by many. It entails, on the one hand, a reference to
a relationship that history failed to smooth over, to trivialize,
between a man and a scientific innovation, and, on the other,
a consensual orthodoxy that imposed itself without having the
ability to convince anyone, as if history, in this case, no longer
had the power to silence "vanquished" objections and frame the
controversy in the victors' terms. As if reference to the "vanquished" in the controversy over quantum mechanics lingered
on, along with the idea of a possible alternative.

In fact, the history of quantum mechanics does not respond
at all to the canon of the history of science. The anomaly is all the
more striking given that, from the point of view of the produc-
tiveness of the relationship between theory and experimentation, of its ability to encompass all fields of physical chemistry,
and the creation of new methods of observation and measurement, it fulfilled its creators' most ambitious hopes. But it is as if
this successful history was also a source of permanent intellectual dissatisfaction, one a multitude of authors have attempted
to counteract.[4] The term "orthodox," a very unfamiliar one for
scientists, concentrates the ambivalence of the situation. For

those who feel that the success of quantum mechanics constitutes its necessary and sufficient source of legitimacy, the term
refers to an interpretation that is external but useful because it
silences the futile problems that others insist on presenting. For
those who refuse to acknowledge the situation and attribute its
longevity to the triumph of a mediocre, instrumental concept,
through which physics has abandoned thought, it sounds, on the
contrary, like a protest and an accusation.

This situation is even stranger when we consider the difference between Niels Bohr's own interpretation and the "orthodox"
interpretation attributed to him, which frequently assimilates
the very distinct ways of thinking of Bohr and Heisenberg. In
fact, these stylistic nuances seem superfluous to anyone who
refers positively to the so-called orthodox interpretation. For
them, this interpretation is reduced to the authorization to
move forward without asking pointless questions. Quantum
formalism would present no problem; it would simply obligate
physicists to abandon the dream that caused them to confuse the
information supplied by their instruments with the attributes of
a given reality. Quantum mechanics would be nothing but a "formal machine" that produces predictions about observable data
on the basis of observed data. The purely instrumental nature
of this knowledge would be a point of no return because what it
asks physicists to renounce would never be anything other than
an illicit dream anyway. Correlatively, the critics of the orthodox
interpretation have no need for stylistic nuance. It is against the
threat of a positivist and instrumentalist reduction of physics
that they defend the vocation of physicists, the faith in the intelligibility of reality that animates them.

However, it is not unanimously held that the "orthodox"
interpretation can be expressed in "positivist" terms, that is,
a general epistemological claim that rejects, in principle, any
attempt at intelligibility that is not limited to the coherent interrelation of observable data. This kind of presentation was even

denounced as the "curse of positivism" by Léon Rosenfeld, Bohr's disciple and spiritual heir.[5] In fact, the rejection of positivism was the one thing Bohr and Werner Heisenberg most strongly agreed on. We must then ask why Bohr and Heisenberg were unable to escape the "curse" in question. Why were they unable to get others to accept what they defined as the obligations of quantum mechanics, becoming instead, and quite unwillingly, for everyone who saw prohibitions rather than obligations, the authors of the "orthodox" interpretation of quantum mechanics? Here, as elsewhere, general ideas, determinism or realism, risk masking the *practical* question tying together the problem of values and the technical definition of problems, the question that helped to consolidate the dissatisfaction of the opponents of the Copenhagen interpretation.

To address this practical question, I'll begin with a contrast. Critics and partisans alike of the Copenhagen school celebrate Einstein's theory of relativity as a model. Bohr and Heisenberg have always said they have faithfully followed the example of their older peer. Why have critics not acknowledged their fidelity? Why did relativity escape the "curse of positivism"? Why, for some, do relativity and quantum mechanics represent a continuously coherent history and, for others (including Einstein), a great history interrupted by betrayal and abandonment?

Bohr and Heisenberg believed they were following Einstein's example. The argument that allowed Einstein to deny that the concept of the absolute simultaneity of two distant events had a physical meaning was based on the fact that no experimental device would ever allow observers moving relative to one another to agree on this simultaneity. And Bohr based his claim that quantum formalism constituted a "complete" theory on the same type of argument: to assign a physical meaning to an element of theoretical description implies that conditions of observation can be defined in relation to which that element can acquire its meaning. Moreover, it was at this point that he "awaited" those who hoped to get "beyond" quantum formalism. He was easily able to demonstrate to them which of their hypotheses brought into existence a "beyond" that, like simultaneity at a distance, can never be subject to any possibility of measurement. Why, then, did Einstein allow himself to be caught in Bohr's trap, and, more crucially, why did he resist the outcome, when he was the one who had invented both the trap and the outcome?

We know that if, for a time, the Viennese positivists who followed Ernst Mach saw in Einstein's theory of relativity the triumph of their claim, Einstein successfully severed any connection to these early allies and relativity escaped the stigma of the "curse of positivism."[6] Relativity was successful in showing that it did not satisfy a general obligation of an epistemological nature, which ordered scientists to eliminate any type of resemblance between their theories and the speculative descriptions adopted by metaphysics. Relativity's obligations were *unique*, associated with the *unique* properties of the speed of light, properties that condemned, in spite of the intuitive evidence, the possibility of providing a meaning to simultaneity that was not relative to the observer. Moreover, those obligations did not result in the abandonment of an "objective" description dictated by the object but in the reinvention of that objectivity. For physicists, relativity is not positivist because it does not question the possibility of a description that is independent of observation. It "merely" obligates us to recognize that that description cannot be spatiotemporal. Objective description has the status of a matrix for all possible spatiotemporal descriptions, which depend on the frame of reference of the one who describes them.

The universe of general relativity is populated with observers who conduct measurements of simple registration, armed with chronometers and standard meters. It is they who are responsible for the transition from an "objective" four-dimensional mathematical description to a physical description in space and time.[7] But this responsibility provides the observers with no more than a subordinate role, defined by the fact that measurements in space and time separate what, in truth, is inseparable. Thus, the world of our observations is certainly only phenomenal, but we are not prisoners of this phenomenal world. We are able to judge it in the name of four-dimensional mathematical truth: according to the terms of this judgment, we know how our space and time observations indirectly testify to the existence of space-time. The presentation of this judgment becomes a sign of glory for a physics that managed to understand the relativity of appearances and uncover the absolute from which they are derived. Laplace's observing demon is dead, but the Queen of Heaven continues to reign as magisterially as ever, her incommunicable attributes now being identified with a mode of mathematical existence that can only be subject to communication, that is to say, measurement, at the price of a redefinition of phenomena. These phenomena resemble the phenomena in Plato's cave, but now it is the physicist who conceives the mathematical truth of the "Sun," placing the distorted shadows in the correct perspective.

Quantum mechanics, on the other hand, centered on Planck's universal constant, appears to preclude the dream of an "objective" description of quantum beings. But it does worse than this. For the formalism that precludes the dream also succeeds in giving rise to it. As I'll show later, it provokes the dream so that it can preclude it later on. In other words, it both "disappoints" and "deceives."

It is "disappointment" that serves as the customary thread in stories about the creation of quantum mechanics. The starting point emphasizes the continuity between classical dynamics and quantum mechanics, and the central role played by the "wave mechanics" associated with Louis de Broglie and Erwin Schrödinger. Wave mechanics purported to be the worthy heir of the Queen of Heaven and appeared to promise physicists an authentic fairy tale, a triumph of mathematical harmony that was able to reconcile the science of masses in motion and the science of light. It was at this point that an abrupt shift occurred: de Broglie and Schrödinger were "stripped" of their creation, the "wave function" whose mode of behavior served as a splendid match to the intelligible reign of the Hamiltonian Queen of Heaven. It is almost as if the unwanted appearance of the wicked fairy Carabosse had transformed the young princess, already blessed by Einstein, into an ugly old crone. For we later learn that the wave function doesn't describe any "physical reality." In fact, the wave function apparently doesn't describe anything at all; it simply allows physicists to predict probabilities, the probability that the measurement of a physical quantity will have a given result. The function itself corresponds to "probability amplitudes."[8]

Einstein's observers brought about a spatial and temporal "localization" whose (mathematical) consequences "caused" distances, durations, forces, and accelerations "to appear." But this localization, the result of location measurements carried out by the observers, is incapable of affecting the "four-dimensional reality" that situates them and allows them to determine what will be observed. On the other hand, the transition from the wave function to observable probabilities establishes a disturbing compromise that makes it impossible to situate "observable reality," to describe it as determined in principle by "physical-mathematical reality." The "reduction of the wave function" to observable probabilities cannot be assimilated to the definition of a pathway to some "higher reality," indifferent to measurement. Something "happens" to what is being measured. The wave function is not "reduced from the observer's point of view" but *actually* reduced; the representation of the quantum system, which is to say, the probabilities resulting from previous measurements or the preparation of the system, are transformed after the measurement. According to the "disappointing" interpretation given by Max Born, the reduction of the wave function should be associated with *knowledge production* in the strong sense, in the sense that what is "known" is transformed by that very production.

In the conventional narrative, this is often the positivist curse that Carabosse utters over the cradle of the wave function that believes itself to be a Hamiltonian princess. This wave function, Carabosse says, will disappoint all your hopes. You will be forced to accept that it does not represent reality but merely the predictive ability that describes the state of our knowledge. Don't be surprised if it is transformed every time it is measured, this is simply the result of the fact that your knowledge and your ability to make predictions change. Physicists were asked to reject any reference to a reality that "existed independently of their production of knowledge."

Such a statement is unquestionably a curse for the physicist. Even Pierre Duhem would have refused to abide by it; for him, "reality" existed as a source of obligations, obligating one to employ tact and skill in the use of theoretical instruments. The difference between Duhem's conception and that which quantum mechanics appears to celebrate in its positivist version is the difference between the practice of a skilled artisan and the fantasy of a push-button device. The artisan does not know the material "in itself," independently of gesture and tool, but her knowledge is much more intimate than anyone else's, through the relevant precision of her gestures and her choice of the appropriate instrument. And it is such distinctions, Duhem claimed, cultivated through intimate knowledge, that can claim to express a "natural classification" that refers to an order that transcends physics. The push-button device, on the other hand, relates to the fantasy of an all-powerful instrument of interrogation, one that no one has refined or adjusted, and which creates knowledge in spite of its complete contempt for the problem posed by what it is we are trying to know.

How is it that positivism, which can customarily intervene only "afterward," in the form of a sempiternal reminder to indifferent physicists, intimating that they should not confuse "observable data" and "access to reality in itself," has been so closely associated with a physical theory? Where did the positivist curse obtain its power?

In fact, the customary narrative strategy is rather curious, for the wave function seems to have come to us out of the blue. The singular practices from which it may have originated are relegated into the background. More specifically, it continues to be presented in terms of its relationship to the ambition that de Broglie and Schrödinger believed it embodied. However, responding to the challenge of a theorization of the atom's behavior was not an ambition de Broglie and Schrödinger associated with the wave function. Such a theorization *existed already*; it was the "first quantum mechanics" initiated by Niels Bohr. De Broglie and Schrödinger's ambition was to bring that first theory into line, to avoid aspects of it they felt were scandalous.

In other words, the customary narrative strategy, which related a hoped-for continuity between Hamiltonian physics and quantum mechanics, and the disappointment of that hope, silenced or attenuated the fact that a rupture had *already* taken place, brought about by the model of the atom proposed by Niels Bohr. The hope that was disappointed by de Broglie and Schrödinger's wave function is not the hope of establishing continuity between "conventional representation" and a representation of the atom, it is the hope of reestablishing a continuity that had already been broken.

To escape the positivist curse, fairy tales, and Carabosse, it might be worthwhile to tell the story somewhat differently, to focus on the novelty of Bohr's model, of the way he redefined the practical relationship of knowledge involving the physicist and the new type of being known as the atom. For the scandal that de Broglie and Schrödinger experienced, the break they had hoped to smooth over, can take on another meaning, one that is fully realistic in Duhem's sense. For, it is because it was able to constrain Bohr to bring about this rupture, to dare to create a scandal, that the atom is indeed "real." The transformations of the theoretical edifice physicists felt obligated to undertake in order to produce its relevant description constitute its claim to an existence that "transcends" physical knowledge.

Bohr's quantum model of the atom introduced electrons arranged in discrete "stationary states" around the nucleus and making "quantum jumps" by which, with no time having elapsed and no space having been crossed in the sense understood by dynamics, those electrons "jumped" from one state to another. This discrete *change of state* could not be described, but it could be defined by its energy cost, which measured the energy distance between orbital "states."

Bohr's atomic model, therefore, reverses the terms with which the question of dynamic motion is presented. Dynamic observables are spatiotemporal; energy is a constructed function, affirming the conservative character of any dynamic temporal evolution, its definition in terms of complete equality and total effect. In Bohr's model, it is the energy cost of the nonspatiotemporal change of state that is directly observable. It corresponds to the quantum of light emitted or absorbed by the atom during the change of state. Moreover, the states themselves are defined only by means of those transitions, as terms of the energy difference measured by light quanta.

In Bohr's model, we cannot observe an atom the way we think we can observe the Moon, by visual identification. We know atoms, in the sense that they are defined by the model of stationary electronic states, only to the extent that they absorb or emit light quanta, that is, the extent to which they interact with their environment. Regarding this situation, Bohr was literally indefatigable, speculating on the novelty of quantum mechanics and the new obligations that followed for the physicist, even on his deathbed.[9] But it is this "nonmechanical" model that Louis de Broglie and Erwin Schrödinger felt to be unacceptable. That is why they directed all their hopes to the construction of a wave equation. The wave equations, when the appropriate limit conditions are used, have a discrete set of solutions. Therefore, they might help eliminate the scandalous notion of an energy that is not a continuous function of position and velocity, or help interpret stationary states no longer in terms of discrete energy levels but on the basis of frequencies compatible with the limit conditions. The scandal of quantum "jumps" could then disappear to the benefit of simple "frequency changes."

Wave mechanics was unable to tame the scandalous singularity of the behavior of the quantum being introduced by Bohr. If we refer to the well-known distinction between "context of discovery" (what the discoverer thought, believed, and hoped) and "context of justification" (what subsequently happened and can be substituted for the tale of discovery), the hopes of de Broglie and Schrödinger should have been forgotten, and the wave function should have been presented as a new way, more powerful and more rigorous from the physical-mathematical point of view, of theoretically acknowledging the new practice to which the atom constrained the physicist. Yet, the customary narration preserved the memory of the hopes and disappointments of the creators of the wave function and is therefore oriented not to the scandal of a reality endowed with the power to refuse to submit to the requirements of a classical spatiotemporal description, but to the disappointment of those who had attempted to restore that submission.

However, that particularity of narrative, which helped make quantum mechanics vulnerable to the "curse of positivism," leads to another particularity, one associated with theoretical language itself. A very unique relationship binds Bohr's first quantum model with the formalism we have inherited. The latter, contrary to the atomic model, does not introduce a "quantum being" endowed with the power to impose new questions, in the positive sense of the term. More specifically, it introduces this power in negative fashion: although endorsing it, it transforms the problem. The quantum formalism created in 1927 promotes the forgetting of obligations associated with the encounter with microscopic reality, to the benefit of an apparently metaphysical problem that has continued to fascinate philosophers ever since—the problem quantum mechanics presents us with of having to *choose* what will become the object of measurement.

3

Niels Bohr's Lesson

How should we think of measurement? According to the positivist approach, measurement has no exterior, cannot be referred to a reality capable of defining "what" is being measured. From this point of view, the impossibility, in the quantum scenario, of representing a reality to which measurement would provide access while remaining autonomous from it simply reflects the fact that physics has finally acknowledged epistemological norms. Another thesis, associated with quantum theory but extendable elsewhere, "explains" the limits of measurement in the quantum scenario, which makes no sense from the positivist point of view. This thesis states that the interaction needed for measurement "perturbs the phenomenon." It would then be the perturbation associated with the operation of measurement that would create the necessity of choosing what to measure. Here, the choice is between those variables that are being measured and those that will be perturbed as a result of the operation. Correlatively, the preparation of a system that allows physicists to predict a physical quantity with certainty, one corresponding to position, for instance, will have as its price a radical uncertainty concerning some other quantity, here the quantity of motion. The perturbation, then, would provide the key to Heisenberg's well-known uncertainty relations, which present us with "com-plementary" variables that singularize quantum formalism, variables that cannot be measured simultaneously, and between which a choice must be made.

This was not Niels Bohr's claim but Werner Heisenberg's. And it is Heisenberg who would marshal the arguments around which controversies concerning the relationship between physical knowledge and reality currently revolve.

The idea of a perturbation implies reference to what is perturbed ("reality as it exists independently of measurement") and, therefore, immediately raises an objection by physicists. Why, they asked, can't the reference to what is characterized by a quantity of motion and a position be preserved, even if the perturbation excludes the possibility that these can be measured simultaneously? However, those who objected were immediately confronted with what might be called the Einsteinian prohibition: physical theory cannot give meaning to something it defines as escaping prediction. In other words, the strength of the argument lies in the way it operates. First it ensnares physicists by convincing them that a common view of reality can be shared, then it strikes by confronting them with the prohibition. The ongoing discussion about "hidden variables" reflects the strength of the argument, which assigns a crucial role to the rules that scientific knowledge should or should not obey. Because the measurement of certain variables results in the impossibility of assigning a well-defined value to other variables, the values of the latter are "hidden." Do we have the "right" to refer to something to which we don't have access, to claim that, prior to the measurement that will perturb them, those variables have a value, even if it remains hidden to us? Or should we eliminate all reference to hidden variables as being devoid of meaning? Can we use Einstein as an example, when he claimed that the simultaneity of two distant events must be recognized as devoid of meaning?

Bohr, however, perceived the obligations that follow from the new quantum mechanics quite differently. The first quantum model of the atom, of which he was the author, unhesitatingly connected contradictory physical elements. Classical mechanics, more specifically, electrodynamics, was both required and denied by the two postulates of the model: electrons were in a stationary orbit around the nucleus, whereas they should have, like every electrically charged body in motion, created a field, that is, gradually dissipated their energy. The amount of light (now, the photon) emitted or absorbed by an atom reflects the individual, nondecomposable, nature of the event constituted by an electron's transition between two stationary states. In other words, Bohr was convinced that the quantum atom constituted a new physical reality, requiring a new physics, releasing the physicist from any obligation toward classical mechanics. Bohr affirmed the "acausal" nature of quantum physics, a statement that should be understood in a strictly technical sense: quantum reality cannot be understood in terms of the causal measurement on which the classical edifice had been built.

And yet, in the second formulation of quantum mechanics of 1927, the two variables of position and motion, which in the classical description exhaustively determine dynamic behavior, retain their central position; and energy, or the Hamiltonian operator, which articulates those variables in classical mechanics, continues to do so, although differently. Quantum formalism, centered on the wave function, is not, like Bohr's model, a "description" that, at the cost of two scandalous postulates, designates the atom as an actor and allows us to indicate which "aspect" of that actor's activity is measured through detection of the photons it absorbs or emits. The formalism, which introduces an equation that affirms the relevance of the categories of Hamiltonian dynamics, is a postulate in every sense of the word. I will return to this later. Here, I want to emphasize the uniqueness of Bohr's position, for he never allowed himself to be influenced by this relationship of resemblance. For Bohr, the equation that connects the evolution of the wave function with the Hamiltonian of the "quantum system" (the Schrödinger equation) has *nothing to do* physically with Hamiltonian physics. The equation does not imply that what is described is *determined* in terms of position and velocity, that those variables are "objective" in the sense that they can be attributed to an object. In quantum mechanics, Bohr insisted, the data about position, velocity, and all other measurable quantities, cannot be attributed to any reality. It is their formal presentation that makes them variables, but their physical meaning relates them to the detection devices used to measure them. Those devices create the possibility of making an observation about a quantum being, but what is observable cannot be attributed to this being while ignoring the device. In other words, for Bohr, measurement of the quantity of motion has no effect on position whatsoever. It reflects the intervention of a detection device that *actualizes* the "quantity of motion" observable.

Bohr retrospectively defined classical dynamics, from Galileo to Hamilton, as determined by an idealization that becomes misleading whenever the physicist addresses the quantum world. What did causal measurement rely on? As we saw in Book II, it was causal measurement that enabled Galileo to create the very concept of instantaneous velocity and, with the rational mechanics of the eighteenth century, gave physicists the ability to measure force. In both cases, the visibility of the body, the possibility of determining where it is, is presupposed. For Galileo especially, the equal sign implies that we know both the initial height and velocity of the moving object: the height that characterizes it at a later moment will then allow us to determine the amount of velocity it has gained or lost since leaving its point of departure, and thus its resulting instantaneous velocity. The idealization arises because the possibility of spatial determination is taken for granted, it can be "seen," and this visibility hides the energy interaction it presupposes. Even in astronomy, we know the Sun's position only through the emission of light. In both cases the idealization that overlooks this interaction is legitimate in that the order of magnitude of the energy interaction required (given by Planck's constant) is negligible compared to the energy associated with the motion of the body being characterized. Detection has indeed taken place, but it has taken place at the level of the retina or the photographic plate. Who would dare to claim that the Sun is in any way affected by the fact that the photons it emits enable us to locate its position? Doesn't it emit them regardless? Irresistibly, we think: it is where it is.

It is this irresistible conclusion that quantum mechanics, according to Bohr, obligates us to resist. Neither position nor quantity of motion answers to the idealization of mere location. To determine the position of a particle is to answer a question we have taken the means to make decidable by the appropriate use of a detection device. And in order to be answered, the respective questions about the localization and evaluation of the quantity of motion imply *logically* incompatible means. The term "logic" is important, for it emphasizes that it is *information*, intelligible information, leading to reasoning, that we demand of our detection devices.[1] Thus, the device that provides information must also produce meaning. And it is what is implied by the production not of data as such but of meaningful information, information about position, that is incompatible with the production of meaningful information about velocity. The first requires a rigid device, which makes energy exchange uncontrollable; the second must introduce causal measurement, measurement of the transfer of energy, which prevents us from rigidly attaching the device and makes position uncontrollable.

Bohr insisted that our measurement devices are part of classical mechanics by definition, and he was often heard expressing what sounded like an arbitrary limit to the ingeniousness of future devices. However, he adhered to the obligations that followed from the project of interrogating quantum beings in terms that allow us to interpret detection, to claim that if a mark is produced on a photographic plate, the quantum being was "there," or that if electrons reflected by a target lose energy, it is because they were the "cause" of a recoil of the target particles, which received energy equivalent to the energy lost by the electrons. In other words, just as the laws of classical electrodynamics were required by Bohr's model of the atom, which happened to contradict them, the classical idealization, made untenable by the finite interaction any detection entails, must still be maintained for the interaction to become a measurement, productive of meaningful information.

The great contrast between perturbation and Heisenberg's uncertainties, on the one hand, and Bohr's production of determinateness, on the other, is related to the respective status each confers upon classical mechanics. From the point of view inspired by Heisenberg, it appears to be something of a lost paradise. The fact that, in the quantum case, the perturbation cannot be neglected (Planck's constant, h, ensures that it will have an order of magnitude comparable to what is being measured) can only inspire a certain nostalgia given that this perturbation has only negative value, creating a kind of slow torment for the physicist. The Bohr interpretation, on the other hand, requires a crucial distinction that mechanics, ever since Galileo, has, according to him, spared the physicist. Causal measurement authorized description and reason to coincide. Between the object defined by measurement and the "measured reality" there was thus no interruption requiring that an observer be implicated or introduced, or obligating the physicist to recall that measurement had to be understood in terms of intervention, mediation, or the production of meaning. Reality was determined "in itself" by the very mode of determination introduced by measurement. For Bohr, the lesson of quantum mechanics was not so much about knowledge as it was about reality itself,

more specifically, the fact that reality did not, either in classical mechanics or in quantum mechanics, dictate its categories
to the physicist. Variables, whether classical or quantum, refer
observables to variables, the meaning of which is the function
of which they are variables, but they never authorize the physicist to speak of a "functional" reality. They merely express the
fact that the physicist has taken the means to *characterize reality as a physical-mathematical function*. For Bohr, with regard
to the requirements of a functional definition, "reality" has to
be referred to as *indeterminate*, independently of the measuring device able to provide a determinate interpretation to an
observation.

Here, it is appropriate to mention the radical distinction
made by Gilles Deleuze between the virtual and its actualization, on the one hand, and the possible and its realization, on
the other. The only thing "missing" from the possible is existence. Indeed, this is what is presupposed by the measurement
device assumed by dynamics: a body can have any possible position or velocity; measurement occasions the transition from the
possible to the real of one of those values. Actualization, on the
other hand, is associated with creation. It implies a change in
kind, not the determination of a preexisting possible. "The virtual possesses the reality of a task to be performed or a problem to be solved: it is the problem which orientates, conditions
and engenders solutions, but these do not resemble the conditions of the problem."[2] In reply to the concept of the problem in
Deleuze we have Bohr's concept of choice. For Bohr, the choice
of a detection device is the choice of the problem by which the
production of determination (or the actualization of the virtual)
assumes meaning.

We are familiar with Einstein's well-known complaint,
"God does not play dice with the universe." But Einstein, like
all the critics of the Copenhagen group, would have been vastly
more satisfied if quantum mechanics had in fact presented us
with a probabilistic description, like that of nuclear decay or

the spontaneous return of an excited atom to its fundamental state.[3] The physicist would be free to believe that another level of description exists, one that "explains" probabilistic laws the way four-dimensional mathematical reality "explains" our observations in space and time. Einstein's opposition to quantum mechanics had less to do with a conflict between deterministic and probabilistic visions of the world than with the question of the relationship between "reality" and "determination."

In 1935 Einstein launched a final attack on the idea that quantum mechanics constitutes a "complete theory." He coauthored an article with Boris Podolsky and Nathan Rosen titled "Can Quantum Mechanical Description of Physical Reality Be Considered Complete?" in which the authors (customarily referred to as EPR) defined the obligation to which any theory claiming to be complete must comply. They write: "Every element of physical reality must have a counterpart in physical theory." The element of physical reality in turn had to satisfy a requirement: "If, without in any way disturbing a system, we can predict with certainty (i.e., with probability equal to unity) the value of a physical quantity, then there exists an element of reality corresponding to that quantity." Commenting on their definition, the authors remarked that it was in agreement not only with classical mechanics but also with the constraints that quantum mechanics claims to adhere to. In other words, EPR do not provide a general argument against quantum mechanics. They offer to construct a critical argument that will find fault with it on the very grounds it has defined for itself. They intend to show that there exist quantum "elements of reality" that have no counterpart in theory.

The expression "without in any way disturbing a system" indicates that the EPR approach requires Heisenberg's perturbation interpretation. It will involve finding a way around Heisenberg's prohibition regarding hidden variables. The EPR article introduces a thought experiment in which a "quantum object" spontaneously breaks down into two subsystems—let's call them A and B—each of which, based on what the experimenter decides, could be subject to measurement. The crucial point is that, because of the unique object that produces A and B, we have information about A and B such that they preexist the measurement they may be subject to. For example, if the single system was a molecule with spin 0, we "know" that the spin value of one of the "particles" must be +1 and the spin of the other −1. Therefore, if we measure the spin of A in a known direction, let's call it x, we can, *without disturbing B in any way,* deduce the value of its spin in that direction. What prevents us, then, from measuring *on subsystem B* another, complementary observable (here, the spin value in direction y, for example), an observable that quantum mechanics states can be given no physical meaning if we give physical meaning to the first. Naturally, this measurement will make it impossible to actually measure the spin of B in the x direction, but this is, in any case, redundant as the value can be determined by measurement of A. And reciprocally, measurement of the spin value of B in the y direction allows us to assign to A the value of the observable that the previous measurement of A had made inaccessible. Therefore, each particle is characterized in itself by two variables that quantum mechanics states cannot have physical meaning simultaneously. Consequently, quantum mechanics is not complete.

Léon Rosenfeld recounted Bohr's perplexity when confronted with this objection.[4] His perplexity arose not from the difficulty of the response but because the EPR experiment presented no problem to him at all. On the contrary, for Bohr it served as an excellent restatement of his own message, a message he himself was unable to contemplate without a certain uneasiness because it could not be reduced to the idea that measurement "disturbs" the measured system. In the EPR case, he stated in his response that it is assumed that at the moment their spin is measured, the two particles are arbitrarily distant from one another and do not interact. The whole question comes down to knowing if we can state that they are separated in any unambiguous sense. It is not enough to say that measuring one does not affect the other. We must also be able to show that it does not affect the possibility of measuring the other. And it is here that Bohr strikes back. If we can use the information derived from the single system from which the two subsystems arise, we do so to the extent that these two subsystems are pertinently defined by their common origin, that is to say, are related to the same experimental device through which the information about the original system has acquired its meaning. Therefore, no matter how far apart the two subsystems are, they continue to form, from the point of view of the unambiguous formulation of the information we have about them, a single system. And in this sense, we cannot say that measurement of one of the two particles allows us to characterize the second while leaving the physicist free to make another measurement of this second particle. In fact, the measurement results in the "destruction" of the single system, in the sense that its singleness could serve as a common reference, leading to a determinate inference, for the two "separate" particles. If measurement can be used to assign a spin to what then becomes the "second" particle, this is another way of saying that this second particle has been subject to measurement: the "first" particle has become an integral part of a measurement device that detects the spin of the second. And the physicist's choice of a device is irreversible; it actualizes this "second particle" to the extent that it is separated and characterized by the measurement performed. A similar measurement made directly on this particle would confirm the value of the observable in question with a probability of one.

In other words, for Bohr the "spatial" separation had no physical meaning in itself, for to the extent that their common origin had to preserve a precise, unambiguous meaning, the two subsystems are in fact not separate from the point of view of an observer. Measurement of the one actualizes the physical meaning that will be given to their shared origin, allowing us to infer the result of a measurement of the other; and it is in this unambiguous sense that it can be considered measurement of the other subsystem. But this actualization destroys any possibility of once again referring to this common origin for other, complementary measurements. Taking "another" measurement of the second subsystem would, *as always,* amount to introducing a device that destroys the meaning of the first measurement and creates another actualization, logically contradicting the first.

Once again, Bohr used the consistency of formalism to defeat those who thought they had detected a weakness in the theory. If EPR were right, quantum mechanics would not be incomplete but wrong. And indeed this was what was at stake when the implications of the formalism in a more recent, EPR-like case were challenged experimentally. As a result, the hypothesis of the kind of "hidden variables" postulated by EPR, which are said to characterize the quantum system in itself, in spite of Heisenberg's prohibition, has since been refuted.[5] But Bohr's answer was not vindicated. It is still seen as confirming critics' darkest suspicions; Bohr "denies" reality, he "prevents" us from referring to what we "know."

Obviously, Bohr never denied that reality "exists," or that it is "knowable." Reality is clearly "knowable," as the existence of quantum mechanics demonstrates. But being "knowable" is something quite different than the possibility of knowledge that critics demand. What they express is nostalgia for that blissful situation where reality itself seems to dictate the categories of its definition. For the strategy adopted by EPR—initially claiming that quantum formalism accepts the obligations of a theory that claims to be complete, and then revealing a situation where it betrays those obligations—continued to define a "reality element" (which had to have a counterpart in theory) in a way that confused the demand for an "independent reality" with the demand that reality be theoretically defined as independent. EPR continued to require that each "element of physical reality" authorize this formidable claim: the possibility of defining that element, of assigning to it a physical and theoretical meaning independently of any practical connection.

It is important to understand that the requirement EPR reproached "quantum reality" (as defined by quantum theory) for failing to satisfy has nothing to do with what other practices require of their "reality." Even less with the knowledge one human can have of another. How can I even dream of knowing the other "in itself," independent of the relation I have with it, independent of our respective abilities to form relationships with one another and with others? The notion that knowledge, in this instance, is dependent on words, on contexts of meaning in which those who know one another participate, is not something to regret. It is, rather, the possibility of knowing in the absence of a relationship that is a nightmare. Should we regret what takes place in a laboratory, where phenomena are effectively staged, purified in such a way that they become experimentally meaningful, acquiring the power to authenticate their representation? No one would dream of imagining that the necessity of a laboratory, of the devices that are used to transform an "empirical" fact, subject to a thousand and one interpretations, into an "experimental" fact, implies the "unknowable" character of reality. Quite the contrary, experimenters are all the more "realist" to the extent that their practice obligates them to fiercely distinguish between "fact" and "artifact," that is, to distinguish between those cases where "reality" has indeed satisfied the requirements that define it as a reliable witness and those where the device has extended the power of interpretation to produce a "false witness" who cannot but confirm that reality. Laboratory practice connects "reality" not to the possibility of predicting without intervening, but to the possibility of an interaction productive of evidence whose meaning can be determined.

From this perspective, Bohr's indeterminacy does not signify unknowability. It reminds us that every determination is productive of a link that carries meaning, creates the ability to make a difference for the one who has the means to determine.

Bohr, therefore, did not give up trying to "know" reality. He remembered that it was only because the bodies interrogated by classical mechanics allow themselves to be presented in terms of the idealization of mere location that, in Galileo's lab and that of his heirs, experimental practice appeared to have simply staged a reality determined in itself and by itself. The reality he wanted to reject was not the one presupposed by experimental practice, or the one each of us presupposes when addressing the world and other humans. It was the reality of the Queen of Heaven, the dream of a reality whose truth could be attained independently of any practice, any question, any relationship. It was the reality that had been preserved by Einstein's general relativity, to the extent that the different points of view situated in space and time, and creating an observable phenomenon in space and time, could be deduced from four-dimensional mathematical truth. It was the reality that everyone who had tried to create a quantum mechanical beyond wished to reestablish. Even though that beyond might assume the seductive and quasi-mystical appearance given to it by a physicist like David Bohm, that of an "undivided, multidimensional, and implicated" totality, based on which we can understand the engendering, or the unfolding of our "explicate order" as secondary, limited, and valid only in certain contexts.[6]

In Book I of *Cosmopolitics*, I addressed the problem of the "vocation" of physicists, which commits them to disqualifying all other knowledge practices. The "vision of the world" to which physicists' faith is addressed situates them in an "elsewhere" from whose vantage point all other practices are judged by their benefits, limitations, and misinterpretation. From this point of view, the "end of the dream" to which Bohr exhorted his colleagues may seem profoundly satisfying and entirely healthy. Those who treat the loss of this dream as something dramatic for thought and assimilate it to the loss of an independent reality, intelligible and describable, thereby translate the polemical definition of "physical" reality to which they adhere, a definition that pits this reality against all the "realities" presented by other practices.[7] Correlatively, the way in which Bohr generalized the obligations entailed for him by quantum mechanics—the idea of "complementarity"—isn't so much fascinating as it is relevant for other practices. For the first time, physics is not presented as a model but as something subject to an obligation, one that can be transposed to other sciences, yes, but as a form of repetition rather than something to be complied with. What is at stake is the singularity of each individual science whenever the complementary affirmation makes itself heard: there is no answer without a question. In other words, each science must undergo, in its own way, the challenge of statements such as: no knowledge content can win its independence from the question that gives it meaning; no question can gain its autonomy from the choice from which it proceeds; no choice can prevent its selective nature from being taken into consideration, can ignore what is excluded from being presented so that what is chosen can present itself.[8]

Nonetheless, the fact is that physicists have not abandoned the dream. We can even say that high-energy physics, when it addresses the mathematical symmetries that characterize its objects rather than behaviors in space-time, has reinvented the dream, which then becomes, as Heisenberg noted, frankly Pythagorean. Physicists no longer require of the interrogated reality that it subject itself "in itself" to the determinations in terms of which we measure it. They address symmetry properties that, *independently of measurement,* characterize the mathematical beings presented by theory. In this way they take as their object the symmetry properties of the wave function even though the reduction of this wave function, which gives meaning to observable properties, destroys those symmetries. It is important to note that the separation thereby instituted between "objective" symmetry and determination by measurement in itself constitutes the negation of the so-called orthodox interpretation. For Bohr, the wave function had no meaning at all independent of its "reduction." The very idea of "posing the problem of the reduction of the wave function," that is, considering the transition from probability amplitudes to determinate probabilities as a problem, was devoid of meaning for him. The fact alone that it was in these terms that the "problem of measurement in quantum mechanics" is ordinarily presented accurately reflects the fact that physicists have not felt in the least bit obligated by the obligations that, for Bohr, follow from the novelty of quantum mechanics.

It is impossible to acknowledge that Bohr was right and denounce those who didn't follow him without transforming my approach into a critical, and thus normative, endeavor. Bohr, like Duhem, failed. But we can use his failure to better understand the practice to which his proposition was addressed.

"Bohr was incomprehensible, his language was obscure, he would think out loud." I don't want to let such statements stand in my way, even though they might be relevant. Not only did people listen to Bohr, intensely, not only did he continuously struggle to express his thought in ever more lucid terms, but he had, in the person of Léon Rosenfeld, a faithful and perfectly intelligible interpreter. Anecdote alone is inadequate. Here, misunderstanding must primarily be understood as a refusal to understand, that is, a refusal to *consent* to understand, as William James would put it. And psychological interpretation is of no more help than anecdotal evidence. It seems to me that the refusal to consent has not been adequately interpreted in terms of physicists' attachment to their dreams of determinism, power, and omniscience. In order to have a chance of discovering an interesting reason for the intellectual dissatisfaction of physicists with the Copenhagen interpretation, we must set aside commonly used slogans such as realism, determinism, positivism. And we must reformulate the "dream" that physicists refused to abandon in more technical terms, using terms that singularize their passion and do not condemn it a priori.

4

Quantum Irony

From now on, the question is no longer one of interpreting quantum formalism but of the problem associated with the formalism itself—what it commits physicists to and what it prohibits them from doing. I would like to show that it's not impossible that the intellectual dissatisfaction of physicists reflects the irony with which quantum novelty was inscribed within the tradition of physics.

This irony began early on, with Bohr's first model of the atom. As we have seen, in order to incorporate the well-defined spectral lines that served as the signature of each chemical element in the periodic table, Bohr had imagined the electrons to be distributed in stationary orbits of well-defined energy. But we have already encountered a version of the notion of a well-defined stationary energy orbit in Book II. The Hamiltonian representation resulted in such "stationary movements" whenever it made use of *cyclical variables.* In this case, each degree of freedom is characterized by an action variable, J, independent over time, the system being represented as a collection of free points that do not interact with one another. Through the work of the mathematician Arnold Sommerfeld and the astronomer Karl Schwarzschild, this representation, which had until then been reserved for use by astronomers, was transposed to Bohr's quantum atom, as if it formed a natural framework in which to formulate the rules of quantification for the series of "orbits" that defined the energy transitions introduced by Bohr to interpret atomic spectra.

The irony here consists in the sudden encounter between the "obscurity" of the atom's acausal behavior and the most luminous form of intelligibility. What is obscure, in the sense of being postulated with no explanatory justification, is the idea of a "quantum stationary state," where a charged body such as an electron is able to remain indefinitely in orbit, and of a "quantum jump" that occurs instantaneously, without any "transition" between two states. Its luminous intelligibility is found in the cyclic Hamiltonian representation. As mentioned earlier, it is the only Hamiltonian representation that does not present physicists with a problem to be solved—the integration of the Hamiltonian equations; rather, it presents the problem in terms that coincide with those of its solution.

In other words, whereas standard dynamics, with its interactions, accelerations, positions, and velocities, was equipped for writing differential equations whose integration—that is, the effective description of the spatiotemporal behavior of the system—had been perceived as a kind of Holy Grail, the atom appeared to require the analog of a cyclical representation, synonymous with the Grail's possession. The atom's stationary states correspond to the way a dynamic system can be represented if and only if the problem of integration has been solved: in terms of "free" modes that are independent of one another. The atom, which had entered physics as an actor in random kinetic events, therefore creates, where its behavior is most obscure, the possibility of a junction with the most elaborate, the most refined, the most singular fiction produced by dynamics: a fictional description that absorbs interactions, and the accelerations they determine, into the definition of the system's constituents, so that each of them reciprocally constitutes a local expression of the overall identity of the system. And correlatively, this fiction acquired the means to claim a kind of privileged relevance because it is immediately confirmed by the energy spectrum that identifies the individual atom.

Where is fiction? Where is reality? In classical dynamics, each stationary state is represented as "alone in the world," its definition expressing the totality of a system itself defined as closed, interacting with no environment. Accordingly, the equations defining the postulated quantum stationary state prohibit us from defining the atom as a system interacting with an environment. This implies that the atom defined by its stationary states is *literally unobservable* since there is no observation without interaction. But if we are, nonetheless, able to know something about the atom, it is that the cyclical representation, in the case of the Bohr atom, does not tell us everything. The quantum atom does indeed "interact" with the world; it emits and absorbs light quanta. In 1916 Einstein showed that the "transitions," the "quantum jumps," if they were to satisfy the laws of Planck's black-body radiation, had to correspond to two types of processes. Either the transition (the emission or absorption of a quantum of light) is caused by an external electromagnetic field, or it is spontaneous—an excited atom spontaneously "falls back" to a state of lower energy while emitting a quantum of light. In other words, the "excited" stationary states have a lifetime, like radioactive nuclei. Whereas Bohr's first postulate creates a connection with the dynamic representation of a world where nothing happens, where nothing can happen, the second, that of the "quantum jump," refers us to kinetics, the science of events.

The two concepts postulated by the Bohr atom—the stationary state and the quantum jump—were in a relationship of reciprocal presupposition. However, they are very differently inscribed in the mathematical physics of the time. The stationary state resulted in the most astonishing reciprocal capture between Hamiltonian dynamics and the atom, a reciprocal capture that short-circuits the question of the resolution of the dynamic problem and asserts its solution, the stationary state being described in the language of integrable dynamic systems. As for the quantum jump, it associated the atom with the active protagonists of kinetics, those whose activity had to be detected, those whose "initiative" had to be recognized—when will it emit a photon? Between the two concepts, no hierarchy should be possible. It is on the basis of the energy differences between the levels, that is, the transitions, that stationary states can be characterized. Logic seems to demand that accepting the effective symmetry between the two postulates means challenging the previously accepted claim that kinetic events correspond to an incomplete dynamic description (see *Cosmopolitics,* Book III). But logic does not rule the history of physics.

The irony of the situation derives from its duality, from the two different ways it can be described. On the one hand, we could say that Maxwell's "Queen of Heaven," whose triumph we followed in Book III, had finally touched ground, where its relevance could only be confirmed by stating, simultaneously and symmetrically, the relevance of what escaped its power. On the other hand, we could say—and this is the path de Broglie and Schrödinger followed—that the only pocket of resistance to dynamic intelligibility was now successfully identified: if they were able to "smooth" the quantum jumps, Hamiltonian harmony would reign over the physical world.

De Broglie and Schrödinger did not succeed, but their ambition still haunts the formalism surrounding the "wave function" defined by the Schrödinger equation, tempting physicists to revive it. This formalism indeed breaks the symmetry of Bohr's two postulates. The definition of stationary states corresponds to a solution of the Schrödinger equation, but the discontinuous quantum jump no longer corresponds to an "event" that would "happen to the atom." It is the "reduction of the wave function," the transition from probability amplitudes to probabilities, that takes the place of the "quantum jump," because, "like it," this transition is associated with the possibility of assigning to a "quantum system" a determinate value of an observable property.

Here we encounter the greater irony, superimposing itself on the first and stabilizing its effects. A form of mathematical wordplay has literally caused the kinetic actor to disappear, to be replaced by the question of the reduction of the wave function associated with "measurement." Bohr's atom was "knowable" only because it emitted and absorbed light, but the absorption or emission event, the condition of measurement, was not defined as a measurement. If we live in a world of color, it is because, according to the terms of Bohr's model, atoms continuously absorb and emit photons. The fate of the majority of photons remains undetermined. Some of them hit our retinas, others strike a photographic plate in a physicist's laboratory and result in a measurement, but the majority of them participate in other interactive adventures that we can only imagine. Yet, after 1927 this "realist" language could no longer be used. The emission and absorption probabilities have physical meaning *only in terms of the measurement device* and, one way or another, the device appears as "responsible" for the transition from probability amplitudes to probabilities. Regarding the atoms that were "free with respect to the physicist," the ones responsible for the bright colors of flowers that attract butterflies, nothing could be said. Spontaneous emission was no longer a property of the atom, through which it could become known; it is the production of knowledge about the atom, the record of a photon of carefully determined frequency, that affects the representation of the atom, that is, reduces the Schrödinger wave function.

Where I have referred to colored flowers and butterflies, Léon Rosenfeld, David Bohm, and Jean-Pierre Vigier confront the question of dinosaurs, that is, the possibility of "knowing" a period of life on Earth when there were no humans, much less physicists, but a period, nonetheless, when the world that quantum mechanics claims to describe resembled our own. Vigier and Rosenfeld knew well, given that they were Marxists, that behind this exchange loomed the menacing shadow of Lenin, who, in *Materialism and Empirio-Criticism*, denounced as idealist and solipsistic those who might be suspected of claiming that "reality" depended on their perception.[1]

ROSENFELD: All that we can say about those Saurians is based on what we can see of them . . . I cannot see the least philosophical difference between descriptions of the state of the world in the secondary epoch and the description of the world as it is today . . .

AYER: . . . I suggest that if Professor Rosenfeld does maintain this position he must do it in a form that would escape Professor Vigier's trouble about the Saurians. He must say that in talking about the Saurians he is talking not about anything he did observe but about something somebody might have observed, had he been there, even if he wasn't there. To make this theory work, you have got to do it in terms of the possibility of making the relevant observations and not in terms of actual observations . . .

ROSENFELD: I am glad that you have mentioned that—this is exactly what I meant. But I didn't mention it explicitly because I thought it was quite obvious.

VIGIER: I was very happy about what Professor Ayer said and I also agree with Professor Rosenfeld, for this is the first time I have heard Professor Rosenfeld disagree with a plain positivistic sentence. I would put the thing in a stronger form: I do not think that things which exist are things which might have been observed. This is where the split comes in a very clear form.

ROSENFELD: I don't say that. Don't continue on that line because I do not say that things only exist in so far as they could have been observed. All the statements we make about the world are necessarily descriptions of a state of affairs, of mind, of material, that an observer might perceive if he were placed in those particular circumstances.

VIGIER: Let us say then we agree that the world exists outside any observer. Did the laws of quantum mechanics apply to the world at a time when there were no observers present?

ROSENFELD: Of course.

VIGIER: OK. If you say then that the laws of quantum mechanics did apply at that time, then the laws of quantum mechanics are real, objective, statistical laws of nature, which have nothing to do with the observer, and are verified whether there are observers or not.

ROSENFELD: No.

VIGIER: You can't change your position and say something two minutes ago and another thing now . . .

Why quote this exchange here, when I've already indicated that it was no longer a question of "interpretation"? Because it exposes what is most often dissimulated in matters of interpretation: a genuine malaise when confronting the role that formalism gives to observation. Vigier's frustration is caused by the technically, rather than deliberately, ironic position of

Rosenfeld. When Rosenfeld states that "things" exist, whether or not they are observed, and that quantum laws apply whether or not there is an observer, Vigier thinks he has won—quantum laws are, therefore, laws of "nature." But then Rosenfeld says no, to Vigier's horror. If Rosenfeld answers "no," it is because Vigier has failed to take into account that the laws of quantum mechanics, like every "objective" statement, apply only to "states of affairs," *not to beings*. And a state of affairs must include reference to a well-defined possibility of stating the affair. The colored flower is a meaningful affair for the butterfly. And when we apply quantum mechanical laws to it, it no longer refers to butterflies but to the photographic plates that allow us to describe its atoms in terms of observable quantum facts.

To speak of irony is to claim a relative sympathy for the victims of irony. If paleontologists can reconstitute a dinosaur from fossilized remains, if molecular biologists are not troubled when they speak of the DNA molecule as if it existed or acted, why should physicists be forbidden from referring to what they investigate as an existent? Should they really say good-bye to any possibility of constructing the representation of an actor to which we could "really" assign responsibility for our measurements of it?

This is the intellectual rebellion that Erwin Schrödinger introduced with his famous parable of the "cat" in the box. A cat is enclosed in a box together with a device that triggers the release of a poison gas when a radioactive particle decays. This means, Schrödinger emphasizes, that the cat "actually" dies when the decay event occurs. However, quantum mechanics does not allow the cat to die until the box is opened and the cat's state—dead or alive—is observed. The cat is, indeed, a measuring device like any other, and quantum mechanics demands that we accept that the particle will never "really" disintegrate, any more than an excited atom will fall back to its fundamental state while releasing a photon, unless measurement has taken place. As a consequence, we must conclude that the "event," here a chain of events ultimately leading to the cat's death, can be postponed indefinitely "until" the conditions of observation have been fulfilled. This means that "until" that time, until the measurement takes place, the cat, like the radioactive particle, will be described as a superposition of probability amplitudes, in other words, as being both "dead and alive" rather than being either "dead" or "alive" with definite probabilities.

Schrödinger's thought experiment has since been assimilated and presented as a scenario in which quantum mechanics is assigned the fascinating power to be able to claim that the cat is indissolubly "dead-alive." In fact, it should instead be thought of as a cry of rebellion, a cry of passion. Quantum mechanics is not complete and cannot be considered complete! Schrödinger was unable to define just "what" quantum mechanics was missing, but to him the Copenhagen position was perverse. We "know very well" that the nucleus decays on its own. We know that the cat dies. We could say that these are "actors" to whom things happen. On behalf of what can we consider as satisfactory a formalism that makes physical measurement the only "actor"? Of course, measurement alone provides the information that has meaning for physicists. But why should they give up inventing the theoretical syntax that would "cause" the disintegrating nucleus, or the excited atom falling back to its fundamental state, to "exist"?

Vigier and Lenin were wrong. The question is not one of idealistic solipsism, which questions the existence of a reality "without us." It is, rather, as Schrödinger's protest vividly expressed, a question of the radical divorce instituted by quantum formalism between reality without us and reality "for us." It is a question of the genuine "claustrophobia" of the physicist, who is suddenly obligated by formalism to no longer refer to the signifying "counterpart" of practices that are productive of meaning. And the term "refer to" is important here. The source of the severe dissatisfaction aroused by quantum formalism is no longer the impossibility of "revealing" a real that, like the Queen of Heaven, would have the power to equate description and reason or, like the four-dimensional mathematical reality of general relativity, the power to designate a truth that is "superior" to our measurements, that gives them their meaning. The fact that reality does not satisfy this requirement is not the problem. The fact that light does not allow us to construct an identity that transcends measurement and explains its "phenomenalization," either as a particle or as a wave, is not the problem. But why does giving up the requirement of a transparent reality, intelligible in itself, why does acknowledging complementarity and the fact that our descriptions include an irreducible reference to measurement, lead us toward a language whose syntax makes measurement the only thing responsible for the world's observability? Cats exist, they observe and die! In the world, meaning is created "without us," and it is created without reference to measurement. The butterfly is guided by the photons emitted by flowers. Why must physicists agree to be trapped in the sudden, stifling abstraction of criteria of meaning that refer to measurement alone? Why must they accept a definition of their practice that obligates them to cut themselves off from the world, imprison them in meanings they, and their measuring devices, bring into existence?

It is the significance of the place known as a "laboratory" that is in question here. As Bruno Latour, in discussing Pasteur, has admirably shown, what is made in a laboratory never really leaves the laboratory. Rather, it is the laboratories that proliferate, and they do so in places—hospitals, the food-processing industries, government agencies, in the case of Pasteur—where the scientific proposition is to become relevant. Wherever a science appears to have "spread," we find that it is the devices, practices, the gestures of sampling and measurement, the rules of interpretation that have been successfully adapted and implanted. Yet, the laboratory of Pasteur's heirs has continued to be transformed with the appearance of new instruments as well as new problems associated with new circumstances. The laboratory certainly creates devices that are both filter and amplifier, purifier and stabilizer, but what it detects and cultivates is also able to alter the practices of detection and culture. In other words, while the biologist never leaves her lab, she is not locked in. In its own way, the innovative and bifurcating lineage of laboratories bears witness to a dynamic of apprenticeship for a world where "something" is encountered that may create new problems that challenge the requirements of detection, measurement, and culture.

We could say, then, that the Pasteurian biologist, although dependent on the laboratory, asserts the existence of the microorganism the way the sunflower asserts the existence of the sun, and Europe, since Columbus, that of America. Of course, the methods of affirmation are distinct, as are the consequences for the other term of the relationship: the sun is apparently indifferent to the capture of its photons by plants; the microorganism experiences new adventures but—and this is what makes it an experimental "factish"—those adventures, once stabilized, have, until now, been "explained" in terms of the properties that have been attributed to it as an "autonomous being."[2] As for "America," its human inhabitants, then gradually its animals, vegetation, water, land, and climate, have been radically redefined by the history that followed their European "discovery." But beyond those considerable differences, the creation of significant connections and the affirmation that "there are worlds," in any case, go hand in hand. On the other hand, the quantum physicist finds herself enclosed in her lab: there she is all-powerful because she determines the measurements to be made, but her measurements seem to prevent her from bearing witness *for* the world, something all other creators of connections can do.

To describe what quantum irony calls impossibility without getting wrapped up in the question of a "reality knowable in itself," of "strong objectivity," the concept of delegation introduced by Bruno Latour is extremely useful, for it integrates, without contradiction, the topics of invention and existence.[3] Delegation is always an invention in the sense that its point of departure is not a general project, oriented toward the grand theme of truth, but one that is partial and partisan. When a "hydraulic door closer" is invented, delegated to automatically close the door behind the visitor to the Center for the History of Science and Technology in La Villette, it is not acting as a golem or a Turing machine supposedly capable of anything we call "thinking." The door closer is "delegated" to closing the door, period. It is not replacing a "human in general," but an employee with a specialized function. This delegation brings into existence an "agent," and this agent is a hybrid being because it testifies both to the interrelations among the nonhuman "actors" it assembles and the human projects that this assemblage realizes. And this agent in turn becomes, for humans, a source of new problems. It can be made to break down if we try to close the door "ourselves" and, if the hydraulic piston is poorly adjusted, it will slam the door in the face of a second visitor unthinkingly following the first. Its mode of existence is constantly being negotiated between two positions: if the device functions as expected, humans will have to adapt; if not and it "malfunctions," it's "back to the lab." For Latour, the technical invention must be understood in terms of delegation, of "having X do Y," through which both users and devices-agents are invented in a continual process of redistribution between what belongs to humans and what is inscribed in the "logic" of the operation of things. This is "sociological" invention in the strong sense, which enabled Latour and Strum to contrast human societies and baboons in terms of the difference between complication and complexity.[4] Baboon society is complex in that the identity of each baboon—what it is willing to risk, hope, anticipate—must be continuously negotiated in real time, with no mark, badge, or word to stabilize the situation. Our societies, on the other hand, are "complicated" in that the majority of our activities satisfy "sociological" prescriptions resulting from successful operations of delegation that indicate the extent to which one can trust a situation, that allow us to determine "who to talk to," how to proceed, or with whom or with what we are dealing.

Obviously, the laboratory device entails the success of an operation of delegation of this kind. It operates on behalf of the scientist whose preoccupations it embodies in turn. But this operation alone is unable to characterize experimental practices; more specifically, it is in their case subject to an additional requirement. This requirement corresponds to the "realism" invented by the experimental sciences. Not only must the experimental device function as a vector for the creation of a meaningful, information-producing relationship between human and nonhuman, but this relationship must be able to undergo a genuine syntactic transformation so that the information produced can be assigned to one of its poles, the meaning of the relationship being interpreted on the basis of that pole. The activity of the delegated agent must have the ability to attest to the properties of an actor "independently" of the experimental project that staged that activity. So Pasteur was able to successfully argue that living actors, microorganisms, were responsible for the activity of fermentation empirically associated with the presence of what were then known as *ferments,* and he was able to do so because the making of *ferments* as his delegated agents included the requirement that the way they would act would constitute a strong argument in favor of such an attribution of responsibility.

A transformation such as this is not limited to science; we do it all the time, for instance, whenever we attribute the responsibility for the answer to a question we ask to the one who answers. But what is unique to experimental practice is that this transformation functions as a critical issue, and as such is subject to demanding requirements and challenges. Of course, no requirement and no challenge can eliminate the sleight of hand involved in the transition from the relationship between the scientist and her agent to the assignment of a property by the scientist to an actor. Pasteur's microorganism is nothing other than the coherent expression of the relationships that Pasteur was able to create from what he addressed. But it is a witness for the specificity of experimental relationships: these are such that, if successful, they will become part of the proof of the existence of a being endowed with specifiable activities that explain that success. The "experimental factish" is constructed, but the way it is constructed implies that it has to resist the accusation of being no more than a construction, and has to satisfy challenges that will test its claim of being endowed with autonomous existence. That is why experimental achievement is an event. Delegation has brought into existence a being that can explain what happens to it in the laboratory and is thus capable of existing *outside* the laboratory.

Pierre Duhem pointed out the necessity of resisting the temptation to forget that physical theory is an instrument. The rational thermodynamics he defended was distinguished from rational mechanics by the impossibility of moving from its theoretical object, "reversible displacement," to a reality that would explain the way it is characterized, that would explain "why," unlike reversible displacement, "it is dissipative." And this example allowed him to claim that even in rational mechanics, a gap must be created between what is subjected to experimentation and measurement and the theoretical representation that is constructed about it. The victory of atoms over thermodynamics allowed physicists to forget Duhem's claim, but it did not really refute it. The victory of atoms is the victory of devices that can be used to assign actors and events the responsibility for their detection, but it remains silent about their "theoretical representation." And when the construction of that representation occurred, in 1927, it transformed the gap into a prohibition. It proclaimed the radical impossibility for theory of justifying experimental practices of delegation liable to bring into existence what it is the physicist addresses, the impossibility of constructing the quantum being responsible for measurement and of speaking of the laboratory as the place where we "encounter" that being and where we learn from it the obligations associated with a meaningful encounter.

In previous chapters I proposed two kinds of "theoretical factish": physicomathematical factishes (for example, the Hamiltonian operator) and enigmatic factishes (especially the increase of entropy). To qualify the formalism postulated in 1927, I now want to introduce a "divinatory factish." For it was indeed a kind of divination that Niels Bohr practiced when he insisted on the fact that the meaning of our measurements—and therefore the definition of the "matters of fact" they allow—are relative to "classical" language, "causal" language, a language that quantum beings do not speak. The wave function *together with its reduction* serves as a kind of divinatory apparatus, enabling us to interpret, to confer meaning, to inscribe within a practice, "messages" that have arrived from a different reality, that physicists are unable to appropriate, and whose ways they are unable to penetrate, much less represent. As Bernard d'Espagnat has said on many occasions, "reality is veiled," but this veil is not a "discovery" that would apply to all humans and would require that they reject what they understand by reality. The veil is the counterpart of the formal divinatory factish of quantum mechanics. More specifically, it reflects the irony of the message that quantum formalism seems to deliver: it is about some reality, but it cannot be deciphered as such without being, at the same time, cut off from its "source," from "reality."

Irony, regardless of the praise showered upon it by

postmodern thought, is apparently not much favored by phys-icists. Some critics of the Copenhagen interpretation have conveyed Schrödinger's protest in a way that expresses their frustration on this point: why can't we, in the quantum case, create a syntax that allows us to move from experimental del-egation to attribution? Their protest does not contradict Bohr's complementarity or deny that the concept to be constructed relates to practices productive of the information it presents. It contests the interruption of the process of construction, without prejudice to the kind of activity we would be led to attribute to quantum entities as actors.

To take but one example, throughout his life Karl Pop-per tried to construct their identity in terms of "propensities," probabilities that owed nothing to our lack of knowledge. For Popper, quantum practice entailed the obligation to accept a concept of physical reality in which properties would become "dispositions" (toward this or that), that is, in which the con-cept of "possibility" would obtain the status of an irreducible category.

However, today, an element of novelty has been added. Now, the question of the construction of quantum actors could become a practical issue. One representative of these new voices is Hans Christian von Baeyer. In *Taming the Atom,* von Baeyer challenged quantum mechanics in a new way. He did not speak as a representative of a persistent intellectual dissatisfaction but on behalf of new devices by means of which the atom had, for several years already, been presented and interrogated.

It is not without interest that the author is the great-grand-son of the great chemist Alfred Baeyer, whose name is associated with a construction that translates the scandalous success of the "transition" leading from delegation to attribution. Baeyer suc-ceeded in representing chemical molecules as commonplace constructions consisting of balls (atoms) connected to one another at specific angles by means of rigid rods. Of course, the rigid rod doesn't tell us anything about a chemical bond any more than the ball tells us about an atom. But Baeyer's model was the first to inform chemists about the nature of the agents they delegate whenever they "make" chemical bodies "react," and therefore the kind of actor the chemist's practice uses and addresses. Could the molecule be conceived as an assemblage in space? Yes, it could. And it's this "yes" that celebrates the success of delegation leading to attribution, the fact that the assemblage satisfies the requirements of the chemist who uses it to imagine possible synthesis operations.[5]

The great-grandson of the man who "caused" molecules capable of explaining their operations to chemists to "exist" tells us of his excitement when he was confronted with the first device that enabled him, not to record the emission or absorption spectrum of the light of a multitude of atoms, but to "see an isolated atom," a "captive" atom, caught in a cavity under vacuum, where it was held motionless by electrical forces and activated by an ultraviolet laser.[6] The laser is turned on: "Right in the middle of the trap, a little star appeared. Tentatively at first, amid the flickering reflections all around, and then with increasing intensity, the mercury atom poured out its light . . . So here it was, an atom in captivity. It was then, as I watched spellbound, that I began to notice that the atom was blinking. At first I thought that this was just part of the general flickering of the screen, but it soon became apparent that the mercury atom was definitely turning off and on, at the rate of several times a second. This was surely the most astonishing thing I had ever seen. Whatever might be the cause of this phenomenon, it was a powerful reminder that atoms are active, dynamic systems, capable of the most intricate internal transformations and convolutions, and not in the least bit like the immutable, eternal kernels of matter the ancients had imagined them to be. Although I understood this difference intellectually, it took the impudent winking of a trapped mercury atom to drive the point home to me in an unforgettable way."[7]

I have chosen a quotation dominated by "subjectivity," for it embodies, more than any theoretical expression, the novelty of contemporary laboratories, where the "microworld" has obtained the means to become perceptible in a variety of ways. In fact, there is no theoretical novelty, in itself, in this becoming perceptible. Von Baeyer's mercury atom "responds" to quantum expectations, and also corresponds to what was envisioned as a thought experiment sixty years ago. And yet, as von Baeyer's text shows, they expose physicists to a new experience, for which quantum formalism had not prepared them. It is only indirectly that quantum mechanics was able to construct the question of the atom on the basis of detection experiments involving multitudes of atoms (electrons, molecules). But it is the atom itself that is, today, in the process of being tamed or domesticated.

In this case, domestication should not be understood to mean something like "good-bye wild freedom, hello sad captivity." Becoming perceptible means that the atom had obtained the means to impose itself as an actor, and an active one at that, flickering on and off "for real." We should view domestication in this case as the beginning of a practice wherein a *domus* is created, a suitable habitat, which helps keep the studied being "alive," helps "influence" it, multiply the connections with it.

The atom of quantum mechanics was a being that had been unable to resist its interactions with its milieu from being absorbed into the theoretical question of measurement. The fact of being able "to see it" by means of a subtle manipulation of those interactions, through the creation of a "habitat" where the atom is able to indicate its individual existence, is less about the capture of the atom than the capture of experimental physicists. Von Baeyer is literally *captivated* by the mercury atom's "impudent wink."

Today, when untold dollars, hours of work, and subtle negotiation among all the laboratory ingredients have resulted in an atom that "displays" quantum jumps, that atom has gained the power to "provoke" questioning of the value judgment that resulted in the destruction of the original symmetry between stationary states and quantum jumps. A captivated von Baeyer protests: this active atom requires that we describe it as such, that is, as an actor. The theoretical syntax that causes quantum jumps to disappear in favor of measurement becomes not only intellectually but practically paradoxical because the laboratory has, by its own means, brought about the syntactic transformation that attributes responsibility for an activity to a being. Through technology the atom has become capable of requiring that we describe it as endowed with that activity. And this technical-syntactic transformation is twofold: it "causes" the atom "to exist" as a "being," comparable to a flower outside the lab, and it does so with respect to an environment that becomes inseparable from its existence. The herbarium or the test tube wherein its biochemical constituents are identified have never been the habitat of a "flower"; its habitat makes it the being of air, light, water, and earth whose existence as an actor is witnessed by those who cultivate it. The "quantum vacuum," laser light, and subtly calibrated electromagnetic fields have become the habitat of the atom. They no longer belong to a detector that "actualizes" a possibility of measurement, but to an "environment" that is an integral part of the atom's existence.

Von Baeyer's flickering atom is not, in itself, capable of refuting the way in which quantum mechanics represents the matter of fact it corresponds to. But it is, literally, impudent, no longer a "veiled" reality. It doesn't dictate how it must be described, but it transforms the laboratory into a "piece of the world," where it exists, like a flower. An enormous amount of knowledge and technology has resulted in providing it the means to claim an existence independent of our knowledge and our technical means of investigation. Correlatively, the question of actualization is decentered. Actualization cannot be reduced to the transition from indeterminacy to known determinacy, that is, to the recording of values that correspond to well-defined observables. Nor can the question of what to actualize be a matter of choice for the physicist. Actualization becomes inseparable from the mode of existence of the "quantum actor" with which new laboratories learn to interrelate.

It is not impossible that in these labs, where a new "culture" of the atom in the strong sense is born, questions can arise that create immediate problems for quantum mechanics. Contrary to the strictly academic experiments that challenged hidden local variables, it is from industrial labs that "new facts" might arise.[8] For the questions raised in these labs are associated with new interests. As von Baeyer explains, issues that IBM is deeply concerned about are based on the answer to the "forbidden" question: how does the flux that crosses a superconductor "really" behave? "What was once a philosophical debate among the high priests of physics is becoming a bread-and-butter issue. Ordinary scientists in the industrial laboratories of the world are starting to wrestle with the meaning of quantum mechanics. Until they understand whether the flux actually exists in a particular superconducting device, they cannot fully explore its practical applications. Technological development, which has profited richly from the well-oiled machinery of quantum theory, has come up against a barrier, making it necessary to reach inside the machine to see what makes it tick."[9]

There's nothing ironic about experimental delegation, but it creates effects that do not lack humor. There is humor, for example, in the creation of a "partial delegate." The delegation is relative to a practice, yes, like Adolf Baeyer's erector set, but, like the erector set, the delegate is also capable of becoming the interlocutor with which the researcher thinks and invents, wondering what "makes it tick." There is humor in the interconnection between "scientific" delegations and "technique": the thought experiments used to challenge Bohr's prohibition against assigning the value of an observable to a system "in itself" have now been resumed "for real." What is at stake is no longer testing the link between formalism and the prohibition. From the point of view of the industrialist, "behavior in itself" loses its epistemological connotation. It is a question of "mobilizing" a being we would like to ask to carry out, "on its own," the operations that interest us. Finally, there is humor in a situation where the "reality in itself" claimed by critics of the Copenhagen interpretation so they might have the right to escape the stifling confinement of their own questions, the right to describe the laboratory as the place where an encounter with the world is organized, appears in laboratories where technical confinement is pushed to its extreme limits: the atom *and its world* are now confined. The laboratory has become capable of opening up to the world to the extent that the physicist can delegate to her apparatus the role of "world creators" for quantum beings that are no longer required to state where they are, or what their velocity or spin are, but are asked "to act as if we are not there," that is, to activate themselves within their "own" surroundings.

Quantum irony has triumphed over its critics. The formalism is "complete" and is capable of trapping all those who undertake to attack it head-on, who try to demonstrate that it is possible to overturn the prohibitions it defines. But "complete" does not imply that it is invulnerable or unchangeable. As Léon Rosenfeld repeatedly pointed out, "complete" means endowed with internal consistency, so that any question it allows finds answers devoid of ambiguity, and so that all the questions it says are impossible actually turn out to be impossible to ask. But everything depends on what we mean by "asking a question." Quantum irony is omnipotent in the sense that it goads its critics into falling into the trap it has set for them. "What question have you decided to ask?" Anyone who accepts the challenge, by that very fact, places herself in the position expected of her. She will, in effect, accept the position of "someone who asks questions," the position that gives measurement a determining role.

The atom that flickers in captivity might indeed inspire a shift in the very relationship between the one who questions and the one who answers. For example, "Where are you?" or "What are your energy states?" could give way to the question "What can we expect of you?" addressed to a potential delegated agent.

However, we need to return to the—heretofore irresistible—question of the ironic trap constituted by the quantum "divinatory factish." For the syntactic torsion that transforms a detector into a condition for the actualization of significant information (the transition from Bohr's first model to the quantum mechanics of 1927) was indeed endorsed by the majority of critics, those who initially sought above all to transform the scope and meaning of that information.

Here, we can use the present to "relativize" this past. If physicists in 1927 had, like von Baeyer, "seen" a mercury atom wink at them impudently, I assume they would never have gotten interested in a formalism whose primary feature is to deny the possibility of referring to that wink as an event, as something that happens and can be detected. Nonetheless, they had indeed "seen" the frequency spectrum of the light emitted by atoms. The radical transformation of theoretical syntax that destroyed the possibility of talking about the atom as "detectable" (that is, as an actor), but only to the extent that it is "detected," may have resolved the difficulties that plagued the Bohr model and resulted in impressive experimental successes. However, it is not enough to explain why it imposed itself as if it represented progress without a price, as if the quantum mechanics of 1927 spoke the "truth" about the model devised in 1913. In other words, quantum formalism, complete and endowed with formidable internal consistency, is also revealing about the value judgment that helped physicists forget the atom-actor of 1913.

The actor of 1913 was a "kinetic" actor. The events that affected it were characterized in terms of their frequencies, and some of them, notably the spontaneous transitions from

an excited state to a state of lower energy, seemed able to lay
claim to an "acausal" initiative. Like the unstable nuclei found
in radioactivity, these excited states were characterized by a
"lifetime," which is another way of stating, positively, that they
could not be subjected to the requirements of a function, of a
representation that would identify the variables used to predict,
control, or determine the electron's "return" to its fundamental state.[10] Certainly, the Bohr-Sommerfeld atom gave an unexpected role to the dynamics of integrable systems, because the
stationary orbits of electrons were characterized very simply
and elegantly by the invariance of the action variable J. But the
Bohr atom didn't obey mechanics. On the other hand, as will be
described later on in greater detail, Schrödinger's wave function
did "obey" a form of mechanics—a modified form—but one that
preserved the essential attributes of Maxwell's Queen of Heaven.
Correlatively, the definition of the observables associated with it
privileged those with a clear mechanical analog (energy levels),
not those that referred to the kinetic actor.

If, as I speculated, the future of quantum mechanics entails
the flickering mercury atom and the involvement of industrial
labs, it may also entail the question of how the privilege conferred upon the "attributes of the Queen of Heaven" played a role
in theoretical practices. The laboratory-as-habitat, because it
"shows" rather than "detects," gives "that which" is detected the
power to insist. But wouldn't kinetic description, initially disqualified as an "incomplete" system of dynamics, then by quantum theory's emphasis on "mechanical" properties also have the
power to insist? Wasn't a measurable property like "lifetime"
also relegated, along with atomic flickering, to the status of an
ectoplasm in the ironic construction of a "matter of fact" that
privileged Hamiltonian facts?[11] To investigate this further, to
measure the theoretical effects brought about by the seductions
of the Queen of Heaven, will require a bit of technical analysis.

5

The Physicists' Double Standard

Once upon a time there was a philosopher. Her name was Nancy Cartwright and, in 1975, she published, like many others, "her" analysis of the problem of measurement in quantum mechanics. This was almost an obligatory step for American philosophers of science and, like many of them, she decided to proceed on the basis of the contrast between the law of evolution to which the Schrödinger wave function was subjected and the reduction of that wave function, which supplies measurement with its meaning. Eight years later, that same philosopher published a book—*How the Laws of Physics Lie*—whose title vividly reflects her discovery that she, and all those who accepted the measurement problem as a problem, had been misled.[1] The last essay in her book, "How the Measurement Problem Is an Artefact of Mathematics," completes the act of rebellion: the problem that critical philosophers and physicists have been struggling with for generations is a false one.

To speak of the "problem" of measurement is to have already abandoned the Copenhagen interpretation for the one proposed in 1932 by the mathematical physicist John von Neumann.[2] As I noted earlier, for Niels Bohr the wave function determined by the Schrödinger equation lacked all physical meaning independently of its "reduction," which actualized a choice of measurement, the wave function then allowing a correspondence to be made between this choice and clearly determined probabilities of observation. The Schrödinger equation, then, has nothing to do with the description of a given physical reality, even though it "resembles" the equations of Hamiltonian mechanics, which cause description and reason to coincide. It is part of the physicist's "divinatory equipment." Von Neumann's analysis, on the contrary, contrasts the determinist and reversible "behavior" the Schrödinger equation defines with the discontinuous and irreversible event associated with its reduction. On the basis of this contrast, he asked: is it possible to expect the equation to provide the secret of its reduction? The answer is no, it's impossible. And it is the consequences of this impossibility that von Neumann's interpretation pushes to the limit. Because any measurement device, including the retina upon which a photon falls, including Schrödinger's cat, could, in principle at least, be described by a wave function governed by the Schrödinger equation, none can be held responsible for its reduction. Then, what is it that makes something a measurement device? Von Neumann saw only one place where his regression operation could be stopped, where what appeared to be responsible for the reduction would not in turn prove to be itself subject to reduction: this would be the consciousness of the observer.

Bohr had spoken of a logic embodied in a concrete device, one whose characterization includes its purpose and, therefore, the way it will be used. His favorite example was that of a stick held in the hand. Depending on whether it is held tightly or loosely in the hand, the stick satisfies two logically incompatible definitions concerning the information it will provide by tactile exploration. Thus, we are forced to choose between two meaning-producing uses.[3] As for von Neumann, he associates the finality of the device with pure conscious choice, arguing that every device, if it is considered independently of consciousness, belongs to the same physical reality as the quantum object itself.

The problem of quantum measurement, as it is usually presented, begins from the position established by von Neumann but tries to escape his conclusions regarding the decisive role consciousness would play in reducing the wave function. Technically, it centers on the contrast between the wave function as a *superposition* of quantum states and a statistical *mixture* of systems, each characterized by a single quantum state (and therefore the possibility of predicting the value of the corresponding observable with probability 1). Measurement on an ensemble of quantum systems results in assigning to each system one, and only one, value of the measured property. The result appears as a mixture of distinct, single-state systems. It is as if the measurement had effected the transition from a "pure case," where all members of the ensemble are described by the same superposition of states, to a "mixture" of systems, each with a separate identity, that is, each described by a single-state wave function corresponding to one, and only one, value of the observable. Can we give a theoretical interpretation to this transition, which cannot be understood in terms of the Schrödinger evolution?

"Quantum mechanics requires a superposition: the philosophical problem is not to replace it by a mixture, but rather to explain why we mistakenly believe that a mixture is called for."[4] This is the way Nancy Cartwright presented the problem in 1975. She indicated how she wanted to see it resolved: the "transition" would be explained away as an illusion. The Schrödinger evolution, whose subject is superposition as such, would govern the situation. To construct this solution, it is necessary to accept, like von Neumann, that the Schrödinger equation has unlimited validity and, in particular, obligates us to define the measurement instrument in terms of superposition. But, contrary to von Neumann's position, the analysis would not be reduced to the axiomatic claim that the atom and the measurement device, whenever abstracted from its aim, are the same thing—just more complicated. The crucial point would be to explain why measurement results *seem* to imply the destruction of superposition, the reduction of the wave function to a mixture of single-state wave functions. Why do we maintain the fiction that we can assign a fixed value to an observable, even a well-constructed one?

From this point on, measurement becomes a physical problem, an interaction whose specificity needs to be defined. As Léon Rosenfeld wrote, "we should represent measurement as an interaction between the observed atomic system and a recording device, an interaction that ultimately leaves some permanent mark unequivocally associated with a definite quantity that characterizes the state of the quantum system."[5] Speaking about a permanent mark points to a way of defining the specificity of the measuring device. It should be such that the measurement interaction can be described in terms of an "evolution toward equilibrium." Therefore, it is to statistical mechanics, a science that articulates equilibrium in the thermodynamic sense with dynamic or quantum description, that we might turn for the secret of the apparent reduction.

Like Cartwright, Rosenfeld referred to the results proposed in 1965 by three Italian physicists, Adriana Daneri, Angelo Loinger, and Giovanni Maria Prosperi. They had shown that when considering measurement, now qualified as "macroscopic," as an evolution toward the equilibrium of the "measuring system," whether the point of departure for this evolution was a "pure case" or a "mixture" made no discernible difference. From the moment the relative proportions of the different systems—each characterized by one, and only one, quantum state—constituting the mixture correspond to the respective probabilities of the possible measurements that can be calculated from the wave function for the pure case, the result of an ensemble of measurements will be the same whether for a pure case or a mixture. The "reduction" of the wave function would then be nothing other than the result of the fact that the evolution of the "measuring system" toward equilibrium allows an approximation that eliminates all information capable of indicating the difference between a superposition governed by the logic of "and" and a statistical ensemble governed by the logic of "or." We are mistaken if we think that, once measurement has taken place, we can assign one, and only one, quantum state to each quantum system—that is, whenever we think that the measurement has transformed the pure case into a mixture, or whenever we think that our "measurement" allows us to associate one, and only one, value to what we measure. But, concluded Cartwright in 1975, this error is of no consequence. It will never lead us to make a prediction that would contradict a possible measurement because our measurements are, by definition, macroscopic, and relative to the state of equilibrium attained by the measuring device.

What happened so that, a few years later, Nancy Cartwright would reject what is in effect a conciliating solution? Why did she suddenly no longer feel "obligated" to give the wave function the power to produce the theoretical understanding of measurement? In the interim she had taken a highly unusual step for a philosopher. She examined not only the literature in which problems of interpretation were discussed, but also the texts that transmit the procedures accepted by the community and from which physicists learn their craft. She came to an astonishing conclusion. When the specialists of quantum mechanics were no longer concerned with the fundamentals but were building the "real" foundations of quantum mechanics, that is, its ability to occupy its corresponding experimental terrain, they no longer felt obligated by the mathematical properties of quantum superposition. More specifically, they used those properties for part of their calculations, but, at a given moment, introduced a judicious approximation that destroyed the superposition.

Worse, Cartwright discovered, quantum theory did not establish the point in the calculations at which the approximation could be introduced. The physicists were guided by experimental data.

In fact, Cartwright found that the "reduction of the wave function" is not even specifically associated with measurement. Physicists "need" the reduction of the wave function not only when they make a measurement but every time they "prepare" a system, whenever, for example, they produce a beam of electrons in a linear accelerator that are supposed to be characterized by a clearly determined quantity of motion. "This kind of situation occurs all the time. In our laboratories we prepare thousands of different states by hundreds of different methods every day. In each case a wave packet is reduced. Measurements, then, are not the only place to look for a failure of the Schrödinger equation. Any successful preparation will do."[6] Moreover, this failure occurs whenever it is a question of accounting for a process that presents the problem of its preparation rather than its detection. For instance, when physicists produce laser radiation, they make use of equations suggesting that transitions "actually" occur, similar to events that are indifferent to whether or not they are being observed. And if those equations, which are kinetic in nature, are to characterize events, the interference terms resulting from the quantum superposition have to disappear, as in the reduction of the wave function. Of course, those terms are necessary for correctly presenting the problem, Cartwright notes, *but their disappearance, through the intervention of an appropriate approximation, is just as necessary* if the problem is to have a solution. In other words, for the vast majority of "experimental confirmations," which add their prestige and authority to quantum laws, there is no theoretical deduction at all of the value of a measurable property that would confirm the measurement. Rather, there is *negotiation* between the way the "laws" present the problem and the type of solution required by the experimental data, between theory and "phenomenology."

The "lie" associated with physical laws indicated by the title of Cartwright's book indicates that she has discovered that, even in classical dynamics, "laws" do not explain phenomenological regularities. In all concrete cases, the law requires additional details that provide it with its relevance for experimental facts, but the crucial nature of those details, which the law needs but does not justify, is "forgotten" whenever we claim that the law "explains" the fact.

Cartwright recognized the affinity of her position with that of Pierre Duhem, for whom theoretical structures were, above all, instruments for ordering and classifying experimental accounts, but she doesn't follow him in his fight against theoretical entities themselves. This accurately reflects the transformation of phenomenological and experimental practice within the century. Atoms, electrons, and protons "really" exist in the sense that they are not simply presented by theory but are agents delegated by practice. When Rutherford subjected a gold leaf to the "radiation" from a radioactive source, there was no possibility of understanding what he obtained other than by introducing entities capable of causing change and undergoing change, that is, no possibility of describing what he did other than as a showering of his target with discrete entities that, under the circumstances, behaved like tiny projectiles. For Cartwright, an entity used in an experimental device to provoke or create the new phenomenon being studied obligates the experimenters, by that very fact, to endow it with existence.[7] "Laws," as Duhem claimed, "save" phenomena. But, Cartwright adds, most likely because, unlike Duhem, her physics is populated with detectors, the entities such laws introduce must be recognized, *independently of those laws*, as integral parts of the phenomenon. And because of those detectors, physicists and electrons coexist in the strong sense, the way the plant coexists with the sun, or Europe, since Columbus, with America.

Cartwright recognized the autonomous existence of what I

call "experimental factishes," beings whose existence experimenters accept because they are able to make them do things in the laboratory. The physicist who manages to delegate "puts into practice" an "agent" and, therefore, also brings into existence an actor, if the agent has the power to present the physicist with problems "of its own." On the other hand, Cartwright feels free to consider as "liars" what I have called "theoretical factishes" (physicomathematical, enigmatic, and divinatory). These intervene whenever theories "judge" experimental practices and refer them to a "reality" that assigns them one role, that of access to its own—now discovered—(theoretical) truth. Of course, there are so many causes to be defended and value judgments encumbering and shaping this assignment that it seems possible to associate it with a merely social mode of existence. But with her notion of the "lie," Cartwright may be overlooking the fact that those "factishes" are themselves capable of posing problems of their own to those who bring them into existence. More specifically, her position makes her vulnerable to the temptation to disqualify such problems for being *false problems.* It might be better to say that it is a question of entirely different types of factishes and of their respective modes of evaluation. Experimental delegation, which seems to be part of a concoction made in a laboratory kitchen, may ultimately interest everyone through the technical devices that continue its practice, while the theoretical factish, which is assumed to interest everyone, which is offered as the focus of the public's (and the philosopher's) fascination, gets its power solely from the passion that singularizes a physics in which Duhem appears as the vanquished.

I want to return to the case of quantum laws, where Cartwright can indeed speak of lies because she has been forced to take sides by the way these laws were presented. For Cartwright, the "philosophical" problem of measurement is a false problem, an "artifact" purely and simply, created entirely from mathematical formalism. Those laws are not limited to "lying";

like all the laws of physics, they invite lies. Not only do they create the temptation to lie, as we have seen, they appear to impose it upon those who refer to them. For, the physicist "at work" who attempts to describe laser radiation or design a del-egation device must use the Schrödinger equation "as if" it had descriptive value, but is also forced to cheat, to have recourse to approximations that enable her to clandestinely deny what the equation asserts openly.

Cartwright introduces a problem that must be understood in terms of obligations. If physicists at work are not obligated by quantum laws, if they feel free to introduce the needed approxi-mations, why should those laws claim an authority that would create obligations for others, for philosophers and for those who are mesmerized by the question of knowing if and when Schrödinger's cat "really dies"? What we hear in the accusa-tion of lying is not the repetition of Schrödinger's "cry" when he insists that the cat really dies. His cry dramatizes the disastrous shortcomings of Cartwright's 1975 solution, for that solution amounts to saying: what "happens" to the cat, a macroscopic measurement instrument, does not matter; what matters is that what we call the cat's "death" is a well-constructed approxima-tion; what matters is that the terms that translate the fact that the "dead" cat is in superposition with a living cat make no mea-surable difference *statistically*. The accusation of lying, rather, follows from the realization that the whole question of the cat dissolves once we turn to what physicists do at work, construct-ing tools and accepting obligations entailed by questions far less "strident" than the question of what happens to the cat.

It is by presenting the contrast between, on the one hand, the "quantum event" physicists at work continue to have need of—especially that "primordial" event, the decay of an excited atom, by which the atom forces us to recognize its nonclassi-cal electronic structure—and, on the other hand, the "simula-crum" to which that event corresponds in post-1927 quantum mechanics, that Cartwright answers Schrödinger's "cry." There is no need to cry about the cat; rather, let us cry about the absurdity of this simulacrum, about the absurdity of the theoretical story of a decaying atom.

"On this story nothing happens. In atomic decay the atom begins in its excited state and the field has no photons in it. Over time the composite atom-plus-field evolves continuously under the Schrödinger equation into a superposition. In one component of the superposition, the atom is still in the excited state and there are no photons present; in the other, the atom is de-excited and the field contains one photon of the appropriate frequency. The atom is neither in its outer orbit nor in its inner orbit, and the photon is neither there in the field traveling away from the atom with the speed of light, nor absent. Over time the probability to 'be found' in the state with an excited atom and no photons decays exponentially. In the limit, as $t \rightarrow \infty$, the probability goes to zero. But only as $t \rightarrow \infty$! On the new-quantum-theory story, never, at any finite time, does an atom emit radiation."[8]

Of course, this cry could be silenced by a physicist who responded to Cartwright that time as presented by the Schrödinger equation has nothing to do with any physical temporality associated with the phenomenon. It's relative, like everything else, to the time of measurement. This answer would be that of an authentic defender of the Copenhagen interpretation, for whom the problem of the "reduction" of the wave function is a false problem because the wave function doesn't correspond to any story, is nothing without its reduction. Quite simply, Cartwright doesn't have the right to use the Schrödinger equation to "narrate" what happens between the initial moment (excited atom) and infinity (photon emitted). But this time, and this is what is interesting about Nancy Cartwright's position, the discussion doesn't end here, with the silence of its protagonists—the "ascetic" silence of Bohr's disciple, and the frustrated silence of the one who "knows very well, but all the same." The trap no longer functions with its two-stage irony, creating the temptation to use the Schrödinger function to "describe," then disallowing the description. Cartwright has entered the physicists' "kitchen," she has read what philosophers don't normally read, the "technical details." She refers to their practices rather than to the "masterpieces" they expose in public.

In order to deny any physical meaning to the Schrödinger equation and relate it to an ascetic encounter with the experimental data that are the sole source of legitimacy in physics, those data must correspond precisely—and here the cases are assimilable to "masterpieces"—to the privileged observables that are the quantum analogs of Hamiltonian variables. But the fecundity of quantum mechanics makes reference to all kinds of situations that offer no such analogy. And it is for such situations that, notwithstanding the Copenhagen interpretation, physicists have to behave "as if" the Schrödinger equation did indeed describe an objective situation. They preserve the syntax of the equation, which prevents them from bringing into existence the atom, the field, or the electron independently of measurement, but use it to introduce agents that they treat as if they existed independently of measurement. It is in the course of this delicate operation that the "kitchen concoction" of physical-mathematical approximation intervenes. It will enable the situation they interrogate to preserve, where it must, the relevant quantum features, and, again where it must, to conveniently ignore others as insignificant. For example: approximation will result in assigning a lifetime or a cross section to a particle, so a kinetic description can be provided, or approximation will allow a clearly defined role to be attributed to an agent. Whenever they need to create an experimental situation where a quantum being is "delegated" to causing events, physicists no longer feel obligated by the asceticism they propose, to the fascination of philosophers. Physicists are playing both sides of the coin.

That's why Nancy Cartwright dares shift her position from that of a philosopher, who comments and tries to understand, to that of a "physicist," in the sense that one must be a physicist, in a technical sense, to be part of the controversy with other physicists. For her, quantum language is not only "deceitful" in the sense that, in general, all theories lie, it is "distorted" in the sense that it is prone to present "bad" observables as significant. We make use of a screen to claim "the electron is here," and the reduced equation seems ready-made to provide us with the probability that it is actually there. By recording a photon "emitted" by an excited atom, we can claim that the atom is now in its stationary state, and in the Schrödinger equation the probability amplitudes corresponding to the different states are effectively in superposition. But we "distort" measurement when we use it this way, making it bear witness to something about which it is, in fact, silent. Measurement never bears witness, Cartwright notes, to a "state" defined in terms of energy, position, momentum, or any other observable to which Schrödinger probability amplitudes for a state would correspond. It witnesses *transitions,* and associates with transitions properties, such as lifetimes, or cross sections (in the case of scattering experiments), that do not belong to the same physical syntax as the observables to which the definition of quantum states corresponds.

Why, Cartwright asks, has this tension gone unnoticed by critical philosophers? "They want to find some way to ensure that a quantum system will possess values for all the classic dynamic quantities—position, momentum, energy, and the like. But this motivation is ill founded. If we want to know what properties are real in a theory, we need to look at what properties play a causal role in the stories the theory tells about the world. This is a point I have been urging throughout this book, *and it makes a crucial difference here.* Judged by this criterion, the classical dynamic quantities fare badly. The static value of classic dynamic variables has no effect; it is only when systems exchange energy or momentum, or some other conserved quantity, that something happens in quantum mechanics."[9]

It is worth remembering that Bohr asserted the "acausal" nature of quantum physics, but he gave the term "cause" the same meaning it has in dynamics, one in which change observes the conservation of cause in effect. When Cartwright speaks of the "causal role" that allows us to refer to a property as real, she emphasizes the concept of a "role." She asks that the physicist who brings experimental devices into existence construct a coherent theoretical language capable of "telling stories" about what happens. In other words, here, the meaning of a cause is found in the practice of delegation, in the staging that simultaneously brings into existence an agent, the story of its production, and that of its action. From this perspective, dynamic "cause" becomes a unique case, which structures "stories" whose common feature is the role they give to the = sign. But these stories, like all others, belong to the process of delegation. Galileo asked the rolling ball to bear witness to the fact that, from the viewpoint of what its motion makes it capable of, the only thing that counts is the change of height that occurs during that motion. In Hamiltonian physics, the extraordinarily simplified nature of the "story of what happens to the ball" is used to make its monotonic nature explicit: there exists a point of view that allows us to state that, no matter where the ball goes, *"nothing" happens to it.*

The tension Cartwright points out between the "causal role" quantum experimental devices confer on energy exchanges, such as the spontaneous transition of an excited atom to its fundamental state (which "causes" a photon to be emitted), on the one hand, and the theoretical syntax of quantum mechanics, focused on variables analogous to those of dynamics, on the other, is even greater than she describes. Take the mathematical simulacrum of atomic decay as Cartwright characterizes it: what the Schrödinger equation "describes" is nothing other than the time evolution of the superposition of the excited-atom-plus-field-without-photon and de-excited-atom-plus-field-with-photon, a photon whose frequency can be measured. But such a simulacrum is already itself a *compromise solution*. If the super-posed probability amplitudes evolve over time, it is only because the initial representation involves an atom described as if it were "perturbed" by an electromagnetic field, that is, with ref-erence to an atom that would be "truly isolated." Such an atom corresponds to a Hamiltonian, H_o, in the Schrödinger equation, and its energy levels will be characterized in terms of *constant* probabilities over time. In other words, it will never decay. The fact that the equation for *the atom perturbed by the field* allows us to define the evolution of an "emission" probability—the prob-ability that a photon will be measured—is therefore directly dependent on the way the problem has been presented. It is because the interaction between atom and field is defined as a disturbance of the atom by the field, and not as a unique sys-tem defined by a Hamiltonian that would include the field, that the probabilities evolve over time. And this choice of defini-tion, even if in general it is the only one practical, is nonetheless "only practical" from the point of view of quantum mechanics. Indeed, from this point of view, "it should be possible" to con-struct the Hamiltonian of the "atom-plus-field" system in such a way that, like the isolated atom, it is represented by a superpo-sition of stationary states each of which evolves individually in what is called *Hilbert space*.[10]

Until now, I have limited myself to claiming that quantum mechanics had extended Hamiltonian dynamics by transform-ing it, but I didn't specify what that transformation consisted of. That I have now introduced the quantum Hamiltonian and Hil-bert space indicates that I need to say (just a little) more about them. In a field such as physics, the telltale sign that questions are gradually gaining in relevance is always the same: it is found in the transition from "ideas" to "technical details," that is, the singular mode of existence of the beings the physicist brings into existence. That's why I have chosen to discuss the approach taken by quantum mechanics in reverse—by not supplying the "necessary technical elements" and moving directly to their "consequences for each and every one of us." I wanted to begin with the "grand problems" and end up by asking, "What game are physicists playing?" The moment has come to describe the strange (official) rules of that game.

6

The Silent Descendant
of the Queen of Heaven

Let there be a quantum system represented by a wave function ψ. This statement, which is extremely commonplace, can refer to the universe as a whole in certain areas of theoretical cosmology as well as to "Schrödinger's-cat-unstable-atom-poison-triggering-device." It's as if the statement were free, presenting no problem of relevance, as if we could require any portion of reality to respond to a quantum description. It is important to recognize that, in his own way, it was Niels Bohr who led quantum physics down this rather strange rhetorical road. Von Neumann's axiomatic interpretation, by which any measuring instrument—in fact, anything (except the comprehending mind)—must be able to be represented by a wave function, followed the path of Bohr's thought experiments. Bohr tried to show that the identity of a measuring instrument inevitably implied a choice, a way of using that instrument, and that the complementary measurements satisfied logically contradictory methods of use. To do so, the physicist had to behave "as if" what was to be the quantum object, represented by a wave function, and what was to be the instrument responsible for the reduction of the wave function *remained perfectly indeterminate until the moment of choice*. The question of the limits of relevance of the wave function, therefore, could not even be raised within the context of Bohr's thought experiments.[1] That is why these experiments helped give the quantum "divinatory factish" a scope that was limited only by the "consciousness" of the observer. I now want to show, referring to von Neumann's axiomatic formulation, which provided quantum formalism with its most purified form, what this unlimited scope means. In writing the simple Greek letter ψ, the physicist makes claims compared to which those of Laplace's demon, who observes the universe as a system of interacting masses, are trivial.[2]

At the center of this axiomatic formulation, which extends and transforms the dynamic Queen of Heaven, we find a fundamental couple, the wave function we have already encountered, and the quantum operator, which "acts" on that function.

The operator is a new being that was invented by quantum mechanics but is not limited to it. So, for today's physicist, every time an evolution equation unequivocally associates the variation over time of a function at a given moment with the value of the function at that moment, the connection determined by the equation can be translated into an "operator/function" formalism. We could then say that the evolution over time of the function f is given by an operator—let's call it G—acting on f, which is then written: $df/dt = Gf$. In general, as its name indicates, an operator "operates" on a function, that is, transforms it into a different function.

The creation of the concept of an operator in quantum mechanics translates, in mathematical form, the difference between the "representation" of a system by a function—which is the case for classical mechanics—and the ability to characterize that system with observable values using a wave function. In quantum mechanics, there is an operator for each observable, but this correspondence has a precise physical meaning only if the wave function on which the operator acts is itself expressed in terms of what are known as "eigenfunctions" (or

"eigen states") of that operator. The particularity of these eigenfunctions is that the operator of which they are eigenfunctions
does not transform them into other functions. The action of the
operator on one of its eigenfunctions yields that same function
simply multiplied by a number. That number is an eigenvalue of
the operator. And it is these eigenvalues, finally, that supply the
measurable values of the observable with which the operator is
associated.

The quantum wave function or, more precisely, the superposition of eigenfunctions by which it can be represented, is
defined as an element of "Hilbert space."[3] For von Neumann,
this Hilbert space, which quantum functions "inhabit," has
become the true subject of quantum theory. We could say that
quantum operators transform *an element of Hilbert space into
another element of Hilbert space.* The way in which they are able to
transform it comprises the rules of the quantum game, the syntax that governs its statements, that is, determines those statements to which meaning can be assigned.

What's the relationship between this syntax and the question of "choice of measurement" that served as Bohr's leitmotiv?
We accept, axiomatically, that a system's "quantum state" is represented by a ψ function in Hilbert space. As such, this function
is physically silent. It has a mathematical existence, but it says
nothing about the possible results of measurements that will
actualize the physical existence of the system. In order for the
function in question to be able to "speak," it must be defined as a
superposition of eigenfunctions of the operator that corresponds
to the proposed measurement. The "choice" of measurement
instrument, therefore, corresponds to the "choice" of representing ψ in terms of this (infinite) set of eigenfunctions rather
than in terms of another family of eigenfunctions.[4] More specifically, the "state" in question will be represented by a superposition of eigenfunctions of the operator, each superposed
eigenfunction being "weighted" by a numerical coefficient.

The set of these numerical coefficients corresponds to the well-known "probability amplitudes" that must be "reduced" so the respective probabilities of the different possible results of a measurement can be calculated. This reduction is obtained by squaring the numerical coefficients associated with the eigenfunctions of the superposition. As a result of this operation, the physicist is able to associate each eigenvalue of the operator corresponding to her measurement with a probability, that is, she is able to determine the respective probability of the different values this measurement will produce.

For each operator chosen, the wave function is represented by a superposition of eigenfunctions (or states), each of which is associated with two properties of interest to physics: the probability amplitude and the eigenvalue. Choice of measurement/choice of operator/choice of representation in terms of a superposition of eigenfunctions in Hilbert space—they are three aspects of one and the same choice. Independent of this choice, the wave function is silent.

And yet, the wave function is not entirely silent, for we know it is governed by the Schrödinger equation, which determines its evolution over time. It is here that we encounter an old acquaintance, the Hamiltonian, which has now become the "Hamiltonian operator." And it is with the definition of the Hamiltonian operator that an axiomatic formulation apparently devoid of "physical meaning" (the four previous paragraphs have only proposed formal definitions) assumes a very precise meaning. Even if, from the ascetic viewpoint of the Copenhagen school, Hilbert space, which shelters the Hamiltonian operator, may indeed have only mathematical significance, nonetheless it bears a message that has the power to lure physicists away from that point of view. How could they resist conferring upon Hilbert space the status of an autonomous "factish"? For it is the direct descendant of Hamiltonian dynamics and, moreover, belongs to the line of the royal family itself. Hilbert space is not limited to claiming the power of equivalence, that is, of governing how the problem of quantum evolution is constructed. The Hilbertian definition of quantum beings, eigenfunctions and operators, implies the ability to characterize any "quantum system" in terms suitable only to integrable systems as understood by Poincaré, that is, an ensemble of autonomous entities evolving over time independently of one another.

In other words, von Neumann's axiomatic formulation elevates the ambiguous irony of quantum formalism to a nearly perverse level. For Hilbert space, where "quantum states" can be represented, irresistibly calls forth a presentation of the situation where the wave function appears as "representing" the quantum system. A "Hilbertian realism," which assimilates Hilbert space to a "physical-mathematical truth" that transcends the phenomenal world the way Einstein's four-dimensional space-time transcends our world, where space and time are separate, could, therefore, almost be said to be dictated to the physicist (and the philosopher). We can now "forget" the ascetic stipulation passed down from Copenhagen: abandon all representation, Hilbert space is purely mathematical; physically, it is nothing without the reduction of the wave function. We can forget it because physicists continue to forget it the way one forgets the "voice of one's conscience"—"Yes, I know, but all the same." Here, my purpose is not to relay this voice and admonish physicists. Rather, I wish to follow the way in which problems that currently fascinate them have been constructed. We must acknowledge that physicists have indeed been "captivated" by Hilbert space and that the way they have used it—in spite of the admonitions from Copenhagen—to state their problems (one example, as we have seen, is the "problem of measurement") have turned it into a physical-mathematical "factish," a physical-mathematical being that is acknowledged to possess an autonomous existence. And we can indeed say that this factish is a "liar." For it pretends to speak the language of quantum reality itself, to supply (Hamiltonian) reasons for the way reality behaves. But physics is not a moral tale, where liars are always punished. The creation of problems is what matters, whatever the starting point.

Let's take a closer look at the Hamiltonian operator, which makes Hilbert space a part of the prestigious lineage of the "cyclical" form of dynamic representation where, as we saw in Book II, power, intelligibility, and beauty converge. In a sense, it is just one operator among many. Its eigenvalues correspond to the measurement of energy, for example, the energy of the different stationary states of the Bohr atom. However, in dynamics, energy is not a property among others. Its dynamic (not thermodynamic) conservation reflects the relationship by which, ever since Galileo and Lagrange, instantaneous states and causal equality mutually define each other. The Hamiltonian dynamic assigns this relationship of mutual definition the authority of a syntactic rule expressing the very identity of the system. The Hamiltonian is at the same time that which every canonical transformation of the representation of the system must maintain as invariant and the key property of the system's evolution equations. Because it is kept invariant during the evolution, this evolution can appear as a simple transformation of representation: the system "is," the point of view from which it is represented "changes." Because the Schrödinger equation links the evolution over time of the wave function with the action of the Hamiltonian operator on that wave function, the Hamiltonian operator inherits this critical role, enabling the identity of the system and evolution over time to communicate systematically.[5] With respect to the question of determining how a quantum system can evolve—more precisely, how the probabilities of the different results of measurements to which it is subject can evolve—it is the superposition of eigenfunctions of the Hamiltonian operator that will provide the answer.

However, as I have stated, the power of the Hamiltonian

operator is greater than that of the classical Hamiltonian. The primary goal of classical dynamics is to construct the equations of motion for a system. In some cases, those equations will be able to be integrated (resolved) and the spatiotemporal defini-tion of trajectory will then be given explicitly. It is in these cases alone that the dynamic system can be given a cyclical represen-tation, can be represented by equations of motion in which each of the system's degrees of freedom evolve independently. In this case, it is no longer energy alone but each of the momenta (or quantity of motion) characterizing those degrees of freedom that become an invariant of motion. But in quantum mechanics the situation is different because motion invariance was the very instrument used in the primordial capture of quantum beings by the Hamiltonian formalism (see chapter 5). Each stationary state of the Bohr atom was already seen to be characterized by an invariant of "electronic motion," the value of the different invariants in this case forming a discrete spectrum rather than being, as in classical dynamics, a function of the initial state of the system (where each degree of freedom of the system can assume any value). And it is this mode of characterization that will be extended and systematized by the definition of the wave function in Hilbert space. In quantum mechanics, to state "let there be a quantum system represented by a wave function ψ" is to assert that this system can be represented as a superposition of stationary "states," each evolving on its own, independently of all the others, over time. And this is so because whenever time evolution is involved, the system will be represented in terms of the eigenfunctions of the Hamiltonian operator.

This, then, is the significance of the Schrödinger equation, which introduces us to the action of the Hamiltonian operator on a wave function. And its scope is shown by its unbridled use by physicists when they claim that Schrödinger's cat—or any-thing at all—is subject to its strictures. For this equation (and the wave function itself) to have meaning, the wave function must be represented as a superposition of eigenstates (or functions) of the Hamiltonian operator. And whenever this is the case, the Schrödinger equation will describe an essentially static situation. The evolution of the wave function in Hilbert space can be compared to the simple, monotonic rotation of the different eigenfunctions of the Hamiltonian it superposes. Consequently, the probabilities corresponding to the eigenvalues of each of those eigenfunctions are, by definition, constant over time.

In other words, the discrete nature of the energy spectrum that characterizes quantum beings has been given an extraordinarily powerful meaning by quantum mechanics. If a system has a discrete energy spectrum—and this property has become universal ever since Planck's constant acquired the status of a universal constant—it *must* be able to be represented in terms of the eigenfunctions of the Hamiltonian operator.[6] Therefore, it must be able to be represented in a way that causes interactions to disappear and allows the system to exhibit periodic behavior in Hilbert space. The benefit of this representation is the answer it gives to the question of evolution over time. For there is one and only one answer: "Nothing is happening."

I want to introduce one last technical term, which will become crucial at the end of Book V, namely, the concept of *spectral representation*. Just as the Hamiltonian of a classical integrable system can be written in terms of variables corresponding to its cyclical representation, the quantum Hamiltonian operator can be defined in terms of its eigenfunctions and eigenvalues. This is the "spectral representation" of the operator. To be able to explicitly write the spectral representation of the Hamiltonian operator of a system is to be able to state a problem using the very terms of what will be the solution of that problem: the problem itself affirms the convergence between power, intelligibility, and beauty. But the spectral representation of the Hamiltonian operator eliminates any possibility of assigning an intrinsic meaning to the concepts physicists need. There is no possibility of defining the lifetime of an excited atom in those rare cases when the spectral representation of a system can be explicitly defined. "Nothing happens," except, as we saw earlier (chapter 5, note 10), the ghostly dissolution of an initially concentrated wave packet, the atom and its discrete energy levels having been eliminated by the spectral representation.

It is fortunate, therefore, that spectral representation is not generally accessible in cases where the physicist knows that "something is happening." In fact, every time physicists' practice leads them to study a quantum situation where things "happen" (the emission or absorption of a photon, collision, acceleration in a field, laser emission), they do not try to provide the situation with a spectral representation where precisely nothing would happen. In such cases, physicists turn to the technique of perturbation. They start from the spectral representation of the "unperturbed" system and "add," in the form of perturbations, the interactions that will be responsible for what "happens." These interactions couple the eigenstates of the unperturbed Hamiltonian, states that are then no longer stationary. Yet, quantum mechanics is categorical. It requires that every "perturbation," every coupling between eigenstates of the Hamiltonian operator, be defined as eliminable, resulting in a new Hamiltonian and new independent eigenstates. The probabilities that vary over time and allow us to construct the physical meaning of properties such as the lifetime of an excited state have, therefore, a meaning that is purely dependent on the fact that physicists choose not to take the requirements of the formalism to their logical conclusion.

Consequently, quantum mechanics imposes the use of a mathematical structure that will never provide intrinsic meaning to observable properties in terms of what Cartwright refers to as the "causal role" of energy exchanges. This role is inappropriate for the inhabitants of Hilbert space. Their presentation, whenever that role is being performed, whenever it's a question, for example, of constructing the kinetic equations that describe laser emission, requires "leaving" that space, a theoretically unjustified operation that, as we have already seen, is realized through the relevant approximations. As Cartwright noted, physicists are forced to "cheat," to introduce approximations that allow them to alter the mathematical structure of the problem. Those approximations clandestinely provide the coupling perturbation with a "truly causal" role, one that cannot be written off by the transformation that would eliminate the coupling by incorporating the perturbation in the Hamiltonian.

More generally, there is no actualization of the quantum object defined in Hilbert space without an "exogenous" intervention, an interaction that is not described in Hilbertian terms. This results in the "theoretical violence" that von Baeyer complained about—the mutilation rather than the negation of practice. For it is this method of definition that relates everything that "happens" to the intervention of measurement, to the recording of a datum by an external device, and which destroys all the words that might describe the practice of delegation. It also results in the double standard of physicists, who bring into existence, more or less clandestinely, something that the mathematical structure of their theory officially denies: atoms interacting with fields, beams of electrons, beings that act, are acted upon, and are affected, not merely measured. The "kitchen" of approximations must resuscitate delegates and actors, whose denial is the price of the "Hamiltonian miracle."

We are now able to understand the problem that confronted critics of quantum mechanics, who allowed themselves to be seduced by the Schrödinger wave function, by the way it extended the harmonious beauty of Hamiltonian physics. For Schrödinger, quantum transitions had been the problem; for these critics, the reduction of the wave function defined by the Schrödinger equation is the problem. But in all cases, it is the particular "narration" associated with experimental practice that has disappeared, to the benefit of the intellectually satisfy-ing prospect of a general extension of the Hamiltonian repre-sentation. However, the question is not whether this extension is impossible. Some physicists, like David Bohm, have con-structed mathematical beings capable of "reproducing" the forecasts of quantum mechanics. Those beings were not taken seriously because their construction came "later" and was therefore incapable of challenging physicists' conviction that quantum mechanics was a reliable guide—*up to and including the "kitchen" to which it obligated them*—that is, up to and including the approximations it suggested. There is no doubt that Hilbert space serves as a "lying factish," but its lie is very *useful* in the sense that it has succeeded in suggesting procedures and rele-vant approximations, through which practice acquires meaning. And it is this usefulness, this relevance of quantum formalism as a guide rather than as a law, that is lost in the attempts to go "beyond" the wave function. Naturally, "it is not impossible" to construct the mathematical expression of a "nonlocal" field that produces "localization" effects, but this expression is extremely complicated. Correlatively, the relevant ways of "leaving Hilbert space" are no longer presented as "plausible" on the basis of such a field. In other words, the attempts to construct a "factish" that does not lie, to the extent that they try to reconstruct the behav-ior of the wave function, confirm the requirements inherited from the triumph of the Queen of Heaven and are committed to further conceal the continuously negotiated character, fecund but conditional, of the alliance between Hamiltonian formalism and experimental practice.

Here, we confront a very unique situation whose unique-ness exacerbates, while transforming, the historical reciprocal-capture between "realism" and "causal measurement." From inception that history has brought about a binary register of evaluation. Galilean rolling balls satisfy the requirements of causal equivalence only to the extent that they manifest the ideal situation of the frictionless fall for which they serve as a (good) approximation. Lagrangian, then Hamiltonian, dynamics have developed a syntax that separates, with increasing radicality, the power of presentation (and the mathematical definition and invention that corresponds to this type of ideal situation) and the "phenomenological" description of the systems we are effectively confronted with. As we have seen, Hilbert space is the direct heir of dynamic inventiveness. The object of classical dynamics was constructed from the fixed point provided by the satisfaction of the causal equivalence between full cause and total effect. But the construction of beings inhabiting Hilbert space does far more than assert and presuppose the satisfaction of a requirement that allows the problem of motion to be formulated and therefore still alludes to the physicist who formulates that problem. This construction brings into existence beings from whom nothing can be demanded, beings that cannot present a problem to anyone.

Like causal requirements, Hilbertian requirements coincide fully, in a way that is both necessary and sufficient, with the exhaustive definition of the beings that satisfy them. That is why those requirements are, at the same time and identically, vectors of obligation, conferring on what they define the power to engage, authorize, or prohibit. Unlike causal requirements, however, Hilbertian requirements do not communicate with a property of the object that may well require idealization but that, nevertheless, makes intelligible a determinate experimental practice. That is why, unlike the classical dynamic object, the quantum object does not appoint a physicist-judge whose categories of interrogation would be those to which the object itself is subjected. The measuring physicist appears, rather, as an *intruder*. She breaks the superb mathematical symmetry of the quantum object, asks questions for which no answer should be able to be given as long as we "remain within Hilbert space."

We might object that, strictly speaking, the classical object described in terms of cyclical variables no more accounts for the possibility of its description than the quantum object described by a superposition of eigenstates of the Hamiltonian operator. Both are literally "unobservable" in the sense that any observation presumes an interaction and in that they are both defined mathematically as devoid of interaction with an external world. But the possibility of constructing a cyclical representation is not something causal equivalence requires. This equivalence enables us to establish dynamic equations but does not guarantee that they can be integrated. Moreover, in the customary formulation, the dynamic system is defined as closed, with no place for the observer. But this definition acknowledges, without dramatic consequences, that this closure is an idealization, neglecting the interactions that enable observation. Such interactions can no longer be neglected in the quantum case, as Bohr untiringly emphasized. And Hilbert space, because it asserts the quantum analog of classical integrability, transforms into an axiom the fact that time evolution cannot be related to interactions. As a result, the very fact that the quantum object cannot communicate with experimental practice becomes part of the definition to which this object obligates us.[7]

There is a connection between the "practical lie" of quantum mechanics and the distinction between causal requirement, on the one hand, and the Hilbertian mode of existence, on the other. The causal requirement concerns the way of defining a problem that corresponds to a situation characterized by measurable variables, and the way the equations expressing that problem are written, whereas, to follow the analogy, the Hilbertian mode of existence doesn't correspond to any problem. This distinction matters because the practice of physicists, classical or quantum, requires the mathematical definition of a problem, one corresponding to the situation they create. This practice is no more excluded by the equations of classical dynamics than it is by kinetic equations, for example, or by the equations of phenomenological physics. Such equations all ask the same question: "What will happen if . . . ?" On the other hand, Hilbert space provides answers that *do not satisfy any question*. It is made to bring about pure affirmation, an affirmation that needs "something else," the measurement that reduces the wave function, in order to become relative affirmation, relating Hilbertian beings to the device that interrogates them. In order to be associated with a problem, Hilbert space must be *violated*.

It may be concluded that physicists' questions and misgivings do not refer primarily to any "grand ideas," to realism, determinism, or any others, but to their own constructions, when read in terms of requirements and obligations. In no way does quantum mechanics challenge the "existence" of a "reality in itself." Rather, what is challenged is the relevance of the requirements of quantum mechanics when expressed in terms of statements that bring quantum objects into being (let there be a quantum system represented by a wave function ψ . . .). But, in answer to this challenge, the definition of requirements that would restore the possibility of defining quantum objects belongs to physical-mathematical inventiveness, and is relative to its historicity. There can be no question of physicists starting from scratch, of contemplating the world with fresh eyes, of escaping a history that does not belong to physics alone. Not only are all physical entities, without which our experimental devices would have no meaning, part of that history, but, in their own way, and even if it's in a way that is slightly deceptive, a large number of the instruments that populate industrial or hospital labs, as well as many commercial products, work hand in glove with Hilbert space and confirm its "relevance," given the approximations that "concoct" it, that force it to have consequences it is unable to justify. Is it possible to negotiate the meaning of this relevance without having to introduce the violation of the lying factish?

But why introduce the problem? Maybe because "grand

ideas" do not lack for importance in the history of mankind. From Schrödinger's cry, claiming that his cat has to be able to die even if no one can observe it, to Vigier's fury, from Einstein's protest to Popper's plea for a "realism of propensities," from Bohr's obsession with Einstein's objections, which troubled him until the time of his death, to Nancy Cartwright's accusation of lying—all of this, it seems to me, represents the expression of a profound dissatisfaction, the feeling that history cannot stop there, left with the irony of a physical-mathematical language whose requirements turn against the intelligibility for whose sake they were conceived.

Can we "civilize" the strange quantum descendant of the Queen of Heaven, bring it to accept practices it could not exist without, bring it to utter words that haven't been coerced? Can the Queen of Heaven herself engender ideas of intelligibility other than those her Hilbertian descendant has so paradoxically satisfied? If this exploration concludes with such questions, it is not accidental. All narration is also always a suggestion, with the twofold risk of captivating those it addresses, or enabling them to condemn the biases of the one who produces it. To explore the question of quantum mechanics presents nearly as many problems as choosing a path and viewpoints for exploring a labyrinth, where paths continuously multiply and bifurcate. Without a compass, losing our way is all but guaranteed, just as following a path that leads to a quantum justification of para-psychology, or the strange doctrine of multiple worlds, or the (strong) anthropic principle will lead us astray. The rejection of dizzying perspectives (see *Cosmopolitics*, Book I) is, however, far from sufficient for explaining the way in which I have pre-sented quantum mechanics and its dissatisfactions. To identify a source of dissatisfaction is never innocent, and the one who does so often conceals in her sleeve the hypothesis of a solution to the problem identified.

I don't intend to make a mystery of what guides me. It is

toward the perspective that situates me that I now turn, toward a response to the challenge of the Queen of Heaven and her Hilbertian descendant, unique in that it experiences no nostalgia for the lost intelligibility of dynamics. From the point of view defended by Ilya Prigogine, dynamics and quantum mechanics are both equally unsatisfactory, for one reduces the difference between past and future to the imperfection of our understanding, and the other to the act of measurement, or the act of awareness. Both condemn any possibility of coherence with the body of other practices, scientific, technological, or cognitive, all of which assume the nonequivalence of past and future, the "arrow of time."

The uniqueness of Prigogine's position stems from the fact that he felt obligated as a physicist by the need for consistency between physics and these other practices. But, ultimately, this obligation forced him to "swim upstream" against the current of history, to that moment in physics when the Queen of Heaven asserted her triumph and made kinetic description an "incomplete theory of dynamics." This is the meaning of the manipulative suggestion I have employed on the reader, the suggestion to allow the moment of perplexity experienced at the conclusion of Book III to extend to the dissatisfaction I, in my own way, have associated with quantum mechanics. There is no answer without a question, and no question that does not translate a choice that precedes it, and whose terms it cannot by itself justify. In this case, the "manipulation" is nothing other than the translation of what has, since I began to take an interest in it, activated my questions about physics.

BOOK V

In the Name of
the Arrow of Time

PRIGOGINE'S CHALLENGE

7

The Arrow of Time

The expression "the arrow of time" does not cause any problem for physicists, but casts some critical philosophers into an abyss of perplexity. It is as if the physicist had taken it upon herself to speak about time "in itself," whereas her time is, and can only be, the time found on a clock or any other device capable of providing a quantitative measurement of change. Time is the number of motion, Aristotle said. From this critical viewpoint, physicists' time has the status of a condition of knowledge. It enables us to study motion but should not be confused with it. Yet, doesn't the fact of discussing the arrow of time manifest this confusion between "time-as-measurement" and "measured motion"?

The critics are obviously correct. The arrow of time is a metaphor. But they are mistaken if they believe that condemning this metaphor will get them very far, for behind the metaphor can be found the problem it points to, a problem criticism cannot make disappear.

Time, as the number of motion, is a general statement that applies to all motions. Using the same chronometer, I can number the movement of a tortoise or a running horse, a ball or the decay of a corpse. In that sense, it is one "external" measurement among others. Readers may recall (see Book II) how "work" was defined by nineteenth-century engineers, who indifferently measured the activity of rats, men, or the rolling ball, with or without friction. Just as, unlike rats and workers, the ideal Galilean ball provides work with an intrinsic theoretical meaning, that same ball binds the relationships between "time" and "motion" that will create the problem of the arrow of time. The beat of the pendulum not only creates the numbers that will be used for the measurement of all other motions, it is itself subject to a temporal law that explains and guarantees the standard it constitutes. In other words, the regularity of pendular motion is not only a condition of measurement (the way the apparent motion of the sun once was). Of course, it does not directly tell us about time. But its telling concerns the articulation of successive instants of motion, which are subject, for the ideal pendulum as for any other dynamic motion, to causal equivalence. The instantaneous state is then able to define past and future states, which are all seen as equivalent. Correlatively, the beat of the pendulum not only gives motion a number, the number itself becomes a function (of gravitational acceleration and the length of the wire). Time-as-measurement has become internal to the object. It enters into the definition of the object, and this definition, in turn, determines the existence of the object in its relationship to time.[1]

We see that the metaphoric possibility of shifting from dynamic evolution over time to time itself, as understood by dynamic evolution, points very precisely to the singularity of the dynamic object, which causes description and reason to coincide. The question of the arrow of time belongs to the field of mathematical physics and is, in that sense, distinct from philosophical questions such as "What is time?" It asks the question of the meaning (and scope) of dynamic symmetry as illustrated by the motion of a frictionless pendulum or the Galilean ball returning to its initial height. Similarly, whenever the literature of physics refers to a phenomenon "going back in time," it is not a speculative statement in questionable taste but an illustration of an essential property of the equations of dynamic evolution, that is, the strict equivalence between two mathematical transformations: the reversal at a given moment of the sign of all velocities in the system and the reversal of the sign of the time parameter. This is what the perfect pendulum realizes, right before our eyes, whenever its velocity changes direction at the end of a cycle. If we consider this end point in the "present," the description of the *future* motion of the "descent" of the perfect pendulum will not be distinct from that which would lead it "back to its past," that is, from a description that led it to the present when transformed by a reversal of the time parameter.

To say that the pendulum goes "back to the past" can hardly be taken literally since nothing in the behavior of the pendulum can serve as a witness to its having accomplished this great deed. In fact, the very definition of the ideal pendulum is that it does not "speak," in whatever sense of the word we choose, no matter how metaphoric. The ideal pendulum entertains no relationship to the world that could be understood as bearing witness. The ability to speak of going "back to the past" is, in effect, strictly equivalent to the claim that its oscillating motion is fully determined by the conservation of cause in effect. In other words, the motion of the ideal pendulum, precisely to the extent that it can challenge the difference between past and future, can cause nothing but itself. It cannot "tell time" to anyone because it is in relation to nothing.

The example of the pendulum reflects, in nearly paradigmatic fashion, the power relationship between two claims of "time-as-measurement": the time the pendulum gives expression to, the standard of measurement, and "dynamic time," introduced by the equations of dynamics. In order for the pendulum to "tell time," it must be supplied with an external observer who "sees" the pendulum, who counts the number of beats, who uses it as a measurement standard—activities that cannot be reduced to dynamic evolutions. By means of the observer who counts the beats of the pendulum or, armed with a chronometer, measures its period, as she might measure the movements of a horse or the gestures of a laborer, the human practice of measurement asserts the difference between past and future. But the dynamic equations for pendular motion introduce the self-determination of the motion of oscillation, the autonomous variation of positions and velocities that allows it to be characterized exhaustively as an *existent* in time, in a way that denies the difference between past and future.

We might conclude that the two modes of existence—that of the observer and that of the pendulum—correspond to two meanings, external and internal, of time-as-measurement. The correlate of this conclusion would be that the mode of existence of the subject who observes obviously has nothing to do with physics and cannot, therefore, present a problem for the physicist. However, like any conclusion that introduces a "subject" confronting the "world," this one gets ahead of itself. It lends itself perfectly well to "high-level" meetings between philosophers and physicists, and irresolvable confrontations where every participant claims to represent an authority that has the power to make determinations for the others. Yet it overlooks the fact that there is no need for a human "subject" for the pendulum to be able to "tell" the time or to bear witness. Any clock will do, and can be integrated, as is the case in any physics laboratory, into an automated device that "traces" the evolution of a property measured over the course of time. The question of the arrow of time arises, in fact, *within* physics, for example, when questions arise about how to describe the (dissipative) interactions that are required to connect the pendulum to the hands of a clock.[2] Do the laws to which the pendulum is subject reveal the truth about the clock?

It won't escape the attention of readers, or anyone familiar with quantum mechanics, that we are very close to the problem of measurement presented by von Neumann. For von Neumann, there is apparently nothing that authorizes the physicist to "stop" quantum description, in terms of the wave function, when interacting with a measurement device. But if he represents the device in quantum terms, in Hilbert space, he is obligated by that representation to recognize that he has lost the ability to characterize a measurement device. At this point, the human observer, endowed with consciousness, is required by von Neumann to define the conditions that make quantum measurement possible. But as we saw in discussing the work of Nancy Cartwright, this approach to measurement, which creates a dramatic conflict between human awareness and "Hilbertian reality," is remarkably artificial. Just as the clock, if it is to function and tell time, *must* break the ideal symmetry of dynamics, the quantum description of any process where time "counts" for physicists *obligates* them to "leave Hilbert space"—which they do so unhesitatingly—that is, to introduce the approximations that will free them from the obligations imposed by the wave function. In both cases we are dealing not with a confrontation between "mankind" and its "object" but with a question produced by physics. It is physicists, not human beings in general, who find themselves facing conflicting obligations: between their obligations vis-à-vis relevant experimental practices and their obligations vis-à-vis a Hilbertian or dynamic definition of "reality."

The parallel between the question of the arrow of time and the problem of quantum measurement extends even further. One way of saying that the clock must violate the theoretical ideal that claims to describe it is to recall that its movement will proceed toward equilibrium if it is not maintained. Similarly, as we saw in Book IV, the interaction of measurement can be compared to an irreversible evolution toward equilibrium, which is translated by the production of the permanent "mark" without which an experimental datum has no meaning. In both cases the infraction can then be considered an approximation. Because of its dissipative interaction with the gear mechanism, the clock's movement approaches, but cannot achieve, the ideal of the pendulum. Similarly, the evolution toward equilibrium that is required by quantum measurement could be compared to the "forgetting" of certain interference terms, the approximation corresponding to this forgetting being justified by the fact that it has no observable consequence for the measured datum.

That is why in the quantum as well as the classical case the need to "violate the law," to break the symmetry of the fundamental equations, far from seeming to challenge the authority of the law, can very easily be concealed. In fact, the approximation seems to confirm the authority of the law. Even more so since it can mask itself behind a perfectly anodyne appearance, and appear fully justified by the insignificance of what is being overlooked. In fact, once the "violation" can be defined as negligible, the game is over. Just as the most negligible "kitchen concoction" is sufficient to enable one to leave Hilbert space and end up with a probabilistic description that is realist in nature, the least approximation of the defined ideal suffices to introduce the arrow of time into a description governed by dynamics. More accurately, in both cases, *any approximation* always results in breaking the symmetry that singularizes the ideal representation. Consequently, constructing the meaning of what experimental practice obligates us to do is trivially simple.

The "double standard" that Nancy Cartwright discovered in quantum mechanics is, therefore, a shared feature of both classical and quantum mechanics. In both cases, the physicist at work bears no resemblance to the physicist as she presents herself to the philosopher or as she presents herself to herself when attempting to reflect on the obligations of her science. In such situations, the obligation attached to the ideal representation takes precedence. Like measurement, the arrow of time is merely a well-constructed fiction. "We"—the physicist and, by

extension, all humans—must accept that the world we characterize as yielding to observation is, first and foremost, defined
by its distance from intelligible reality, dynamic or Hilbertian,
of which it is merely an approximate translation. On the other
hand, at work the physicist requires only that theory be relevant.
Whenever it is a question of preparing and predicting observable processes and properties (lifetime, cross section, and so
on) that imply a difference between past and future, the physicist "negotiates," without the least qualm, the transition between
"fundamental" theory and the relevant representation, and uses
this successful negotiation as an experimental confirmation of
the theory.

This also results in the double status of the meaning assigned
to observable "dissipative" properties. Dissipation is both
required, because it characterizes the majority of experimental situations that legitimize theory, and disqualified, because
it is considered relative to the approximation resulting in the
construction of their theoretical meaning. Dissipative properties are thus included for, if they did not, the theoretical scope
of theory would be restricted to a handful of special cases and
the theory could not claim to be "fundamental." But the fact that
they conflict with theoretical syntax is not taken into account.
The double standard by which physics manages the "conflict
among obligations" that inhabits it is characterized by the pro-
motion—depending on the situation—of two divergent obligations. The first is the obligation of experimental relevance:
theory does indeed account for what we observe. The second is
the theoretical obligation, more specifically, a reference to the
betrayal of this obligation. It would be only if physicists do not
take the obligations of theory as far as they can go that they may
be tempted to assign a physical meaning to observable "dissipative" properties.

This "double standard" defines the problematic space
wherein the question of the arrow of time *actually* takes place

today. Anyone who attempts to argue that the way phenomena bear witness to a "dissipative" temporality should be understood as a source of obligation runs up against the fact that, currently, these phenomena are not, in themselves, likely to challenge the relevance of the "fundamental" equations because it is on the basis of those equations that they are described. That is why the burden of proof falls entirely on the challenger. She must prove that the means she proposes so that theory is able to comprehend this temporality are not the by-product of a betrayal of the obligations imposed by that theory, that they cannot be reduced to the surreptitious introduction of an approximation into the description, the arbitrary rejection of a possibility of measurement or discrimination provided for by that description. If she cannot, she will have confirmed that, in effect, the theoretical account of the dissipative phenomenon she proposed is relative to the approximations that allow it to be constructed.

There is nothing logical or necessary about this challenge. Quite the contrary. Logic, had it been allowed to function as such, as a constraint with which the judgments of physicists had to comply, would suffice to tip the balance of judgment in the opposite direction. It would lead us to describe quite differently the conflict among the obligations that inhabit physics. The obligation of experimental relevance marks the distinction between a physical theory and a revelation that appears out of nowhere. Theory, therefore, is not logically empowered to disqualify, because of its own obligations, what is the very source of its authority. To return to the question of the relationship between the pendulum and the clock, if the clock is the measurement device by which we were able to assign a law to the motion of the pendulum, that law cannot claim to reduce the dissipative nature of collisions and frictional forces to a simple approximate description, for it is because of those collisions and frictional forces that the pendulum is connected to the device that moves the hands and becomes part of a clock. Logic would

be sufficient to condemn the vicious circle entailed by a judgment that attributes the responsibility for the operation of time
measurement presumed by every description to the imperfect
nature of that description. But the requirements of logic must
bow to the facts. Physics is a history, and the fact is that at the
beginning of the twentieth century, that history experienced a
genuine form of reciprocal capture that produced the respective identities of what we refer to as dissipative phenomena and
level of fundamental description. While the description of phenomena, interactions, and dissipative events was considered an
"incomplete" description from the point of view of theory, theory was given a power of extension that nothing should be able
to limit.

In asserting the inseparability of the wave function and its
reduction, the Copenhagen interpretation attempted to foster
a solution that satisfied logic. The quantum being could not be
described, would have no assignable identity, independently
of what conditioned experimental access to it. However, logic
and the critical philosophers that relied on it are the only ones
fully satisfied with this solution. Yes, it creates perspectives that
can be described in grandiose terms. Heisenberg spoke of the
person who, examining the sand on a beach, studies the tracks
whose origins he questions, finally realizing that they are his
own footprints. Here, the observer is the only rationale for what
he observes, which means reducing the damp sand to general
properties that may be taken for granted. The world, like the
sand capable of receiving an impression, is essentially silent.
The answers we obtain are determined by the questions we have
decided to ask. This expresses the pathos of a situation wherein
physics, in the very act by which it claims for its equations what
temporal symmetry warrants—an intelligible perfection from
which nothing escapes—understands that those equations prevent it from assigning any intelligibility to the world it thought it
was addressing.

No working physicist will ever fully acknowledge this critical lesson. Physicists accept the idea that we may not be able to conceive reality independently of what makes it observable. But is it necessary—to continue the parallel that serves as my guide—to assign to the collisions and dissipative shocks that make time measurable the noble but opaque status of a "condition" for there to be an "access" without which the ideal motion of the pendulum would have no observable meaning? Is it wrong to believe that those collisions and frictions are physical processes to which we certainly delegate the role of "connector" between pendulum and clock hands, but which happen "regardless," the way an excited atom emits light regardless of whether this emission is a condition for assigning stationary states to the atom?

In fact, Niels Bohr, was no different than other physicists in believing that the quantum world was not mute. But he was haunted by the need to assert the novelty of that world, its irreducibility to the phenomenal world concerning which the arguments of mechanics are constructed. And his way of asserting it was to insist that we could never dream of "inhabiting" that world, of constructing a "rational" account of it, identifying the intrinsic reasons explaining what we observe, the way we believed we could construct the reasons explaining the Galilean world. The quantum world will never reveal its "reasons"; in fact, it forces us to recognize the idealization behind the idea that the world of mechanical phenomena was able to assert its own reasons. That is why, in my discussion of Bohr's position, I introduced the concept of the "divinatory factish." "Divinatory" here should be understood literally. Even though the physicist constructs the message according to her own questions, it is on signs that have arrived from *another* world, and not on some amorphous noise onto which she would project her own meanings, that the physicist acts.

Yet, it is pointless to defend Bohr's proposition, to acknowledge that it has nothing to do with the impoverished idea of a silent world to which a ventriloquist physics would lend its own language. The history of physics does not put us in a position to defend what it has set aside because its "decisions" are inseparable from the creation of new problems and new beings. Just as the history of physics shifts from "dissipation" to "incomplete description" via "reciprocal capture," it is clear that it did not take the path Bohr suggested. More specifically, it bowed "in passing" to the inevitable character of those propositions, only to quickly sidestep them. Today, Hilbert space is presented as the worthy heir of Hamiltonian dynamic space, that is, as separate from the "reduction of the wave function" that was supposedly inseparable from it. Physicists "inhabit" that space, ask it questions, use it to introduce problems "as if" it supplied the reasons defining the quantum world. The very fact of introducing "approximations" into a Hilbertian formulation reflects the fact that this formulation has been recognized as a *description.* Only a description introduces the problem of what will be kept and what will be ignored. In no case does a divinatory apparatus authorize such a choice.

Are we to believe that this survival of reciprocal capture reveals an irresistible psychological "penchant" on the part of physicists? Does the defeat of Niels Bohr, like that of Pierre Duhem before him, force us to treat the physicist's "faith" in the possibility of a "vision of the physical world" as an inevitable ingredient of her practice? That would be somewhat premature. For history can be expressed quite differently: opportunity makes the thief. Let's return one last time to the analogy between the quantum question and the relationship between the pendulum and the clock. If it is possible, at the cost of plausible approximations, to derive the description of collisions and frictions from the theoretical understanding of the perfect pendulum, why shouldn't we do so? If Hilbertian reality as such is able, at the cost of plausible approximations, to account for the theoretical meaning of observable properties and processes that are not part of the divinatory economy of quantum formalism, why deny it what that ability allows it to claim? Why deny that this Hilbertian reality has the right, as such, to obligate, to define what the physicist deals with? In other words, although there is indeed a question of "decision making" in the history of physics, this does not refer us either to a subject and its penchants or to a onetime event but to a process of practical invention. What "makes" the decision is the effective possibility of creating relevant extensions to reciprocal capture, which stretch the ability of the theoretical formulation to guide and interpret experimental exploration as far as possible. Like the "thief" produced by "opportunity," the physicist is primarily the product of this process. It is not her psychological expectations that are satisfied, however. She is the spokesperson of the satisfaction produced by the operation of extension.

There is no a priori requirement such that the object of a "decision" *in* the history of physics cannot be reenacted *by* the history of physics. Nothing requires it, however, and what I have just advanced allows us to predict that the opposition to such a move will be stiff. For the physicist-actors in this story are anything but impartial arbitrators. Of course, the nonphysicist might be sensitive to the claim that neither Hamiltonian dynamics nor quantum physics has a "logical" right to equate the asymmetry between past and present that characterizes dissipative processes with the breakdown of symmetry in their own equations, or to judge this breakdown of symmetry in terms of an approximation. The same argument is either dismissed by physicists or, at best, receives an indifferent "yes, I know, but all the same." I know that, in terms of logic, you are correct, but the fact is that the symmetry-breaking approximations have effectively allowed for the characterization of a very satisfactory ensemble of "dissipative" processes. As Einstein said about Mach, logical criticism has never been useful except to kill the vermin that live off physics, never to promote its progress. In other words, the mandate of whoever wishes to "reenact" the question of the arrow of time, the status of the asymmetry characterizing "dissipative" processes, requires the invention of means to demonstrate that the "progress of physics" entails that these questions be taken into account.

It can no longer be a question of unilaterally, that is, in an ad hoc manner, adding an opaque dissipative factor to dynamics, which would ensure asymmetric temporal behavior whenever experimentally required, simply to resolve the problem. To accept this addition "out of nowhere" would imply that physicists consider their strategy of approximation to be a bitter setback that must be corrected at all costs and by any means, even ad hoc. In fact, the very opposite is true. The challenge is to turn a situation generally felt to be one of the triumphs of physics into something unsatisfactory.

The decisions made throughout the history of physics cannot be redone in the name of right or logic, but require a process of invention that, in order to alter history, must become part of the history being altered. And what has to be invented is none other than an appeals procedure, one that provides de facto ratification of reciprocal capture but challenges its modalities. In order to have a chance to convince physicists, this process must begin from a situation they judge satisfactory, one that refers to dynamics and quantum mechanics as providing a "complete theory," which rightfully holds the key to dissipative processes. Correlatively, the question of dissipation must be framed in terms of the *breaking* of the symmetry of the equations of dynamics and Hilbert space. It is then, and only then, that the judgment associating this breaking of symmetry with approximation can be appealed. In other words, the "reasons" justifying an arrow of time that cannot be reduced to an approximation must be constructed in such a way that the symmetric equations themselves supply the justification for breaking the symmetry they assert. Naturally, such reasons must be recognized, along with those that are made explicit by the equations, as "objective reasons" stemming from the *object*. For the proposed solution cannot appear as a simple version that would differ only rhetorically from what physicists are already doing when they introduce relevant "approximations." It is the phenomenon itself, as represented in dynamic or Hilbertian terms, that must be recognized as imposing the *obligation* that the time symmetry of the equations that represent it be broken, that must disallow the assertion that its time-broken symmetry can be explained in terms of an approximation. Finally—although physicists might be divided about the need for this last condition—it would be good if the recognition of the specific obligations to which the arrow of time corresponds were accompanied by the identification of new experimental possibilities. Such identification would endorse the superior relevance of the proposed solution in everyone's eyes. Progress could then resume its normal and moral course: as usual, the scene will be one in which nature itself distinguishes among its interpreters and narration can solidify into a tale of progress that disqualifies error.

It is at this point that the narrative choices for my stories need to disclose their rationale. For the requirements I have just described impersonally are the very same that were discovered, constructed, and accepted by Ilya Prigogine.[3] Can we say that these requirements have been honored and that the statements resulting from Prigogine's work will be confirmed by his colleagues? In the French edition of *Cosmopolitics,* I wrote: "No one can answer this question. Will the existence of doubt and opposition one day be considered 'merely sociological' in the sense that it would cast light on the 'habits' of physicists, the 'prestige' of physical laws, the 'ideal' of perfection associated with those laws, or the overwhelming 'authority' of the principles of symmetry? Or, rather, will Prigogine's effort be seen as sterile, futile, or 'unfortunately condemned by some hidden defect'?" "History," I wrote, "hesitates, without the slightest assurance that the future verdict will be fair." Today what has truly won is indifference—physicists have chosen not to care. Strangely enough, a new concept is now taken for granted, that of quantum "decoherence." This states that the slightest perturbation to a quantum system is sufficient to destroy the superposition that characterizes it—a serious concern for those who envisage the possibility of a quantum computer, and an easy way to ascribe quantum "histories" to a universe defined in terms of a wave function. This concept, however, is merely a name for what physicists already knew, which is that the slightest approximation was sufficient to obtain a "realist" characterization of quantum actors. It may be a slap in the face to the generations of philosophers who meditated upon the "measurement problem" but it also shows that physicists are pragmatic. If there is a "true" need, they accept what they previously rejected, and the principles they proclaimed in order to justify that rejection are put to rest. It may well be that Prigogine's efforts will be associated with the memory of a useless, and desperately complicated, project and the misplaced ambition to transform decoherence into a matter of principle, which is now known to be a property of quantum systems. What is certain is that texts such as this will not reverse the course of events. For physicists, too many things are at play for an "incompetent" intervention to be considered anything more than presumptuous chatter.

However, the situation is different from the point of view of the questions that guide my exploration of physics, those that deal with the consistency of our practices of knowledge. Here, the only thing that counts is the fact that hesitation was possible. In light of the requirements I have just detailed, we can understand how a potential verdict that claimed these requirements had been satisfied would have celebrated the coming into existence of a new physical-mathematical factish, able to claim that it spells out "nature's reasons" better than its predecessors. By means of this verdict nature could be considered to have appointed a suitable representative. However, the very possibility of hesitation introduces the question of what we can demand of phenomena that we represent and the obligations associated with the ambition to represent them. Of course, it is not a matter of challenging the importance of the verdict, the importance of the invention of a new definition of "nature's reasons." Rather, it is a question of emphasizing that the invention of this new definition is precisely what was required by physicists who inherited the reciprocal capture that marks the history of physics. The notion of "nature's reasons" itself does not refer to a quest that would transcend the practices of physicists. It is inherent in the requirements imposed upon those who suggest that they abandon their double standard: they must be forced to do so by "nature itself."

And the challenges are great. For the physical-mathematical beings that might claim the ability to impose obligations that entail the need to take dissipative phenomena into account will not be met with sympathy, as the vectors of a long hoped-for solution. They will have to overcome a certain amount of skepticism. Questions will arise about whether the claim with which they identify hasn't been extorted from them; about whether the obligation they bring into existence does not in fact imply that they forfeit reference to an exact definition, one that is possible in principle. And they will be asked if they are capable of "keeping up," of having the behavior expected in all situations in which they might appear. Can they match the performance of their Hamiltonian rival whenever the conditions of the encounter have been defined? If they fail to satisfy all of these requirements, they cannot hope to gain recognition as representative of an "autonomous reality" that itself stipulates how it should be described.

Whatever verdict history reserves for Prigogine's proposal, its interest will remain to confer on physics an irreducible, historical—that is, path-dependent—dimension, more specifically, to make what might be called the "Galilean miracle" an active protagonist in this history. Physicists who are heirs to that miracle may maintain that they are authorized to claim that the problem they formulate about reality is true to the "problem" that reality itself continuously produces and resolves from moment to moment. In that sense, what physicists call the "fundamental laws" (of physics or nature) support their claim to be the *spokespersons* of reality. But the *achievement* those laws bear witness to does not transcend the singularity of the tradition that has produced such a claim. For the possibility of maintaining it serves as a defining condition, selecting which features of reality can have a legitimate spokesperson. Indeed, this is the miracle that Prigogine has been called upon to repeat: the construction of a physical-mathematical representation that must be able to claim the strange ability to "bring into existence" a reality endowed with the power to dictate its own (and new) conditions of representation, and thus capable of maintaining the claim that it transcends all the "realities" with which our other practices engage. That physicists can require the repetition of this "miracle" as a condition for taking the arrow of time seriously reflects the fact that this pretense to transcendence, a vector of fascination, is not legitimized by their achievements because it follows from a definition of achievement that is strictly inherent in their tradition.

Some may consider the following narrative to be dry, others will find it stimulating, and there are those who may even find it offensive. To trace the "technical" invention of problems and instruments is not to present, as I have done with Hamiltonian dynamics or Hilbert space, a stable physical-mathematical "being," whose behavior we wish to grasp. If Prigogine had been successful in his endeavor, the last step of this narrative alone would survive, which would supply the premises for a functional presentation of the new being. Everything else would be relegated to the status of a scaffolding that can be forgotten once construction is complete, with some elements being redefined as applications of the new theory, interesting only for those who wish to use them within a specific field. The history staged by my narrative has behaved in just this way: each of its steps has redefined the problem by making the shortcomings of the previous step explicit. In this way, each step incorporates the preceding step, but also allows it to be disqualified, to show how it was unable to provide a satisfactory solution to the (new) problem. Consequently, this narrative will be dry for those who perceive no more than an incomprehensible labyrinth, populated with strange beings who suddenly appear grimacing before us, only to disappear at once, as in a haunted house in an amusement park. However, it will be stimulating for those who recognize the passion that split this labyrinth, the tireless effort that created these beings, and hoped to see them fulfill their intended task, but was prepared to destroy them and incorporate some of the attributes they had managed to sustain in new beings, containing new ingredients, let loose across a modified problematic landscape. Finally, it will certainly be offensive for those who maintain, for one reason or another, that a pipe dream produced what tried to come into existence but was destined for failure, and see in its meanderings proof that, in any event, the path would lead nowhere.

8

Boltzmann's Successor

"To us the only hope for obtaining a general theory of non-equilibrium processes seems to be to reformulate the entire problem in a more systematic way on a purely mechanical basis."[1]
The primary function of this quote, which dates back to 1962,
is to dissipate a recurrent misunderstanding. We should not
understand "general theory of non-equilibrium processes" to
mean the so-called theory of "dissipative structures" in the field
of "far-from-equilibrium" thermodynamics, for which Prigogine later became famous. Often, Prigogine's work on the "arrow
of time" is treated as an extension of his work on thermodynamics. But the misunderstanding is not entirely innocent, for it
is used to level the accusation that he suffered from "delusions
of grandeur." After Prigogine had shown that irreversible processes, which produce entropy, can, at the macroscopic scale,
be the origin of new regimes of activity that are impossible at or
near equilibrium, it seemed as if he had convinced himself that
no one should be allowed to ignore irreversibility and dissipation. Supposedly, this is the reason he embarked on his overly
ambitious attack on dynamics and quantum mechanics. Naturally, such statements are often accompanied by a slight sneer.
A thermodynamicist has no business "playing with the big boys"
or attacking the "fundamental problems of physics."

In fact, the man who, in 1962, revealed his ambition to construct a "general theory of non-equilibrium processes" did not do so as a specialist in thermodynamics, the heir to Lagrange, Carnot, Clausius, and Duhem, but in statistical mechanics, and as an heir to Lagrange, Hamilton, Maxwell, Boltzmann, and Gibbs. And he articulated the problem at the exact moment when dissipative physical-chemical processes were being interpreted in terms of dynamics: when Boltzmann's initial ambition of constructing a theory for the irreversible approach to equilibrium was transformed into an interpretation that simply eliminated the problem presented by irreversibility.

In Book III I described how Boltzmann had constructed what he hoped would be a "purely mechanical" representation of the simplest nonequilibrium process, one in which a dilute gas reaches thermal equilibrium. The gas is then represented as a population of elastic particles whose respective velocities change because of collisions. Boltzmann ended up with an "integral-differential" equation that represented the manner in which the collisions cause the statistical distribution of velocities in the population to evolve over time. Through this equation he was able to construct his well-known $\mathcal{H}$-theorem, which he hoped would provide a rigorous mechanical interpretation of the growth of entropy. As we saw earlier, his interpretation was unable to resist the paradoxes of Loschmidt and Zermelo, namely, the obligations imposed by the Hamiltonian trajectories that Loschmidt and Zermelo made explicit for the first time. Boltzmann was forced to recognize that his theorem did not describe the *impossibility* of an evolution that would lead to a spontaneous decrease in entropy, thereby contravening the second law of thermodynamics, but only its *improbability*.

Therefore, it is the association between probability and entropy that was retained as the essential lesson from Boltzmann's work. So what became of the evolution equation for the distribution of velocities that had served as the basis for the expectation that a purely mechanical derivation of the growth of entropy could be found? It was deprived of any authority, even though it was still useful for interpreting the phenomenological properties of gases in kinetic terms. The collisions it introduced could no longer pretend to "explain" the irreversible evolution of a gas toward equilibrium. Those collisions had appeared to Boltzmann to be responsible for that irreversible evolution, but only because of the approximations concealed in his reasoning. When the Queen of Heaven, the Hamiltonian definition of dynamic trajectories, claims the full scope of her powers, collisions are "smoothed" and are unable to produce any effects other than those produced by the other dynamic interactions.

Therefore, the assimilation of the kinetic description to an incomplete theory of dynamics was made possible because of a judgment about "kinetic collisions," as introduced by Boltzmann. Physicists concluded that such collisions, which were used to construct a model for the growth of entropy, were merely phenomenological, a simple approximation of "collisions in a rigorous dynamic sense," that is, devoid of any possibility of giving meaning to the arrow of time.

Let's return to Prigogine's goal of 1962, the construction of a "general theory of non-equilibrium processes," that is, a theory that extends beyond the scope of validity of the Boltzmann equation (dilute gases). In the context I have just recalled, that goal assumes a somewhat paradoxical guise. For, reduction of the kinetic collision to a simple approximation of a dynamic collision has one apparently obvious consequence: the only "general" theory possible must take as its subject collision in the dynamic sense and, in doing so, transforms the description of nonequilibrium processes into a form of dynamic evolution like any other. If Boltzmann's equation introduces a "mechanism" whose role derives from an approximation, its limitations should present no theoretical problem. We should not hope for, or demand, a more general theory. It is normal that a physicist would limit herself to refining, on a case-by-case basis, the most relevant technique of approximation for a kinetic interpretation of the phenomenological evolution she is studying.

And yet, in 1962, Prigogine and his colleagues held the opposite position. Their argument was based on the experimental relevance of the kinetic description: "There is certainly a profound physical meaning in the Boltzmann equation, as is borne out by its remarkable agreement with experiment in the calculation of transport coefficients for dilute gases."[2] It is this meaning that must be made explicit if it is to be extended to "the enormous range of experimental conditions in which transport or relaxation phenomena are now being studied. Starting from low-temperature transport processes in liquid helium and mounting to high-temperature processes in fully ionized plasmas, the range of energies covers ten powers of ten!"[3] The ambition to address what remains "one of the fundamental problems of physics"—"time, so closely related to irreversibility"—is formulated briefly, but there is no question (at this stage) of reviving the confrontation between the mechanical model of entropy, Boltzmann's $\mathcal{H}$ function, and the "attributes of the Queen of Heaven."[4] The problem is limited to the contrast between the theory of equilibrium—which is general and can be used to define the equilibrium of any physical medium from liquid helium to plasmas—and the great difficulty, once we leave the domain of dilute gases, of constructing a kinetic description of the processes that lead to equilibrium.

In Book III I referred to the growth of entropy as an "enigmatic factish." Entropy, whose growth appears to characterize irreversible change, in fact defines only the equilibrium state that is the conclusion of that evolution. The increase in entropy itself does not deliver its physical meaning in thermodynamic terms. Could kinetics supply that meaning? This was Boltzmann's hope and, in 1962, it was Prigogine's as well. He didn't (yet) try to explore the obligations associated with irreversibility where dynamic descriptions are involved, but only to construct the "profound physical meaning" of the obligations associated with a relevant kinetic description of the various irreversible processes. What Boltzmann succeeded in doing with dilute gases, Prigogine wanted to do in other milieus, which, not being dilute, require that interactions other than purely kinetic collisions be taken into account.

Prigogine, when considering the relative modesty of his initial ambition, often stated that he was fortunate that, at the time, he didn't "dare" try to imagine where those questions might lead. In fact, the terrain he explored was both reassuring and crucial. Reassuring, because it limited the scope of the question of irreversibility to the field of statistical mechanics alone, without asking the question of the latter's status. Crucial, because the requirements of a general theory of irreversible processes would stage the scene in such a way as to make explicit the obligations of such a theory. Those obligations would constitute a stable referent, a "pole" serving as a compass during Prigogine's quest. This would subsequently enable him to identify physical-mathematical developments that "could lead in the right direction," which could be put to use in studying irreversibility. What some have seen as opportunism reflects this "on the watch" position. From statistical mechanics to chaotic systems, and to today's "rigged Hilbert spaces," the point of view remains the same: irreversibility implies obligations.[5] What will continue to evolve, however, are the status and scope of those obligations, that is, the twofold determination of who those obligations will affect and the objects likely to impose them.

So, in 1962 the program was limited by two stable references whose stability Prigogine did not challenge. One was the Boltzmann nonequilibrium equation, which represented the (irreversible) evolution in time of the distribution of velocities under the influence of collisions, which is only valid for dilute gases. The other was the general theory of equilibrium in statistical mechanics, which represented the probability distribution, ρ, associated with Gibbs, and its definition at equilibrium, ρ_{eq}, which is constant over time. Let's begin with the Boltzmann equation and see how its limits of validity are interpreted.

The existence of those limits were rather dramatically exhibited by Loschmidt's thought experiment with the reversal of velocities. When the direction of all the velocities of all the particles in a gas evolving toward equilibrium is instantaneously reversed, the evolution resulting from the new initial state should, according to the equations of dynamics, lead the system away from equilibrium just as spontaneously as it approached it previously. Why wasn't this evolution predicted by Boltzmann's equation? Because Boltzmann's argument was based solely on the distribution of velocities. His equation introduced types of collisions characterized by the velocities of particles as they "enter" into collision and as they "exit" those collisions, and it assumed that the relative frequencies of the different types of collisions depended solely on the distribution of velocities among the population of particles. In short, the fewer the number of particles with velocity v_1, the fewer the number of collisions involving a particle with that velocity. This seems to be common sense. However, Boltzmann's assumption, in order to be valid, assumes that the information about the positions of particles at a given moment has no relevance for calculating the frequencies of the different types of collision at that moment. Yet, *after* the velocities are reversed, the description of the gas as it moves away from equilibrium must make use of those positions. The initial state of the evolution away from equilibrium implies the definition of the position and velocity of each particle. Any small modification of the position of a particle will lead the gas to quickly resume its "normal" behavior toward equilibrium. In other words, the positions and velocities acquire equivalent roles in the description of the evolution, the very roles they have in a dynamic description. The collisions that occur during the abnormal evolution cannot, therefore, be predicted solely on the basis of the velocity distribution function.

It was to accommodate the singularity of this "abnormal" evolution that physicists since Boltzmann have introduced "correlations" among particles. This evolution leads the system away from equilibrium because the collisions that produce it "repeat" the past. That is why Boltzmann's hypothesis that the positions of particles have no relevance for the evolution of the velocity distribution function, now known as the "molecular chaos hypothesis," in this case ceases to be valid. Collisions are produced between correlated particles.

Correlation is a physical-mathematical being liable to disconcert readers who seek to retain intact a distinction between what they accept as belonging to physical reality (forces of interaction, trajectories) and what we associate with the questions we ask and the instruments we invent to ask them. To which of these two categories do correlations belong?

If the solution to the problem were explicitly constructed, if the variation over time of the position of each particle in the gas were described explicitly, there would be no place for correlations, which translate the limits of approximation associated with the hypothesis of molecular chaos. Correlations, therefore, do not appear able to lay claim to the "reality" of trajectories and are even explicitly condemned by the way the integration of the equations of dynamics is defined. Recall that a system whose equations have been integrated can also be described in such a way that it becomes isomorphic with an ensemble of independently evolving "modes." In that case neither the concept of collision, nor, a fortiori, the concept of correlation retains any meaning. The existence of correlations appears to be purely relative to the fact that we don't know the trajectories, that is, to questions that *our ignorance and not the system itself obligates us to ask*.

And yet correlations in statistical mechanics belong to a realist syntax. More specifically, their status is comparable to that of the collisions themselves, whose definition they complete. Boltzmann defined collisions in terms of the velocities of particles that "enter" into collisions and by the velocities of particles that "exit" collisions. The complete kinetic definition, which can be used to incorporate the "abnormal" evolution brought about by the reversal of velocities, must take into account the fact that every collision creates correlations between particles. It then becomes possible to say that the instantaneous reversal of all velocities transforms the "post-collision" correlations, created by collisions during the normal evolution of the system, into "pre-collision" correlations that, this time, *determine* which collisions will take place.

What, then, is the status of the collisions and correlations that helped make Boltzmann's failure intelligible? I would suggest assigning them the status of a *problem*. The problem isn't objective in the sense that it would be dictated by the object, or subjective in the sense that it would designate the subject in terms of what it does or does not know. It belongs to the category of relations in the sense that relations precede the solution that distributes what belongs to subject and object, respectively. Loschmidt's reasoning belongs to the category of solutions. It assumes there are perfectly determined trajectories for particles and asks us to "see" collisions "undo" what other collisions have done. It can then restate Boltzmann's velocity distributions as an approximation, that is, it can assign temporal symmetry to the object and the $\mathcal{H}$-theorem to the subject. Collisions and correlations enable the kinetic problem to resist the dynamic solution that denies any objective meaning to Boltzmann's $\mathcal{H}$-theorem. They help "slow down" Loschmidt's reasoning process before it rushes toward a solution. A problematic space can be constructed in which the collisions introduced by Boltzmann retain a privileged relationship with the evolution of the gas toward equilibrium, providing that this relationship incorporates the *question* of correlations "before" and "after" the collision. Boltzmann had assumed that the position of the particles didn't matter. Loschmidt created a situation in which, as in dynamics, they play a role equivalent to velocities. The question of correlations transforms the confrontation of the two solutions into a problem. The problem now becomes: when and to what extent do positions "count"?

From this point of view, normal evolution, to which the $\mathcal{H}$-theorem corresponds, and the abnormal evolution introduced by Loschmidt appear as *extreme cases*. One is associated with the dilute nature of the gas, the other with the specific features of the initial state. In the first case, correlations "don't count." A collision can, of course, be defined as "creating" *post-collision* correlations, but these have no effect on the frequency of the collisions that follow. However, they will play a determining role if a reversal of velocities transforms them into *pre-collision correlations*. We are then presented with the second case, where collisions are *determined* by the correlations that characterize the initial state (resulting from the reversal of velocities). And in this case, the "abnormal" nature of these collisions is expressed by the fact that they are defined as "destructive" of pre-collision correlations.[6]

The problem of correlations is obviously central for the nonequilibrium statistical mechanics Prigogine proposed in 1962. For it is these correlations that allowed the space that lies between the two extreme cases to be inhabited, the case in which kinetic collisions suffice and the case in which all dynamic properties are required. In an ionized plasma or dense gas, we must take into account the interactions determined by the relative positions of the "actors" in the space. Even during so-called normal evolution toward equilibrium, kinetic reasoning cannot then be limited to collisions alone. Correlations as well must also be taken into account. For the evolution leading irreversibly toward equilibrium to be generally intelligible, the collision must escape the categories of "normality" (which creates correlations) and "abnormality" (which destroys correlations) and be capable of being described in general as both creating and destroying correlations. It must be an actor in the *correlation flow*. The *general* theory of nonequilibrium processes can then pose the general problem of determining how, in each case, pre-collision correlations are "destroyed" by the evolution and what role post-collision correlations play in this evolution before equilibrium is attained.

The distinction between pre- and post-collision correlations appears here as an obligation. The "correlation flow" is, by definition, temporally directed. As such, it is not able to impose the arrow of time on dynamics. Its function is to allow the general problem of irreversible evolution as such to be formulated. And this problem is not organized by the dynamic conservation of cause in effect. The correlation flow is expressed in terms of creation and destruction, not in terms of conservation. The problem of irreversibility as articulated by correlation flow is focused on the question of "roles." What plays a role in such a situation? What must be taken into account?

Let's turn now to the second stable reference: the general theory of equilibrium formulated by William Gibbs. This theory defined the "state" reached, in one way or another, by *any* irreversible evolution that takes place in an isolated system or one that is at equilibrium with its environment. Here, the formulation of the problem fully supports the abandonment of Boltzmann's project, which was to provide a precise theoretical meaning to processes associated with an increase in entropy. This meant that models describing these processes at the molecular level should have the ability to overcome the limitation of thermodynamics, which does not concern evolutions but the definition of equilibrium states, corresponding to maximum entropy or to the extreme of some other thermodynamic

potential. The only role statistical mechanics plays in Gibb's definition is to make explicit, in a general way, the relation between
this thermodynamic definition and its probabilistic interpretation, which affirms the omnipotence of dynamic theory. But it
cannot be used, any more than thermodynamics can, to define
anything other than the equilibrium state, which has become a
state of maximum probability.

Gibbs starts with a Hamiltonian representation of the system, one in which positions and velocities play equivalent roles
and where the concept of a collision, which problematizes that
equivalence, therefore, plays no role at all. If we had an exact
understanding of the system, we could describe it as a unique
being. More technically, it could be represented as a unique
point in "phase space," a space possessing as many dimensions
as the system has degrees of freedom (the position and velocity of each particle). However, in the case of a large system of N
particles, this exact representation is inaccessible. Therefore,
the system will be represented by an *ensemble:* the ensemble of
all the dynamic systems compatible with the knowledge we have
of the system under consideration. This ensemble will correspond to a "cloud" of points in phase space. The evolution of the
system, if it were perfectly known, would be represented by one
and only one trajectory in phase space. It will be represented by
the evolution of this cloud, more accurately, by the evolution of
the "density," or "probability" function, ρ, which describes the
density of points in the cloud in each region of phase space.

Time evolution of the density defined by Gibbs is part of
dynamics. Even if we do not have an exact dynamic description of the system, the very fact that the density is defined over
the phase space forces us to consider that each of the systems
in the ensemble is governed by the Hamiltonian equations
and, therefore, has one and only one trajectory, which is perfectly determined even if we are unable to calculate it explicitly.
Gibbs's representation, therefore, can be considered something

of a *solution*. It presumes the existence of the trajectory, which means that the evolution of density over time must be conservative: each point belonging to the ensemble remains one and only one point. The cloud of points behaves like an incompressible fluid, which can change shape but whose "volume," the number of points it includes, must remain constant in phase space.[7] This is known as the "Liouville theorem."

What does this mean for the description of equilibrium? The essence, the goal, the touchstone of the equilibrium statistical mechanics is that the thermodynamic properties that describe a physical system at equilibrium must be able to be represented on the basis of the average value of dynamic properties. Statistical mechanics must create a bridge between the dynamic description of unobservable dynamic actors and observable macroscopic properties. In the case of an isolated system, whose energy is defined as invariant of evolution, the density, ρ_{eq}, from which the average values can be calculated, corresponds to an ensemble known as the "microcanonical" ensemble, which comprises all possible states of the system characterized by the same energy (or Hamiltonian).[8] In other words, the description of equilibrium implies that all states compatible with the *given* value of the amount of energy in the system have an *equal probability* of representing that system.

But what of the evolution that leads toward equilibrium? How can the evolution over time of ρ result, at ρ_{eq}, in an extension of the initial ensemble to an ensemble comprising all the states characterized by the same energy? There is a radical contrast between the Liouville theorem, which subjects density to a form of conservative evolution, and evolution toward equilibrium, which implies that, in one way or another, *all* the points compatible with the overall definition of the system (in terms of energy or temperature) *acquire* an equal chance of representing it.[9]

There exists a solution of great simplicity, which involves

abandoning any reference to a "purely dynamic" evolution. It is sufficient to introduce a *coarse-grained* description, whose technical definition would be that of a "mean local statistical description per block." This description does not try to define phase space in terms of points but partitions it into "blocks," or volumes of arbitrary but finite size. Regardless of granularity, each small volume corresponds to a given characterization being applied to an infinite number of distinct points—distinct from the dynamic point of view. And every time a point belonging to the ensemble representative of the system enters a "block," *all the points in that block are considered to belong to that ensemble.* As a result, dynamic evolution is no longer governed by the Liouville theorem. What should be described, from the viewpoint of dynamics, as a dispersal of representative points throughout space, coarse-grained evolution describes as a continuous extension of the ensemble, a kind of block-by-block epidemic propagation. The behavior, ρ, is no longer that of an incompressible fluid but that of a fluid that expands continuously until it reaches a maximum corresponding to ρ_{eq}. Coarse graining does the needed trick.

That irreversibility is obtained by the introduction of an approximation confirms the interpretation that assigns it merely probabilistic status. There is, however, one small problem, which signals that a genuine coup has taken place: the temporality of the expansion of the density function becomes purely relative to the approximation, to the "granularity" chosen. This is unimportant for the description of equilibrium, which corresponds to the end point of evolution and is thus independent of the time required to reach that end point. However, this negates any possibility of making sense of "dissipative" properties, such as cross section and relaxation time, which quantitatively characterize the temporality of the approach to equilibrium and have well-determined experimental values. As we have seen, it is precisely the agreement between the predictions of the

Boltzmann equation and the experimental value of such properties for dilute gases that was used as an argument by Prigogine to claim that this equation "should have" a profound meaning. The theory of Gibbs ensembles, however, doesn't predict or preclude anything. It involves the physicist in determining—carefully—the "correct granularity," the proper approximation, needed to "save" the experimental properties that characterize the evolution toward equilibrium.

In 1962, Prigogine adopted a problematic space constructed from two classic, and incompatible, "solutions" of the problem of the evolution toward equilibrium. One was taken from Boltzmann and made use of collisions; the other, from Gibbs, denied that collisions played any particular role at all and implied a connection between evolution toward equilibrium and "coarse-grained" evolution. He demanded from the first that it provide a "profound physical meaning" for what the second defined simply as extrinsic (choice of granularity).

Prigogine started from the time evolution of density, ρ, governed by the so-called Liouville equation, which implies the ability to address both the classical and the quantum cases. At the time, statistical mechanics had been extended to quantum systems, and the Liouville equation, derived from dynamics, could be formulated in quantum terms. There is nothing surprising about this; it simply reflects the Hamiltonian heritage of quantum mechanics. In fact, the formalism adopted by Prigogine is a formalism of operators. Just as the time evolution of the quantum wave function corresponds to the application of the Hamiltonian operator, H, on this function, the time evolution of density, ρ, whether classical or quantum, corresponds to the application of the Liouville operator, L, on ρ. Except for some important differences, the Liouville equation can be used to describe both the problem of the classical evolution of an ensemble of dynamic trajectories and that of the quantum evolution of an ensemble of eigenstates of the Hamiltonian operator.[10] And in both cases, we come to the same conclusion: evolution is "conservative." A classical point remains a point. A quantum eigenstate remains an eigenstate. No classical or quantum system, if governed by the Liouville equation, can achieve equilibrium through the process of dissipative evolution.

How, then, can we express the problem of evolution toward equilibrium in phase space in such a way that the above impossibility doesn't have the final word? Is it possible to "slow down" the logical transition that leads to this conclusion? It was the discovery made in 1956 (together with Robert Brout) of an answer to this question that served as the starting point for Prigogine's work. Brout and Prigogine found that it is impossible to slow the transition of the problem toward its conclusion as long as the Liouville equation, or that of Hamilton or Schrödinger, explicitly express the *integrable* character of the system they represent. For, in this case, they describe an evolution that can be decomposed in terms of independent modes and, what's more, characterized by their periodic behavior. Not only is evolution toward equilibrium impossible by definition, the very concepts of collision and correlation, which imply coupling among the different degrees of freedom of the system, are excluded. It is possible, however, to formulate the *problem* of dynamic evolution by leaving open the question of the integrable nature of the equations that determine it. This is how Poincaré had formulated the problem of integrability: a given system was represented with reference to an integrable system, described in terms of "modes." Each of these modes corresponded to a degree of freedom of the system and evolved independently, without interacting with other modes. As for the divergence between the reference system and the system described, it was represented by a coupling term between these modes, which are, therefore, no longer independent. It is this representation that went on to be generalized. The problem of evolution toward equilibrium would be expressed by distinguishing a "free" term in the density function, where the components are deprived of correlations, and a coupling term that includes pre-collision and post-collision correlations. The problem of "correlation flow" can then be expressed.

In the formulation proposed in 1962, the Liouville operator, L, is not applied to density, ρ, but to the two components of that density, ρ_o, which represents the "correlation vacuum," and ρ_c, which represents the "remainder," that is, the ensemble of correlations determined by the initial preparation or brought about by dynamic evolution. Obviously, the two components evolve interdependently, and it is the nature of this interdependence that allows us to redefine the two "stable references" in terms of which the problem of the general theory of nonequilibrium processes might be expressed.

The equilibrium state assumes a very precise meaning here: it is the state in which collisions and the correlations that collisions continuously create *no longer have any effect on observable properties* and, therefore, on the density, which is now constant over time. This means that, in this state, the component, ρ_c, which continues to evolve because the flow of correlations does not stop, no longer plays a role in the observable outcome, no longer contributes to the definition of observable properties. The definition of density can then be reduced to the description of the "correlation vacuum" state. It is then a question of determining how the system *evolves toward* a state represented only by this correlation vacuum, ρ_o. Here, the interdependence of the time evolution of ρ_o and ρ_c becomes central. Except for the particular case of equilibrium, the component ρ_c influences the evolution of ρ_o. Before the system reaches equilibrium, this evolution is thus different from the evolution corresponding to a system without correlations. Thus, describing the approach to equilibrium becomes a question of describing the disappearance of this influence. The formal distinction between the two components, ρ_o and ρ_c, can be used to expose the implications and obligations associated with the evolution toward equilibrium.

How can we now describe the flow of correlations that expresses the interdependence of ρ_o and ρ_c? It is at this point that supporters of dynamics turn away in disgust, because Prigogine suggests a veritable formal alchemy, which, although it does indeed provide a formal expression of the Liouville equation, replaces this limpid reflection of the elegant simplicity of dynamics with a succession of shifting reflections, decomposing and recomposing an image out of contributions whose identity will be redefined from moment to moment. Four new types of operators are defined, which allow us to specify how the correlation flow contributes to the respective evolution of ρ_o and ρ_c. A characteristic of the evolution of ρ_o is that it maintains the correlation vacuum. To it there corresponds a "collision" operator, which expresses vacuum-to-vacuum transitions (passing through correlated states), and a "destruction" operator (the destruction of initial, that is, pre-collisional, correlations). The evolution of ρ_c expresses the "creation" of correlations (from the vacuum) and their "propagation." Correlations propagate through collisions. If a collision occurs between two uncorrelated particles, a binary correlation is created. But a new collision with one of the two particles resulting from the first creates tertiary correlations between the three particles, and so on.

The crux of the alchemy in question is, as I have indicated, to slow down the transition between the dynamic problem and its solution, that is, to break down the problem by creating "dwelling" spaces for the beings that are required by the characterization of the collision event or, more specifically, of the chronology of successive collision events leading to equilibrium. This chronology helps clarify what is meant by the approach to equilibrium, but it is not capable of imposing this approach in opposition to the dynamic solution, which denies it in principle. Just as correlation flow implied the obligation to distinguish between pre- and post-collision correlations, the definition of operators for the collision, creation, destruction, or propagation of correlations implies the obligation to determine a direction for time. Anyone who rejects this obligation, which falls outside the field of dynamics, will consider Prigogine's approach to be begging the question.

To provide a more intuitive image of Prigogine's "chronological" representation, we could say that his goal is not to follow the evolution of positions and velocities that define this system but to study the evolution over time of relations that define the problem presented by the evolution of the system. In other words, instead of being formulated once and for all, the problem of the identity of the system evolves over time. Obviously, the relations that define this identity must not be confused with interactions. Physical interactions such as gravity are neither created nor destroyed, they exist. The only thing that changes over time is the intensity of their effect. In contrast, correlations and the relations they build up "normally" have no influence on the evolution of density, but in the case of "time-reversed" evolutions their effect is dominant. In 1962, such a "chronological" representation was not seen as being irreducible to a dynamic solution, to a solution that would annul the need to take correlations into account, because it would treat evolution in terms of the equivalence between cause and effect. But this representation constructed a problematic space in which the physical meaning of the approximations used by physicists and the conditions of validity of those approximations could be identified. Approximation no longer referred to a generalized "lack of knowledge." It assumed an "intrinsic" sense, based on a differentiation that intrinsically characterizes a system and allows certain terms to be considered insignificant in comparison to others.

In particular, the reasons for the validity, and the limits of validity, of the Boltzmann equation are perfectly explicit. Its domain of validity, the description of dilute gases, corresponds to conditions for which the evolution of ρ_o can be considered *autonomous*: the influence of the component ρ_c on ρ_o is negligible at any moment. Correlations, which collisions naturally continue to create, are irreversibly "propagated" but do not retroact, do not influence the frequency of collisions. The correlation vacuum at time t then depends solely on the correlation vacuum at the initial moment. That is why, as we have seen, the case of dilute gases helps distinguish two types of collisions, those that belong to the "normal" evolution of a system, where correlations create correlations that will have no observable effect, and "reverse" collisions, belonging to the evolution resulting from a state prepared by reversing velocities, which are defined as destroying correlations.

There is another important result, which is that in the long time limit, it is possible, for any system, to write a separate evolution for ρ_o. For this limit, the evolution of ρ_o no longer depends on ρ_c. In the long time limit, every system reaches equilibrium. Therefore, Prigogine's 1962 proposal can be used to specify what has "become insignificant" once equilibrium has been reached. But, except for the case where the validity conditions of the Boltzmann equation (the molecular chaos hypothesis) are satisfied, it does not seek to define the equivalent of an $\mathcal{H}$-theorem, that is, a general definition of entropy valid throughout the evolution of the system. Such a definition should resist the Loschmidt challenge: the defined entropy should increase even during the system's "abnormal" evolution resulting from the reversal of velocity. In 1962, however, this result was not (yet) being sought. The evolution operators allowed for a "general theory" of nonequilibrium processes in the sense that they could be used to specify the problems imposed by each of those processes. But they could not be used to extract a more general lesson, to claim how, in one way or another, all those processes helped increase the entropy of the universe.

9

Boltzmann's Heir

In 1962, Ilya Prigogine was Boltzmann's successor. Like him, he accepted the verdict concerning the second law of thermodynamics and wasn't trying to challenge the power relations that imposed such a verdict. Statistical mechanics is subordinate to dynamics (and quantum mechanics). It cannot challenge it. After 1970, Prigogine became Boltzmann's heir in the sense that he intended to resume the fight that Boltzmann had abandoned. For the sake of the second law of thermodynamics, our conception of dynamics *must be changed.* It is no longer a question of providing the necessary approximations with physical intelligibility but of demonstrating the need to "enlarge" dynamics. In 1962 it was a question of "slowing down" the problem in order to distinguish its various components. Now it was a question of challenging the very formulation of the problem.

This transformation began with a physical-mathematical development that appeared in 1969, the transition from the use of correlation flows to the study of the *dynamics of correlations.*[1] What made this transition possible was a new physical-mathematical result: it is possible to exactly decompose the evolution of ρ given by the Liouville equation into *subdynamics,* each of which was subject to its own law, in such a way that one of those subdynamics represents the evolution responsible for the final establishment of equilibrium.[2] This particular subdynamics will then be able to serve as the "thermodynamic" representation of the system. The remaining subdynamics are referred to as "nonthermodynamic," for what they describe may dominate the evolution of ρ for a given time but is fated to be "forgotten" at equilibrium. They can then be assimilated to "contingency," whose effects may be spectacular but are transitory. Dumbly, obstinately, irresistibly, the system evolves toward equilibrium.

The separation into subdynamics proposed in 1969 corresponds to a principle of classification that is valid *at every moment*, not only in the long time limit. The "macroscopic" or "thermodynamic" definition of the system is no longer the result of an approximation, no matter how well established. It now appears as a form of stratification generated by dynamics itself. Both the definition of the macroscopic level and the dissipative evolution now claim to correspond to a dynamic (that is, intrinsic) property: the increasing insensitivity of the evolution of ρ to nonthermodynamic subdynamics. This makes possible a formal definition of systems for which the distinction between "macroscopic level" and microscopic description (dynamic or quantum) independent of any approximation may be defined: a necessary (and sufficient) condition is that the equation determining their evolution allows the transformation that leads to the definition of the separate "thermodynamic" subdynamics. An integrable system in the Poincaré sense does not allow such a transformation—no classification into distinct subdynamics is produced. The system is and remains fully sensitive to all the details of all its states; it preserves the entire memory of its own past.

Before further describing this new way of presenting the problem, I'd like to comment on its implications and, especially, ask about the possibility of its being recognized as an "event," that is, as authorizing a new story. This story would relate that

Galilean physics had invented the first physical-mathematical representation that incorporated time into the very definition of its object: the instantaneous state of any "Galilean body" is defined by the equivalence between the past that produced this state and the future it will produce. And the event associated with the classification into subdynamics would reflect the possibility of a different way of incorporating time into the definition of the object. It is in relation to the future, that is, the equilibrium toward which the system evolves, that the evolving entities will be defined. Correlatively, equivalence is no longer a pivot. The "future" will, on the contrary, be defined by its relative insensitivity to certain aspects of the present.

It is important to bear in mind that the results formulated by Prigogine and his team are formal in the sense that they demonstrate the possibility of an operation but provide no general recipe, applicable in every case, for completing the operation. There are a small number of special cases, the "shared assets" of specialists in statistical mechanics that, like Friedrich's model in quantum mechanics, constitute an effective field of experimentation. These can be exactly described according to different formal approaches and, therefore, allow those approaches to be compared. But these cases "prove" nothing, they are the terrain for what can be called "experimentation with functions," in a sense that relates doubly to Gilles Deleuze. For, on the one hand, Deleuze introduces a radical distinction between the concepts created by philosophy and the functions that are the product of science. However, he also defines the practice of philosophy as "experimental," in that philosophy experiments with the way the concepts it creates resolve or help resolve a problem.[3] The concept of experimenting with functions, therefore, specifically refers to the work of physical-mathematical invention, where the question is asked of what a new type of physical-mathematical function can do, what it requires, and the obligations it entails. To the extent this inventive activity does not involve the construction of new matters of fact to which a function would refer, but the possibilities for redefining well-known "cases," it is centered not on the "can do," on the power of description, but on "how to describe."

Alluding to the Deleuzian definition of philosophical concepts to singularize the question of "how to describe" indicates that this question, which relates to Deleuze's notion of "functions," is likely to resonate with the conceptual problems of philosophy. But this singularity is correlated with another concerning the processes of scientific innovation, namely, the difference between what is, and what is not, recognized as a scientific "event." Like philosophy, scientific innovation must take place "among friends"—friends of functions, not friends of concepts. In this case, only those physicists who maintain a "sympathetic" relationship to the problem of the irreversible approach toward equilibrium can, at this stage, find the proposed solution of interest. And the only ones who maintain such a relationship are those who are dissatisfied with the hierarchy of physics, that is, the approximations that ensure the transition from the "fundamental" level to the "dissipative macroscopic" level. For those satisfied with this hierarchy, Prigogine's formalism "provides nothing new" because it doesn't enable them to create new "matters of fact" presenting new properties, merely to obtain known properties in a different way. And in fact, they could also claim that all of Prigogine's formalism assumes a redefinition of density, whose precondition is that time has a fixed direction. But, from the viewpoint of dynamics, this is an artifact, a hypothesis that has been added without justification. Consequently, the stratification that reveals the dynamic significance of dissipative evolution is largely an act of prestidigitation because the equations of dynamics have not been deprived of their power to deny any physical meaning to the difference between past and future.

Yet, for there to be innovation in the scientific sense,

(nearly) all "competent" scientists, and certainly not only "friends," must be *forced* to recognize the superiority of the new proposal. Here we have a typical example of the way the history of the sciences interweaves distinct temporalities. Time, long and weighty, is here part of the definition of what is satisfactory and what is problematic, and this definition is always immediately connected with the ecology of practices, in this case, their hierarchical organization. If the inability of dynamics to account for irreversible evolution had handicapped physics from an ecological point of view, the 1969–70 hypothesis might have become an event, might have constituted an innovation. In our story it is a step, not the end of the story.

Let us return now to the early seventies, when Prigogine took a step that, in a different story, might have been decisive. For it was now an "enlargement of mechanics" that Prigogine tried to interest his colleagues in, an enlargement such that the symmetrical equations of classical and quantum evolution would now correspond only to a *particular* representation. This representation is, de facto, the only one usable in the particular case of integrable systems (or, in the quantum case, of systems that can be described in terms of eigenstates of the Hamiltonian). But in the "general case," described by the field of statistical mechanics, where the concept of thermodynamic equilibrium has a meaning, this representation loses its exclusive character.

The idea of enlarging mechanics shows that Prigogine recognized that irreversibility implies questioning the most stable assumptions of dynamics. As I attempted to show in connection with the definition of a "state" (see *Cosmopolitics*, Book II), dynamics is a highly coherent edifice: accepting one element as self-evident may obligate us to accept all of them. Correlatively, rejecting one of its consequences may result in questioning the entire edifice. It is here that the radicalization of Prigogine's approach becomes most evident—the shift from the question of the *solution* of the equations of evolution to the question of the equations themselves. It is no longer a matter of slowing down the process of solution so as to open a space in which it can be negotiated but of redefining the problem in such a way that it leads "directly" to the sought-for solution, that is, a generalized form of the $\mathcal{H}$-theorem.

This is the challenge of what Prigogine and his colleagues called the *physical representation* of dynamic actors. Dynamics in the customary sense involves interacting entities, and those interactions, identified by the instantaneous equivalence of cause and effect, are by definition indifferent to the direction of time. The very definition of "entities" and interactions will now be at stake as they themselves will be asked to determine how the distinction between past and future is defined, thereby affirming a world in which equilibrium belongs to the future.

The reader may recall that interacting entities in classical dynamics admit a privileged mode of representation, which can be actualized when the dynamic system is integrable: a cyclical representation in which interactions have been incorporated in the definition of the entity and where the system can thus be represented in terms of the independent evolution of each of its degrees of freedom. The Schrödinger equation refers to the same kind of ideal, but this time, it is the eigenstates of the Hamiltonian operator that correspond to the privileged representation. In both cases the ideal (or spectral) representation is defined once and for all: entities defined as independent, not engaging in mutual interactions, preserve this definition throughout the process of dynamic (or quantum) evolution.

In contrast, the "entities" defined by the "physical" representation (broken time symmetry) have an *evolutionary identity*. The fixed point from which the representation is constructed is not, like the = sign, given at each moment and conserved from moment to moment. It is located in the future. "Physical" entities are defined as evolving toward the identity they will have at equilibrium (if they are part of an isolated system). At equilibrium, when the system is defined by average properties only, these "physical entities" will mutually ignore one another. This mutual ignorance reflects both the result of an evolution and the fixed point from which the evolutionary identity of the particles can be defined. Over the course of time, and irreversibly, the identity of the entities—their relative differences, what they are sensitive to—is continuously redefined, and it is this redefinition that the equations of the dynamics of "broken symmetry" describe.

This new statement of the problem is reflected in a new definition of the vacuum of correlation, now written $^P\rho_o$ (where ^{P} is the "physical representation"). In effect, the definition of the vacuum of correlation at each instant constitutes the definition of the identity, at that instant, of the entities framed by the physical representation.

The principle of transformation leading to the "physical representation" can be expressed rather simply.[4] Unlike the relationship of complete equivalence among all the "canonical" representations that can be applied to a Hamiltonian system, the representation resulting from this transformation is equivalent to the dynamic representation only for what concerns macroscopic properties, those we can observe and which correspond to an ensemble description of the system. This transformation actualizes the existence of a form of "selection" that the physical representation will attribute to irreversible evolution itself. It is based on the property of "separability" of the "thermodynamic representation of the system" established in 1969, which formalized the possibility of a classification that distinguishes between those correlations that contribute to the definition of the state of equilibrium and all others. This separability has now been incorporated into the definition of the new "vacuum of correlations," $^P\rho_o$, in such a way that its evolution describes the evolution of the identity of "physical entities."[5]

The definition of the vacuum of correlation in its physical

representation, therefore, includes, at each moment, those correlations associated with the "thermodynamic" subdynamic that
represents the component of evolution leading to equilibrium.
All the correlations associated with other subdynamics, which
will be "forgotten" at equilibrium but contribute to the evolution
toward equilibrium, especially the pre-collision correlations
associated with the initial preparation of the system, belong
to the complementary component $^P\rho_c$. These are not included
in the definition of entities that evolve toward equilibrium but
could be said to belong to the contingency of the path actually
followed in moving toward equilibrium. By definition, they stop
contributing to the description of the system when equilibrium
has been obtained.

In proposing this physical representation, Prigogine fulfills
Boltzmann's project. He is able to formulate an $\mathcal{H}$-theorem that
is valid not only for the restricted field of dilute gases, but for all
domains in which the macroscopic level has an intrinsic definition. The entropy defined by the $\mathcal{H}$-theorem for dilute gases
"benefited" from the fact that, in this case, the evolution of the
correlation vacuum in the customary representation already satisfies an independent evolution equation. The entropy defined
by the new $\mathcal{H}$-theorem within the framework of the "physical
representation" expresses the generalization of this situation. It
has the properties required by any kinetic model of thermodynamic entropy: it reaches a maximum whenever it is a function
of the correlation vacuum alone, that is, when all the correlations associated primarily with the particular nature of the initial preparation have stopped contributing to the observable
evolution of the system. Thus defined, the increase of entropy
is no longer "merely probable," *it is no longer contradicted by the
"abnormal" evolution resulting from the reversal of velocities.* In fact,
the opposite is true. It makes explicit what the reversal of velocities dissimulated. In the customary dynamic representation, the
reversal of velocities appears unproblematic, a straightforward

implementation of the time symmetry of the equations for which pendular movement provides an immediate illustration. The intuitive analogy with a pendulum that appears to "return to its past" while its velocity is spontaneously reversed is destroyed by physical representation. Whenever the pendulum "takes off again" in the opposite direction, it is certainly the "same" pendulum, governed by the "same" equations, whereas in the physical representation, the operation of reversal entails a radical redefinition of the dynamic identity of the system and its constituent entities.

The physical representation proposed by Prigogine can be used to illustrate what the reversal of velocities "does." It is not the "same" system governed by the "same" equations of evolution, which seems to return to its past, it is an *entirely different system that continues its evolution.* For the definition of a state in this representation no longer refers to an instant but entails a relation to a past that has already sorted correlations and to a future equilibrium in which that sorting will be completed. Its definition answers the question: how can the present contribute to the future? The reversal of velocities radically transforms, although in transitory fashion, the answer to this question because it restores to a past that is already "past" the power to contribute to the future. The correlations produced by collisions, which had been integrated in the correlation vacuum, that is, in the definition of physical entities, are transformed into "pre-collision" correlations belonging to $^P\rho_c$. This transformation is expressed as an entropy "leap," a discontinuous variation in entropy, which corresponds to the redefinition of both $^P\rho_o$ and $^P\rho_c$. If the system were approaching equilibrium, the entropy would suddenly decrease, which does not contradict the second law because it is not spontaneous but imposed from outside the system. After the reversal, the entropy of the "new" system again increases, whereas the "new" initial conditions determining this evolution are gradually "forgotten."

The physical representation, therefore, can be used to express what Boltzmann had been unable to make explicit in light of Loschmidt's objection: the difference between the "normal" state and the state resulting from a reversal of velocities. The probabilistic interpretation is unable to clarify this difference. Relative to an instantaneous configuration, probability can be used to explain the privilege of the state of equilibrium: it is the most probable state because the overwhelming majority of the a priori possible dynamic states of the system correspond to its (macroscopic) characterization. But probability is incapable of expressing the difference between two kinds of nonequilibrium state: states whose evolution leads the system toward equilibrium and states whose evolution distances the system from equilibrium. Both states are similarly "improbable," while from the point of view of irreversibility they should be radically distinct. The situation is made even more unsatisfactory because the probabilistic interpretation implies that for each state of the first kind there corresponds a state of the second, which means that both kinds of state have *the same (im)probability.* Indeed, for each dynamic state resulting in the first kind of nonequilibrium state, whose future is equilibrium, there corresponds, by the reversal of velocities, a dynamic state of the second kind, whose future leads away from equilibrium. Consequently, the probabilistic interpretation does not help us to understand why evolution toward equilibrium is privileged, nor why the first kind of "normal" nonequilibrium state is easy to prepare, whereas the definition of the second kind of nonequilibrium state implies having recourse to a theoretical argument such as "what if the velocities of all particles were simultaneously reversed?" It is incapable because it is dependent on dynamics, which states that all dynamic states are equivalent.

We can then say that, in 1972, Prigogine gave the problem of irreversibility a scope that, had it been accepted, would have implied a "rewriting" of the history of physics. This involved

narrating Galilean equivalence and subsequent developments
that resulted in the classical and quantum Hamiltonian for-
malisms in a way that distinguishes what the customary narrative confuses: the two meanings of "first." Naturally, the
Galilean equivalence is "first," not only historically but also as a
de facto technical reference. In the new narrative, systems that
can undergo dissipative evolution continue to be defined on
the basis of their *contrast* with systems that empower Galilean
equivalence. It is on the basis of such systems that the language
of physics, as it poses the problem of irreversibility, is constructed. More specifically, it is on that basis that the language
defining irreversibility *as a problem* is constructed, because,
in phenomenological physics, chemistry, and in all kinetic
descriptions, irreversibility is assumed *unproblematically*. But
Prigogine intended to deny this primacy its relationship with
any claim to being "fundamental," from which everything else
would be derived by approximation. Galilean equivalence is not
the Ariadne's thread among dynamic phenomena that Leibniz
claimed. It does not have the benefit of designating the point
of view from which the labyrinth of these phenomena would be
exposed to secure exploration as guaranteed on the basis of first
principles. Or, if we wish to make use of a Platonic-Maxwellian
reference, to designate the contrast between the cave in which
appearances reign and the truth of the Queen of Heaven.

To the two distinct meanings we can assign to the "primacy"
of symmetrical description there correspond two distinct meanings of the term "emergence." Prigogine rejected the first: irreversibility does not "emerge" from reversibility as a property of
particular physical systems. On the contrary, it is the motions
that allow us to claim the equivalence of past and future—motion
along a smooth, inclined plane, a pendulum or an ideal spring,
the periodic movement of the planets—that retroactively appear
to be particular and *not representative*. However, he accepts
the second meaning: the dynamic equations with "broken"

symmetry "emerge" from the description provided by classical dynamics.

This second meaning is acceptable because emergence, here, expresses the historical inscription of irreversibility or dissipation as *problems* for post-Galilean physicists. Words themselves reflect this inscription: irreversibility denies reversibility and dissipation denies conservation. The "negation" follows from the fact that the language of dynamics asserts symmetry and the conservation of cause in effect because it has been constructed on the basis of exceptionally simple situations in which Galilean equivalence can be confirmed. From then on, the possibility of describing dissipative evolution *must* be expressed as a breaking of symmetry, and irreversibility seems to emerge "conditionally," if and only if the system satisfies certain conditions (the possibility of constructing the physical representation depends on the possibility of defining a "macroscopic level," a possibility determined by a precise "dissipative criterion").

The construction of this "emergence" as a response to a problem imposed by history has an immediate consequence: the breaking of the symmetry of the equations of dynamics brought about by the "physical representation" of particles must, by definition, result in *two classes* of symmetry-broken equations, one corresponding to evolutions that situate equilibrium in the future, and the other to evolutions that situate it in the past. The breaking of symmetry must be accompanied by a statement that denies any physical meaning to the second class of equations. Objections are frequently raised about the need for such a statement: the exclusion of the second class is ad hoc, it cannot be justified; we have returned to the initial situation after a pointlessly complicated detour, and so on. Boltzmann was forced to yield to Loschmidt's objection: his "mechanical" model of irreversibility depended only on the selection of particular initial states corresponding to the hypothesis of molecular chaos. Other initial states should unquestionably cause evolutions corresponding to "antithermodynamic" behavior, where the system spontaneously evolves away from equilibrium. In Prigogine's proposal, such selection takes place with respect to the equations, not the initial states. But isn't this proposition equally arbitrary?

There is one significant difference, however. In dynamics, the question of initial states and the question of the equations of evolution do not have the same status. The initial state is, by definition, *whatever* the physicist or his ideal alter ego, Maxwell's demon, defines it to be. The definition of this state is, therefore, "free"—it relates to the "preparing subject" or the contingency. As for the equations of evolution, these are given, and govern *any* initial state regardless. In this context, any limitation on the definition of states is automatically interpreted as affecting the ability to describe or prepare, but incapable of restricting the demon with 10^{23} arms, capable of simultaneously reversing the velocities of all the particles in a system. In contrast, the selection imposed by the "physical representation" affects the equations supposedly shared by the demon and the physicist. Therefore, the demon is not "free" with respect to the definition that imposes this selection. More technically, the state resulting from the reversal of velocities, the state whose definition expresses the demon's sovereign freedom and omnipotence, no longer challenges the selection that places equilibrium in the future. The meaning and the effects of this reversal, which express what the "subject" in the presence of the "object" can do, are contained in the relationship to the future incorporated in the equations of evolution with broken symmetry. In other words, the selective claim "we live in a world where equilibrium belongs to the future" can be said to hold pride of place with respect to any particular description, for no dynamic evolution, no matter how demonic, will be able to contradict it.

No one would dream of contesting the fact that the formulation of two distinct classes of equations of evolution with broken symmetry is simply the mathematically inevitable consequence of the symmetry of the initial equations. When symmetry is broken, it is always broken in two. It is the "initial" symmetry that creates the necessity of selecting and, if we accept Prigogine's proposition, we could also say that, in so doing, this necessity transforms the meaning of dynamic symmetry in turn. For, this symmetry does not express some ideal intelligibility but betrays the dependence of ordinary dynamics on the particular cases that have allowed it to escape the need to define how past and future differ. That is why Prigogine can claim that his selective statement that equilibrium belongs to the future precedes any physical description, that it is not part of what physics should demonstrate, explain, or justify, but what physics must accept as a starting point.

If we were to bring out the philosophical "heavy artillery," we might be tempted to refer Prigogine's selective claim to a transcendental order, but its transcendental character, to the extent that it refers to the breaking of symmetry and the necessity of choice, is clearly relative to the history of physics. The difference between past and future finds its expression in dynamic terms in the selection of one class of equations and the exclusion of another. But the question this selection seems to impose—why this class rather than another—*does not have to be asked.* It concerns a mathematical artifact determined by the definition of the difference between past and future as the breaking of symmetry. What the selection selects is something that physics literally *does not have the right* to question because its experimental practice assumes it. The selective statement that "equilibrium belongs to the future," far from being arbitrary, would correct the arbitrary, the contingent particularity of the history of dynamics that took no account of the obligations associated with a coherent physical description. It would, therefore, eliminate a possibility, that of an evolution that would contradict the second law, which is nothing other than an artifact, the artificially maintained extension of a syntax that expressed the now unrepresentative singularity of the first objects of dynamics.

By 1972, Prigogine could claim that the requirements of

his program were satisfied. He had defined the general class of dynamic systems for which the concept of thermodynamic equilibrium has a meaning. He had shown that the second law does not require a probabilistic approximation for dynamic description, but belongs to another type of dynamic description authorized by this class of systems. But did he define a new kind of physical-mathematical "factish" capable of "repeating" the Galilean miracle? Was he capable of supporting the claim that his equations with broken time symmetry owe nothing to the physicist's standpoint? In fact, Prigogine adopted a position he compared to Niels Bohr's complementarity principle. The physicist can choose to accept the customary representation given by dynamics, where the very concept of correlation is superfluous. This representation, the only relevant one for integrable systems, can be referred to for any dynamic system, even if it is practically useless when describing most of them. But if physicists choose to continue to refer to it, they won't obtain an exact definition of equilibrium or any of the experimental properties that characterize the approach to equilibrium. If these physicists take an interest in equilibrium and the approach to equilibrium, they will have to explicitly incorporate the arrow of time into the dynamic representation, which they can now do in an exact manner. The "factish" does not, then, define the way nature explains itself. It expresses the fact that this nature imposes a choice concerning the way it will be explained. This means that what we have referred to as the great achievement of dynamics, the coincidence between "description" and "reason," is not reproduced. More precisely, the way this coincidence is achieved no longer refers reason to nature "in itself." Reason is situated, it cannot be dissociated from the physicist's question.

As we saw in Book IV, complementarity is not really popular among physicists. However, unlike Bohr's proposition, there is nothing ironic about Prigogine's. The two representations he puts forth as complementary do not entrap physicists, do not expose them to the temptation of defining a "matter of fact"; they remind them that matters of fact have meaning only with regard to a well-defined possibility of observation. Here, the two "complementary" representations refer to two distinct ways of relating to a dynamic system, of formulating the problem of its behavior. But each of them authorizes the physicist to claim that there is nothing arbitrary about the way the dynamic system is represented: the system will satisfy whatever requirements its chosen representation defines. Thus, there exists a possible history in which Prigogine's proposition would have been seen as enlarging and stabilizing that of Bohr.

However, that history is not ours. In our history, the complementarity suggested by Prigogine has a weakness that is not theoretical but historical: for the majority of physicists, his "either, or" cannot claim to possess the incontrovertible character of Bohr's complementarity argument, which refers to a necessary choice among the quantum analogues of Hamiltonian variables. Why abandon symmetrical representation when it allows us to define equilibrium in terms of an approximation? Why not leave the approach to equilibrium to the field of "phenomenology," to the construction of relations that are experimentally relevant? Why give it the power of redefining the scope and meaning of the "attributes of the Queen of Heaven," of forcing her to assume the diminished status of being only one of two complementary options? In accepting the transformation of representation suggested by Prigogine to replace approximation and phenomenology with exact definition, aren't we using a sledgehammer to swat a fly?[6]

One exception was Léon Rosenfeld. It was Rosenfeld's influence that led Bohr to explicitly associate macroscopic measurement with the permanent, which is to say irreversible, production of a mark. As we saw in Book IV, Rosenfeld was also interested in the work of Daneri, Loinger, and Prosperi, who associated measurement with the evolution toward equilibrium.[7] During the last years of his life, Rosenfeld collaborated with Prigogine and his team, whom he considered—and coming from Rosenfeld, the compliment carries some weight—the author of a "non-trivial extension" of quantum formalism.[8] For Rosenfeld, Prigogine's result was deeply satisfying. It confirmed Bohr's claims that the wave function and the Schrödinger equation have no physical significance independent of measurement and that the definition of observables and macroscopic description were inseparable. But this result removed the suspicion of arbitrariness created by Bohr's rhetoric. It is not the free choice of the observer that defines what will be a measurement device and what will be a quantum system. To function as a measurement device, a system must accommodate the definition of a "macroscopic level," must be able to evolve toward an equilibrium that will be, since it is a measurement device, associated with the production of a "mark." There is nothing arbitrary about the observables that correspond to the definition of the device; they are defined by the type of equilibrium to which this "thermodynamic" system (that is, one that satisfies Prigogine's dissipative criterion) is susceptible. So, for Rosenfeld, the Copenhagen interpretation was freed of any hint of "idealism." Measurement is an intrinsic part of physics and not the decision of a human being. Its inevitable role in quantum mechanics can from now on be expressed without any appeal to human subjectivity. Quantum mechanics is unique in that it must rely on *two* complementary dynamic representations, one, explicit, expressed in terms of the wave function, the other, assumed implicitly by the reduction of the wave function, expressed in terms of broken time symmetry evolution.

The various ways of resolving questions of interpretation do not interest physicists uniformly, and their interest, except when it is constrained by an innovation that offers new opportunities for relating theory to experiment, is largely determined by a form of aesthetic-historical evaluation. The idea that a subordinate science like statistical mechanics, focused on the study of systems with N degrees of freedom, about which we obviously cannot expect an exact definition in terms of trajectories or wave functions, might provide the answer to one of the "mysteries" of quantum mechanics is, for those uninterested in the question of the arrow of time, rather unpalatable. That same problem is addressed by the grandiose hypothesis of von Neumann and Wigner, for whom the observing consciousness determines the reduction of the wave function, and the even more grandiose hypothesis of "multiple worlds" of Everett and Wheeler, for whom every "reduction" corresponds to the fact that the entire universe is multiplied into as many universes as superposed states composing the wave function before its reduction. These hypotheses are consistent with the "revolutionary" aesthetic of twentieth-century physics, whereas the solution offered by Prigogine contradicted that aesthetic. Doesn't his solution assume that we acknowledge a dimension of the observable world that has been denied in "revolutionary" fashion by the probabilistic interpretation of entropy?

Prigogine's quest, therefore, has continued. If it is to be accepted by physicists, the arrow of time mustn't require that they acknowledge that there is a "profound physical meaning" associated with the observables characterizing dissipative evolution toward equilibrium. It must win them over in spite of their refusal to recognize this meaning. To convince those who accept no obligations other than those imposed by dynamic description, it is dynamic description itself that must impose a transformation of representation, one that will also confer meaning on the arrow of time. What is needed is a factish that repeats the Galilean miracle.

10

The Obligations of Chaos

Every dynamic system is defined by a set of equations that, no matter how complicated the system, can assume the same "canonical" form. Whether or not we can explicitly write the solution of those equations, the trajectory of the system, we are required to say that every dynamic system is characterized by one and only one trajectory. As we saw in *Cosmopolitics*, Book III, Pierre Duhem had already fought against this notion, which gives the concept of trajectory unconditional validity, independent of the contingent fact that we may (or may not) be able to write an explicit equation for that trajectory. For Duhem, dynamic trajectory was a method of mathematical representation, a construction that entailed the possibility of determining, based on the value of an ensemble of measurable properties, the evolution of the value of those properties over time. As such, it will always have a mathematical meaning, although it might lose its physical meaning. In order to have a physical meaning, an additional condition is required: the possibility of deduction must be robust with respect to approximation. Duhem referred to the mathematical definition of "pathological" trajectories. For example, two neighboring trajectories, initially arbitrarily close, will always ultimately diverge, that is, will correspond to qualitatively different behaviors. According to Duhem, in the case of such pathologies, the physicist must conclude that her equations do a poor job of expressing the problem given that the "mathematical deduction" they authorize is "forever unutilizable," regardless of the precision with which the practical data can be determined in any conceivable future.[1]

Some seventy years after Duhem, Prigogine would reprise the same type of argument, but in a different problematic landscape. The pathological beings presented by Duhem have become legion and there is no longer anything pathological about them. They now form a well-defined class of systems, the "chaotic" systems. Their "nonrobust" character in the face of approximation is now referred to as "sensitivity to initial conditions," and this sensitivity can be characterized by a number, the Lyapunov exponent, which must have a positive value.

Strictly speaking, the contemporary notion of chaos is not at all specific to the equations of dynamics. It is applicable to all deterministic equations of evolution, whether they are dissipative or conservative. In the public's memory, it is also associated with a computer simulation by Edward Norton Lorenz involving meteorological evolution, a form of dissipative evolution. The equations used by Lorenz were perfectly deterministic and corresponded to a very simple meteorological model. But Lorenz found that radically divergent evolutions resulted even when using initial conditions that were so similar that their difference was not displayed by a computer. The "butterfly effect" is the parabola of the practical consequences of the divergent evolutions determined by those highly simplified equations. Long-term meteorological forecasting is impossible; the growing precision of measurement or the increasing power of computers will not indefinitely increase its scope over time because, in the long-time limit, any difference, no matter how insignificant—a butterfly fluttering its wings thousands of miles away—can "count." Of course, in itself the beating of a butterfly's wings is not "responsible" for anything, it is not the "cause" of the divergence. Rather, it symbolizes, hyperbolically, the sensitivity of the evolution to initial conditions, the way in which the evolution of two chaotic systems, initially arbitrarily similar, will produce behaviors that, over the long term, will become as different as if their initial conditions had been randomly selected.

As such, the butterfly effect can appear as a limit to forecasting efforts, whether meteorological or those associated with the large-scale computer models often used in economics. It can also become a resource for renewed metaphors for games of chance and necessity, renewed in the sense that unpredictable behavior and rigorous adherence to causal explanations are now quite comfortable together. But for mathematicians it symbolizes a problem, not a result. More specifically, it introduces a field of research that addresses the new questions required by chaotic behavior and the obligations they entail for the practice of simulation. That practice is now simultaneously crucial and traitorous—crucial because the equations for chaotic systems cannot be integrated exactly and must be simulated, traitorous because, in the case of chaotic systems, no individual simulation is representative.

How can the topology of the landscape of solutions be characterized? How can the transition from the fact of unpredictability to the disclosure of a regularity underlying that unpredictability be made? Defining a chaotic system by a positive value of at least one of its "Lyapunov exponents" is one of those new characterizations. The Lyapunov exponent doesn't characterize individual behavior but an ensemble of behaviors evolving from distinct initial conditions, arbitrarily close to one another, or, to put it differently, it characterizes the topology of the space where those different behaviors will be represented as so many trajectories. Having a positive Lyapunov exponent means that in the corresponding dimension of that space, the *distance* between two initially neighboring trajectories will increase, or expand, exponentially with time (a negative exponent means that the corresponding distance will decrease, or contract, exponentially with time).

The question of chaos is a question that cuts across scientific practice both negatively and positively. Negatively, it condemns the hope of progress tied to the development of equations associated with increasingly detailed descriptions, for which increasingly powerful computers would compute increasingly illuminating results. Positively, it introduces a new relevance for mathematics, one that no longer gives pride of place to physics. The invention of new questions, new distinctions, new criteria would no longer be primarily tied to the question of processes that *explain* a behavior, because chaos marks the limits of the power such explanations provide. It would result in the creation of a "nomadic aesthetic," wherein the relevance of mathematics would become local and circumstantial. That is why some authors see in it the promise of a mathematical invention of the question of forms.[2] The wave that forms on the surface of an ocean does not imply the theorization of the ocean as a whole but organizes its own questions. The ocean thus loses its status as a "cause," for which the wave would be the consequence. Rather, it is defined solely in terms of the conditions that must be satisfied for the wave to be produced. And milieus other than the ocean can provide meaning to those same conditions, in such a way that the "wave theorist" will be able to travel wherever the mathematical object "wave" can be actualized.

Conversely, protagonists from the most diverse fields of science are inclined, for better and for worse, to grasp the new mathematical questions and objects offered by chaos in order to propose a transformation of the questions, modes of judgment, and priorities in their field. And among those protagonists we now find Ilya Prigogine, who challenged chaos to provide new arguments in favor of the arrow of time. Chaos, if it is to be captured and redefined in the quest for a theory of irreversibility, cannot be dissipative chaos. Prigogine, who is best known for his work on dissipative systems (operating far from equilibrium), was not interested in nomadic chaos, which has become a common reference for far-from-equilibrium thermodynamic systems and many other models, from metabolism to the economy. He would devote his efforts to the kind of chaos that could be defined by equations that are symmetrical over time, the kind of chaos, therefore, that may be able to provide reasons for breaking the symmetry of such equations. And in so doing, he would take an interest in a class of systems that had been the subject of a body of work already considered classical in statistical mechanics, work done primarily by the Russian school of mathematical physics led by Kolmogorov, Arnold, and Moser.

Although the work of the Russian school led to the definition of systems currently referred to as chaotic, this was not in fact their goal. They were focused on studying dynamic systems that Poincaré's theorem defined as "nonintegrable" (see Book II). The Russian mathematicians were the first to see in Poincaré's "resonances" something other than an obstacle blocking the road that led from the problem of dynamics to its solution. They studied the problematic landscape of qualitatively differentiated dynamic behaviors brought about by those resonances according to their density in phase space. This meant that they were investigating a spectrum of complicated behaviors where, based on an initial state, a (Hamiltonian) system could behave regularly or randomly. From this point of view, the periodic behavior of integrable systems is nothing more than a particular case, defining one extreme of the spectrum, the case for which all behaviors are regular. At the other extreme we find the so-called K-flow Hamiltonian systems (K for Kolmogorov). Regardless of the region, finite but arbitrarily small, of phase space, that region will always include points that bring about "random" behavior. K-flows have the property of being "sensitive to initial conditions" that will later be associated with chaos.

From the point of view of Gibbsian statistical mechanics, the uniqueness of these systems is that the points belonging to the ensemble representing a given system, even if they are initially concentrated in an arbitrarily small cell in phase space, will, after a certain amount of time, be found "anywhere" (assuming the conservation of energy, of course). This dispersion helps explain why, in this case, the least approximation in tracking the dynamic system (the finest "granularity") can transform the Hamiltonian description into a description of evolution toward equilibrium, where every point in the macrocanonical ensemble has, when dispersion is complete, an equivalent probability of representing the system.

Establishing a precise connection between the definition of "collisional" systems capable of "physical representation" and the purely dynamic definition of, now chaotic, "K-flows" was doubly interesting for Prigogine. It meant "escaping" a field defined by Boltzmann's kinetic approach, a field felt to be unworthy by most physicists for justifying the challenge to "pure dynamics." And it meant entering the prestigious field of pure dynamics "by the right door," the one that might lead to the abandonment of the identification between dynamic behavior and the integrable systems model. Maybe this abandonment would enable him to establish a "purely dynamic" meaning for the concepts of collision and correlation, concepts that were merely "phenomenological" for specialists in dynamics given that such concepts lost their meaning whenever the integrable systems model was invoked. Couldn't the connection between irreversibility and collisions, which in itself served, from the point of view of those specialists, as a motive for disqualification, be retranslated in terms of obligations determined by "pure" dynamics?

Additionally, if the general definition of dynamic behavior and the integrable systems model could be radically disassociated, the interpretation in "purely dynamic" terms of collisions and the correlations they produce or destroy might cease to be "optional," coexisting with the possible that denies it. Irreversibility would no longer hang on the choice of not using, without any theoretical justification, a mode of description that has the power to eliminate it. Prigogine's program, therefore, became greatly radicalized. For now it was the construction of a post-Galilean factish that was in play. The approach to equilibrium and, especially, the collisions introduced by kinetic descriptions would have to become capable of supplying the dynamic "reasons" for why they cannot be eliminated for the sake of representation in terms of cyclical variables, that is to say, in terms of entities that evolve independently of one another.[3]

And yet the above summary is merely an outline. The ingredients are there, but we haven't yet learned how to combine them, to explore the obligations entailed by dynamic chaos if adequately presented. It was to a model created purposely to illustrate the expansion and contraction of distances characterized by the Lyapunov exponents of chaotic systems that Prigogine and his team now turned. The *baker's transformation*, transparent in that the distance between question and answer is minimal, would serve as a pure field of experimentation for formulating problems and, in this case, as a field of confrontation between the concept of trajectory and that of evolution toward equilibrium.

The baker's transformation is what is referred to as a *map*. Unlike the equations of dynamics, it doesn't present the problem of integration, for it is the entire "space" that undergoes the transformation, and its effects can be monitored on every point in that space. A series of baker's transformations causes every point in a square of surface one to undergo an "evolution" that has the two essential characteristics of a dynamic trajectory: that evolution is deterministic and reversible. However, this evolution is "chaotic" in the sense that, in one of the two directions defined by the square, the distances between neighboring points contract, whereas, in the other they expand (the Lyapunov exponent is positive). We could say that the baker's transformation is a nondynamic model of K-flows for, after a series of transformations, any ensemble of points initially concentrated in a small region of the space will be found scattered throughout the entire space.

The baker's transformation has been frequently described. I will limit myself to pointing out that it can be broken down into two steps. First, the square space is "flattened," without changing its area, into a rectangle (distances along the horizontal coordinate are expanded by a factor of 2, while those along the vertical coordinate are contracted by a factor of 2). The two halves of the rectangle are then stacked on one another to re-create the square. As a result, depending on whether they were in the left half or the right half of the initial square, the points experience two different outcomes. For points in the left half, the value of their horizontal coordinate has doubled, while the value of their vertical coordinate has been reduced by half. For points in the right half, the value of the horizontal coordinate has doubled *and* decreased by 1, the length of the side of the square, while the value of the vertical coordinate has been decreased by half *and* increased by ½, the length of the half-side (or the height of the rectangle). In other words, depending on whether the expansion results in the coordinate of a point having a value greater than that of the dimension of the square, two divergent outcomes are possible. This divergence is illustrated by re-forming the square, but it is not dependent on the specificity of the model. It is inevitable for any transformation that introduces an expansion or contraction of coordinates in a space that is conserved. And it is responsible for the eventual fragmentation of any region of space: no matter how small, any region of space will eventually end up being cut in two, and those two fragments in turn, sooner or later . . .

As such, the baker's transformation allows for a simple and

rhetorically powerful continuation (in fact, a reinvention) of Duhem's argument against trajectory. Regardless of the initial precision with which the position of a "representative point" is defined in baker's space, we can calculate the number of transformations after which this initial information becomes insignificant, after which, in spite of that information, we will have to assign all the points in this space the same probability of representing the system. After which, therefore, we will be in a situation that defines equilibrium in statistical mechanics. The increase in imprecision, however initially small, that gradually invades the definition of points, can almost be "visualized" when the coordinates of the points are defined in binary terms (a succession of 0s and 1s). Every transformation can then be represented as a "shift": for each point, the first decimal of the "expanding" (horizontal) coordinate becomes the first decimal of the "contracting" (vertical) coordinate. The remaining ensemble of expanding decimals shifts up one row while the contracting decimals shift down one row.[4] Because a point must be defined with finite precision, corresponding to a finite number of binary decimals—n, for example—from the first transformation we can "see" the precision of the definition of the expanding coordinate decrease as it is reduced to $n-1$ decimals: a decimal of unknown value has been "called in" by the shift. After n transformations, the expanding coordinate is completely indeterminate. After the $2n+1th$ transformation, the representative point can be "anywhere" in the square. Prigogine concluded that, here, the notion of a trajectory corresponds to an *illegitimate idealization*.

The reason advanced by Prigogine in claiming that trajectory cannot resist the chaos test was initially formulated as follows: we have always known that the precision of our measurements is finite, unlike the mathematical possibility of defining "one and only one point." This difference between physical representation and mathematics was of no importance for integrable dynamic systems, but it becomes crucial for chaotic systems. In the first case, mathematical knowledge appeared to be an ideal that it was possible to approach on a continuous basis. But chaos institutes a qualitative difference between increasingly precise knowledge and the knowledge of "actually infinite" precision. For only the latter provides trajectory with a well-defined meaning for chaotic systems. Chaos puts the physicist's back against the wall. In order to preserve the reference to trajectory, the physicist must refer to a knowledge that is inaccessible by definition. Not that of a demon, whose knowledge may be arbitrarily precise but is still similar to our own. What is needed to preserve trajectories is a knowledge that is independent of measurement, one classical metaphysics attributed to god alone. If the physicist refuses to make the transition from the demon to god, from a knowledge associated with practice, no matter how perfect, to one that is associated with a reality "in itself," or that is contemplated not from a godlike point of view but from god's point of view, she has to admit that chaotic systems obligate her to reject the concept of trajectory.

However, this is merely an argument, not a physical-mathematical construction. Here, evolution toward equilibrium is not intrinsic, has no intrinsic temporality. The number of transformations after which the system is at equilibrium does, of course, incorporate the intrinsic dynamics of the system (the 2 in $2n + 1$). But it also introduces the number n, which characterizes the "granularity" of our knowledge. This is only the first step, the sign that the baker's transformation is suitable for "experimenting with functions" that allow us to put forth a "true" approach to equilibrium, that is, a perfectly intrinsic description of this approach, one that doesn't make reference to the contingent granularity of our knowledge.

Here, I'll limit myself to presenting only those results (obtained in the early 1980s) that best express the "bridge" Prigogine now endeavored to construct between a physical representation based on the dynamics of correlations and "chaotic" dynamics. This involved the construction of a baker's model of the second law of thermodynamics based on a redefinition of both the "states" and the "equations of evolution" that together break the "time symmetry" of the baker's transformation.

Naturally, the baker's transformation does not, strictly speaking, define an equation of evolution. This explains my use of scare quotes. However, it does supply an analog for dynamic evolution. Not only is it deterministic—a point resulting in a point—but it is accompanied by a reverse transformation, where the vertical coordinate expands and the horizontal coordinate contracts. In the mathematical idealization of infinite precision, where a state is represented by one and only one point, this reverse transformation allows for the equivalent of a return to the initial state, analogous to the reversal of Loschmidt velocities: points scattered throughout the space by a series of baker's transformations, if subjected to a series of reverse transformations, gradually come together to again form the initial cell. In the representation with finite precision, reversal is of course no longer possible, but the situation is still not satisfactory. A series of reverse transformations can lead the system toward "equilibrium" as well as a series of direct transformations. As is always the case whenever a coarse-grain procedure for the evolution of density, ρ, in phase space is used, equilibrium can certainly be defined, but the connection between equilibrium and the "future" cannot.

This time the question is presented directly, with no escape possible. Prigogine reached the point that led to Boltzmann's failure, and needed to determine how the new ingredient found in chaos could change the situation. The baker's transformation seems to call for a description in terms of irreversible dispersion. How can this be expressed without introducing approximations, that is, without silencing one of the intrinsic properties of that transformation, the fact that a point results in one and only one point? In other words, how can he confront the contradiction between irreversible evolution and the baker's version of the Liouville theorem, which calls for the conservation of the measurement of an ensemble, that is, the number of representative points in that ensemble? How, then, can he "prevent" a series of reverse transformations from reconcentrating the points in an ensemble that a series of direct transformations has dispersed? It is this sort of experimentation with functions that the baker's transformation lends itself to and it is here that physical-mathematical inventiveness can demonstrate its power, its ability to overcome contradictions. For the chaotic nature of the baker's transformation can be used to bring into existence mathematical beings capable of being substituted for points, insofar as they are part of a representative ensemble, without the need to introduce approximation. Approximation always introduces grain of finite size, regardless of its fineness. These beings, however, function like points, they occupy no "surface" (technically they have measure zero). But they nevertheless allow the ensembles of which they are a part to escape the consequences of the Liouville theorem.

In the space defined by the baker's transformation, this strange being, which now represents the state of a (chaotic) system, is none other than a horizontal or vertical line or "fiber." A line without thickness does not define a surface: therefore, like the point, it has measure zero. And yet, if a horizontal line is subjected to a succession of baker's transformations, it will be multiplied until, in the long time limit, it covers the entire baker's space. A vertical line will gradually be reduced to a point when the number of transformations tends toward infinity. No longer defined in terms of points but in terms of dilating and contracting fibers, the ensembles subjected to the baker's transformation are able to confirm the irreducibility of their dispersive behavior. They can claim this behavior owes nothing to the human choice of "granularity": the "surface" (or measure) of a fiber being zero, it can, like the point, claim the ideal of infinite precision. But this version of the ideal is now robust with respect to approximation because the dilating fiber shares, along with every finite region that satisfies the definition of finite precision, the dispersal process that is the baker's analogue of an evolution toward equilibrium in the future.

Like physical representation ten years earlier, the description using fibers entails a definition of the state expressed in terms of broken time symmetry. As we have seen, although the definition of the instantaneous dynamic state is indifferent to the reversal of velocity, this reversal radically transforms the state it affects if that state is defined by its "physical representation." Similarly, in the case of the baker's transformation, the behavior of fibers radically changes, depending on whether they are subjected to direct or reverse transformation. If subjected to a reverse transformation, the horizontal dilating fibers contract and the vertical contracting fibers dilate. Like the physical representation, the baker's transformation imposes a correlated definition of transformations and states. And the baker's analog of the second law of thermodynamics as a *selection principle* can be precisely formulated, in the sense that it specifically excludes the possibility of a particular type of state. Whenever we are dealing with a series of direct transformations (which lead to equilibrium in the future), states represented by the contracting fiber are physically excluded as they would correspond to the initial state of a system that spontaneously and *indefinitely* moves away from equilibrium. And the description based on the concept of physical representation accurately reflects the unique nature of this type of state. It is the only state where the reference to equilibrium cannot be defined, which also means that it is the only one whose physical representation cannot be constructed. The kinetic analog would be the initial preparation of a system, described in terms of pre-collision correlations, that, rather than being transitory, "consumed" over time, *permanently dominate* the evolution of the system.[5] Using the second law of thermodynamics as a selection principle means that some values of density, ρ, must be excluded, condemned as corresponding to states that are physically *unrealizable*. In other words, the various states defined by dynamics as being equivalent, indifferently realizable based on the free choice of the preparer, are *no longer of equal value.*

The legitimacy of excluding a category determined by initial preparatory conditions can be defended on the basis of the same principle that allowed Prigogine to characterize the trajectory of the baker's transformation as an illegitimate idealization. The definition of such conditions cannot resist approximation. Whereas evolution toward equilibrium brought about by a dilating fiber is robust, the evolution of a contracting fiber is not. In the baker's case as in kinetics, the slightest lack of precision is sufficient to reestablish the norm of an evolution that leads, after an arbitrarily long period of time, to equilibrium.[6] And the exclusion, moreover, is "propagated by dynamics": starting from some admissible state, a system will never spontaneously evolve toward an excluded state.

Prigogine's new argument, with its focus on the baker's transformation, is simultaneously strong and weak. It is strong because he can now engage directly with specialists in the theory of dynamics. There is no further need to ask that irreversibility be taken seriously, it must be recognized once a representation can be constructed that is both robust with respect to approximation and intrinsic, that is, independent of the degree of approximation. Expanding and contracting fibers satisfy these requirements. They provide a sort of "intrinsic granularity" independent of the choice of approximation. This means that Prigogine no longer needs to depend on "sympathy," not for himself, of course, but for his struggle for the second law. He can argue a more general cause, in this case a rationality that is finally relieved of its "irrational" reference to knowledge of infinite precision, inaccessible in principle, and finally free to accept the obligation without which mathematical representation has no physical meaning, a rationality finally capable of expressing this obligation by inventing a reference that is stripped of all arbitrariness and is robust with respect to approximation.

It is weak for at least two reasons. First, the "limited rationality" that has become Prigogine's champion clashes, as did Duhem's claims before him, with the kind of realism that characterizes dynamics. As I have pointed out several times previously, this realism is based on the fact that the dynamic object appears to dictate the way in which it must be described. Trajectory is then no longer a physical-mathematical *construction,* as Prigogine implies when he argues for a different construction, robust with respect to approximation. It is considered to be the "truth" of the behavior of the system it describes. If, in the case of a chaotic system, we cannot access this truth, it nonetheless exists because it is produced over time *by the system itself.* Reference to "divine" knowledge to which the physicist does not have access is insufficient for disqualifying this realism. Such inaccessibility isn't a problem here because the physicist does not conceive of her obligations in terms of knowledge. Because she maintains a reference to trajectory, she feels obligated by the very mode of existence of the system "itself." The system is, at every instant, "truly" in one and only one state, represented by one and only one point. Any attempt to make her forget this "evidence" is merely a sophistic strategy designed so she will betray her obligations. Against what one could almost say is a heartfelt argument, Prigogine can certainly respond by introducing other realist obligations: the description of broken symmetry is richer because it provides an intrinsic sense to the properties that characterize the evolution toward equilibrium, properties whose meaning we were, until then, able to construct only by approximation. But the second weakness of his argument now arises: the baker's transformation yields no observable property because it is nothing more than a geometric model.

In order for this argument to be taken seriously, whatever has been realized through its use should apply to systems endowed with physical meaning. And in this case, Prigogine must confront an additional difficulty: in quantum mechanics chaos, as defined by "distances" whose time evolution is characterized by the Lyapunov exponent, has no clear definition.

The "realism" of the trajectory cannot be vanquished by argument. It belongs to the register of those convictions that can, whenever appropriate, hide behind a "yes, I know, but all the same." Nonetheless, this realism could be avoided, transformed into an outdated conviction, if the interest of specialists were engaged. But in order for this engagement to occur, it must be fed with problems that the baker's transformation cannot produce. We must return to Hamiltonian physics and Hilbert space and subject them to the test of chaos. It is to this work that Prigogine devoted the last years of his life.

11

The Laws of Chaos?

This time it was Poincaré's theorem that defined the scope of the problem, for it was this theorem that was most likely to contest the generalization of the model of integrable dynamic systems. And the obstacle it placed in front of that generalization was as valid for classical dynamics as it was for quantum mechanics. As we saw in Book II, in classical dynamics Poincaré's theorem embodies the impossibility, for the majority of dynamic systems, of constructing a "cyclical" representation, synonymous with integrability, by "extending" the description of an integrable system through the calculation of perturbations. In quantum mechanics the situation is somewhat different. The beings that inhabit Hilbert space not only satisfy causal requirements, the reversible equality of cause in effect, their definition asserts the quantum equivalent of classical integrability. The Hamiltonian operator for a finite quantum system, one that is characterized by a *discrete spectrum,* can, by definition, be given a "spectral representation," a representation of the Hamiltonian operator in terms of eigenfunctions of that operator and its eigenvalues. Poincaré's theorem is an obstacle to the solution of a problem (of integration) that does not seem to arise. Yet, as we shall see, there exists one case where the difference between the two situations disappears. This is the case of "large systems," characterized by an infinite number of degrees of freedom in classical dynamics or by a *continuous,* rather than discrete, spectrum in quantum mechanics.

In quantum mechanics, continuous-spectrum systems are physically very important but always calculated using approximation. They are important because they correspond to all situations where "something happens" in the quantum world, for example, when we transition from an isolated atom, described in terms of stationary states, to the excited atom interacting with a field and characterized by a lifetime. The fact that approximation must be used to provide meaning to this "something happens" has, as readers may recall (see Book IV), aroused the criticism of Nancy Cartwright: doesn't the approximation that allows us to "leave" Hilbert space to describe the "causal" actor reflect the irrelevance of the static observables defined on the basis of that space? But the crucial point now is that, for the ensemble of such situations, the Hilbertian guarantee according to which "there exists a spectral representation" disappears. In fact, specialists of quantum mechanics are well aware that the definition of Hilbert space ceases to be consistent once they are dealing with a system with a continuous spectrum. More than forty years ago, the notion of a *rigged Hilbert space* was formulated to respond to this problem.[1] Here, "rigged" implies that Hilbert space had to be "completed," "framed" by two additional spaces capable of accommodating, as their legitimate inhabitants, the new kinds of actors introduced by continuous spectra. Hilbert space is thus "rigged," the way one might rig a ship before it sails across the broad ocean, far from the tranquil shores along which it first traveled. But the status of these generalized spaces is well expressed in Leslie E. Ballentine's *Quantum Mechanics.*[2] In the beginning of the book (pp. 16–19), the concept is presented and the author's summary concludes with the following: "These two examples suffice to show that rigged Hilbert space seems to be a more natural mathematical setting for quantum mechanics than is Hilbert space." But in the pages that follow, this "natural setting" does not seem to entail any obligations at all: rigged spaces are only mentioned, and briefly, twice more in Ballentine's book. For it is approximation that now takes over and allows us to "escape" Hilbert space when necessary.

For Hilbert space to cease being the "natural setting" for large quantum systems is exactly what Prigogine needed. If the obligations carried by the arrow of time or the ensemble of properties with broken time symmetry that characterize those large quantum systems are to be expressed in the form of new questions addressed to such systems, these must not, by definition, be subject to Hilbertian requirements. A space offering the possibility of construction must be opened between the statement of the problem to which their behavior responds and the definition of that behavior.

In Hamiltonian dynamics, the situation is equally interesting. What happens at the limits of large systems? In fact, the entire economy of difference and similarity, assumed by the representation of a "nonintegrable" system as a "perturbed" integrable system, is overturned. Let us assume some strictly periodic behavior, that of an integrable system, for example: the transition to the limit (number of degrees of freedom tending toward infinity) will have no effect on it. Now, take the case of a "nonintegrable" system, represented as a "perturbed" integrable system in the sense that its degrees of freedom are coupled. In this case, regardless of the value of the coupling, no matter how weak, the transition to the limit of the large system is expressed by the appearance of chaotic behavior. In classical dynamics, the KAM theory explored a diversified range of dynamic behaviors, limited at two extremes: integrable periodic behavior, on the one hand, and chaotic behavior, on the other. For large systems, the spectrum is reduced to the two extremes, which no longer have the same status. The integrable system is merely an

"isolated" case, and it is nonrobust because the slightest perturbation, the slightest coupling is sufficient to tip its description
into the general class of chaotic behaviors.

The fact that in both the classical and the quantum case the
transition to the limit of large systems can be expressed as the
loss of the reference model based on periodic behavior means
that "large systems" could indeed serve as the landscape where
the meaning of Poincaré's theorem changes, that is, becomes a
crucial matter of obligation. It would not be simply an obstacle
to the construction of a solution based on problematic equations, but would assume a positive meaning through which the
appearance of observable properties with broken symmetry
would become "self-explanatory."

Prigogine would call all of these classical and quantum systems *large Poincaré systems*, which makes explicit what he believes
to be their function. It is such systems that would become the
reference countermodel capable of assigning its dynamic meaning
to the arrow of time. It is such systems that would obligate physicists to use something other than approximation when trying
to account for them. However, just as it is not sufficient to designate the "sensitivity to initial conditions" of chaotic systems
to overcome determinist reference, it is not sufficient to identify the obstacle Poincaré's theorem presents to the procedure
of integration. Prigogine had to transform those "large Poincaré
systems" into beings endowed with positive properties; he had
to define them in terms of the requirements they had to satisfy
and the obligations they had to impose.

Two things are certain. To the extent that a bridge must be
made to the statistical mechanics of equilibrium, it is at the level
of an ensemble description that large Poincaré systems should
allow a positive meaning to be assigned to Poincaré's theorem.
And, in those cases that correspond to the approach to equilibrium, this should ensure that the requirements associated
with that approach are satisfied. In other words, an ensemble

representation must "tame" chaos, must turn it into an explanation for *regular,* entropy-increasing behavior.

In physics, the "transition to the limit of large systems" is an operation that is both necessary and difficult. It is necessary because all the objects of statistical mechanics are "large systems": a gas, a liquid, a solid only assume meaning at the "thermodynamic limit" (when the number N of particles and the volume V of the system both tend to infinity while their ratio remains constant). In quantum mechanics, as we saw earlier, the problems that introduce an interaction with a field (with a continuous spectrum) also correspond to large systems. But the operation is difficult in the sense that several properties must go to infinity at the same time (N and V for the thermodynamic limit). When it is not a question, as it is in the statistical mechanics of equilibrium, of "building a bridge" between two known banks but of exploring the possibilities of a formalism, the limit transitions, if they are to be reliable, require extremely complex calculations, which, here, must be carried out under suspicion: somewhere in the process of calculation, haven't some pieces of "information," some possibilities, been eliminated whose loss would explain the new properties appearing "at the limit"? If this were the case, these would be no more than artifacts of the procedure.[3] Of course, one can reply that we "know" that new properties must arise: are the distinct properties of the gas, liquid, or solid our artifacts? But there is a risk that this argument may fail to convince, for the hierarchy of physics has done its job. The critic is not interested in "common" phenomenological properties, she wishes only to remind us, in expressing her suspicions, of the difference between a rigorous "fundamental level" and an approximate phenomenological description.

The difficulty of an operation is not in itself an argument, for the evaluation of that difficulty depends on the value we assign to what the operation targets. Thus, "superstring" theory, which is being promoted as the key to a possible unification of physical interactions, requires calculations of unparalleled length and sophistication. A certain uneasiness has prevailed among thousands of young physicists who had to confront a lifetime of scientific work devoted to such arid calculations.[4] But since unification is the Holy Grail of contemporary physics, the price had to be paid. With respect to the "transition to large systems," however, this price is an obstacle, for the transition can apparently be carried out by means of approximations that are easily controllable and reliable once they are guided by the phenomenological properties we know we are looking for. That is why, in Book IV, I emphasized that if the "Hilbertian factish" was a liar, its lie was interesting, for it was lending itself to the appropriate approximations that served as a bridge between Hilbertian properties and the phenomenological properties physicists need. The problem posed to Prigogine, if he was to interest his colleagues in the class of "large Poincaré systems," was to make the possibility of avoiding approximation and giving rigorous physical-mathematical meaning to the transition to "large systems" interesting in their eyes—independently of the question of the arrow of time, which *for them* was not an argument.

At this point, something new intervenes: "large systems" can be used to construct a method of description that can lay claim to the beauty and intelligibility that give the Queen of Heaven and her Hilbertian descendant their true power of seduction. The new ingredient is a theorem by David Ruelle (1986). According to this theorem, the systems characterized by determinist chaos can have *several distinct spectral representations.*[5]

How can such a theorem change the nature of the problem? Recall that the customary spectral representation, a representation of the Hamiltonian in terms of its eigenfunctions and eigenvalues, is the "ideal" representation once we make use of a formalism involving operators. The evolution thus represented is as synonymous with beauty and intelligibility as cyclic representation in classical dynamics. In both cases, the relationship between problem and solution becomes limpid: the problem appears to dictate the mode of description appropriate to its truth. The possibility of *different* spectral representations indicates the possibility of reinventing, for chaotic systems, this privileged relationship between problem and solution that is characteristic of integrable systems. This time, the equations do not define a system that is then found to have chaotic behavior, they formulate the problem of behavior *as chaotic*.

The definition of a spectral representation is the Grail of mathematical physics; it obliterates the difference between the register of the problem and that of its solution because the problematic definition of the object and the solution to the problem, that is, the formulation of its law of evolution are expressed in the very same language. We could just as easily change the metaphor and say that, for Prigogine, the possibility of defining a spectral representation with broken temporal symmetry would shift the grounds of the struggle from a war of insurrection, where the rebel recruits allies who, in spite of their prowess as fighters, are considered barbarians by those in power, to an attempt to win power by means of a legitimate claim: the rebel finds he is the heir of the power he is challenging, a direct descendant of the great tradition of physical-mathematical invention. Like all members of the royal line, the man who had previously "followed" the phenomenological properties of matter, would have succeeded in "getting the object to speak." Here we find Deleuze and Guattari's distinction between itinerant and royal sciences: "However refined or rigorous, 'approximate knowledge' is still dependent on sensitive and sensible evaluations that pose more problems than they solve: problematic is still its only mode. What belongs, on the contrary, to royal science, its theorematic or axiomatic power, is to isolate all operations from the conditions of intuition, making them true intrinsic concepts, or 'categories.'"[6]

It is with regard to the baker's transformation that Prigogine will define the new object of his quest. The redefinition of this object in terms of contracting and expanding fibers would be reformulated in terms of spectral representations: along with the "customary" spectral representation, corresponding to a description in terms of individual trajectories, there appear two new spectral representations that are probabilistic, that is, defined for ensembles. One represents regular evolution toward an equilibrium located in the future, the other toward an equilibrium located in the past. From this a decisive claim can be constructed: the two representations with broken symmetry *cannot be reduced* to the customary representation. It is impossible, even at the limit of perfect information, to "return" to an individual formulation described in terms of trajectories. This impossibility is crucial, for it implies that the new characterization is not simply another way of presenting the older characterization. And it is this impossibility that generally arouses suspicion. If it is impossible to return to the "ideal" description, it is because information must have been lost. Yet, in the new case, the suspicion is excluded: the mathematics itself contravenes this reduction.

Here I want to give a very brief description of what otherwise might appear to be magic. Let's say that everything now revolves around the distinction between "regular" and "functional" functions, also known as "distributions" or singular functions. Such creatures are now well known in mathematics, although they remain fairly exotic in physics. Yet, it was a physicist, Dirac, who created the first of them, Dirac's "delta," to help solve a quantum problem (which involved the continuous spectrum of a field). Dirac's δ has the strange property of being zero everywhere except at one point where it is infinitely large. It was on the basis of the δ function that the new mathematics of functional analysis developed. And it is the fact that the δ function must participate in the presentation of continuous spectrum systems, even though it does not inhabit Hilbert space, that is the origin of the conception of rigged Hilbert spaces.

The question of irreducible spectral representations was thus the result of the intervention of "functions" that originated in quantum mechanics but have since been endowed with sophisticated new mathematical features in order to satisfy the obligations of mathematical consistency. What they brought with them was a dramatic restriction: an operator that implied a singular function could not be applied to another singular function.

This restriction has a spectacular consequence for the definition of what physicists conceive as the "ideal case," one that is perfectly determined, where a description in terms of ensembles is reduced to the description of an individual system. The density ρ becomes zero everywhere except at one point. This means that from the viewpoint of the theory of ensembles, the object associated with the "individual state" corresponds to a Dirac δ function. The point is no longer an intuitive object but a singular limit case, and the primary target of the mathematical restriction. And what this restriction affects are the new spectral representations that can be defined for chaotic systems, for their evolution operator incorporates a singular function. Such operators cannot, therefore, be applied to an individual system, which also corresponds to a unique function. The description of the "state" of a system can no longer be reduced to a point, but only to an appropriate ensemble.[7] The ensemble description no longer has as its ideal limit the description of an individual system. The probabilities it introduces have the ability to assert that they are *irreducible* and not relative to some contingent lack of information.

For Prigogine, such a result was crucial, for it gave autonomy to the theory of ensembles, without which it is impossible to construct a meaning for the approach to equilibrium. But from the customary point of view of the physicist, it implies a radical change of perspective. If the theory of ensembles were to become the site where obligations that have no meaning in Hamiltonian dynamics or Schrödingerian quantum mechanics could be formulated, it could no longer be defined as a simple auxiliary formalism, intended to handle situations for which we do not have perfect information. Correlatively, the nature of a trajectory or a wave function would become problematic. According to the conventional point of view, it is a question of primary concepts, corresponding, in the case of trajectory, to an intuition of motion. But in terms of the new perspective, the primary concept is the ensemble. Trajectory and wave function become derivative concepts: trajectory can be redefined as a unique object, localized, yes, but constructed in terms of *delocalized objects.*[8] The very term "delocalized" implies that the ensemble has become the fundamental concept, for it is the ensemble that can be said to be localized *or* delocalized. Location is no longer a primary attribute of members of the ensemble but a singular property of the ensemble as such, presenting very specific types of problems.

For chaotic systems trajectory and wave function would now be defined as the result of constructive interference between delocalized plane waves, which means they would be defined in terms of a representation, the object of which is an interference situation. And it is in terms of this interference situation that Poincaré's theorem can assume a positive value. "Poincaré resonances" no longer serve as an obstacle to integration, but are used to explain the destruction of the interference that now corresponds to the concept of location.

Prigogine showed that to construct a probabilistic spectral representation with broken time symmetry, the resonances must correspond to "persistent interactions."[9] This last condition is trivially satisfied for all kinetic systems—it expresses the need to speak of collisions in the plural rather than of a collision as an isolated event. But the condition had to be made explicit in the quantum case because some very important situations did not satisfy it. The atom that spontaneously returns to its fundamental state can be represented outside of Hilbert space, and it is then described by a law of exponential "de-excitation," but in this case the spectral representation with broken time symmetry brings wave functions onto the scenes, that is, representations of the individual atom, not a probabilistic representation.[10] The de-excited atom implies a breaking of time symmetry but gives no meaning to the evolution toward equilibrium.

Like the 1972 "physical representation" whose results it confirms, the "irreducibly probabilistic spectral representation" incorporates the breaking of time symmetry in its equations of evolution (that is, in the definition of its evolution operator).[11] As in 1972, this new definition of evolution brings with it a new definition of state. What changes, however, is the rationale for this new definition. It is no longer a question of taking the "sorting" of correlations seriously, the fact that correlations become insignificant at equilibrium. The 1972 definition of "physical entities" was already nonlocal. Those entities were defined with reference to a future equilibrium where they could be described as uncorrelated (the specificity of initial conditions was forgotten). But the nonlocal character of "states" to which the spectral representation is addressed is now associated with the purely dynamic and nonlocal concept of resonance. Consequently, the nonlocal definition of state has the means to claim its relevance in purely dynamic terms, without having to make use of an explicit reference to the state of equilibrium.

Correlatively, the abandonment of the ideal of the description of an individual system, the "localized" description, is self-explanatory. It is the "local" definition of state, in terms of a "point" in phase space, that becomes the impoverished definition, an *artificially impoverished* one, because it excludes any reference to the topology of that space. This had no effect on integrable systems, whose phase space is structurally trivial. But when the "erratic" behavior characteristic of large Poincaré systems dominates, reference to the topology of the phase space, which the nonlocal definition incorporates, becomes a part of the problem and the point loses its status as a perfect definition of the state of a system. That is why Prigogine did not hesitate to claim, in the name of his representation, the most prestigious term in physics: this representation enabled him to formulate the *laws of chaos*.

Starting from the modest offer to take the experimental relevance of Boltzmann's $\mathcal{H}$-theorem seriously, Prigogine now presented the laws of chaos to his colleagues. Chaos, as the site of a new kind of spectral representation with broken time symmetry, would take its place within the illustrious history of physical-mathematical invention, with the invention of beings who have the unique ability to define how they should be addressed. For, when correctly addressed, description coincides exactly with explanation. And the first heroic deed of this new "factish" would be to break the connection this invention had until then respected, at least in the field of classical physics, between "reality" as it is able to explain itself and "reality" in the "intuitive," "visualizable" sense that the Hamiltonian trajectory still seemed able to identify. From the viewpoint of this intuitive reality, one may well wonder what is meant by the fact that the spectral representation of chaotic systems can be constructed only for ensembles. As a first approximation, the situation might appear somewhat similar to Brownian motion, the old symbol of the triumph of kinetic theory over the laws of thermodynamics. The trajectory of a Brownian particle is "erratic," completely irregular, but the corresponding probabilistic description is perfectly regular: it characterizes the "diffusion" over time of the probability of localizing the particle based on its initial position. However, unlike Brownian motion, the erratic behavior of a "large Poincaré system" does not appear to be the symptom of underlying thermal activity. There is no "reason" for the loss of relevance of the explanations that determine the behavior of the individual system. A new type of explanation has been invented, together with a new type of being that dictates those explanations. In other words, if the Hamiltonian trajectory ceases to be a legitimate object, if the probabilistic behavior of the system alone is the object of a law of evolution over time, it is because what mathematical physics calls an "object," whether a trajectory or an ensemble, never had anything to do with a vision in the first place, in the sense that vision is presented as being detached from any form of practice.

From the viewpoint of this new "realism," quantum mechanics loses its status as a "privileged site" for philosophical reflection, a site where the critical thinker might hope to find herself on common ground with physics, for here "at last" physicists were forced to "think," forced to consider the limits of their knowledge. Now, with Prigogine's hypothesis, the function could again "think" for the physicist. Which confirms Nancy Cartwright's argument about the problem of measurement: measurement, to the extent it had been reserved for a dialogue with philosophers, turned out to be powerless when confronted with the real problem, a *practical* problem corresponding to the fact that physicists had to "cheat," had to introduce approximations that enabled them to surreptitiously escape Hilbert space, so they could "talk about" the experimental schemes in which a given entity played a "role."

Nonetheless, the laws of chaos have different implications for classical and quantum physics. The abandonment of classical trajectory, its redefinition, in the case of chaotic systems, as a singular object within the theory of ensembles, embodies, as I have said, a reinvention that asserts the difference between the physical-mathematical object and what intuition offers us as immediately intelligible. In quantum mechanics, however, this difference has, of course, already been asserted. It coincides with quantum formalism itself. It is the very specific structure of that formalism, with the duality between the wave function and its reduction, which is now situated within a more general context in which it retroactively acquires new meaning.

Retroactively, the structure of quantum formalism responds to a problem that has less to do with quantum beings than with what has been required of them. If an atom cannot be said to have emitted a photon except to the extent that a sensitive plate has recorded that photon, it is because "events," the evolutions during which "something happens," such as the emission of a photon, have been used to construct the representation of an object to which "nothing can happen," the stationary atom. They have been "instrumentalized" as a means to access a physical-mathematical being that denies such a possibility. Within the context of quantum formalism, the emitted photons do not express the emission but the energy difference among *stationary* states. Once the specific character of this object construction has been made explicit, which Nancy Cartwright has done, the question of measurement loses its critical, epistemological dimension, capable of justifying the selective intervention of the philosopher. But as soon as this "something happens" acquires, together with "large Poincaré systems" and their spectral representation, the ability to define how it should be addressed, and the ability to refer Hilbertian language to a specific case, one in which localized beings, described by a wave function, remain local, the question of measurement may appear to have been *resolved*. Because it is necessary for "something to happen" in order for something to be observed, the observation, which provides access to the quantum world, always implies a "large Poincaré system." And this "something happens" must, as a result, be characterized in terms of probabilities. Probabilities are not then relative to a "reduction" that would, one way or another, testify to the intrusion of the instrument into an unobservable world, defined in terms of probability amplitudes. The instrument is not intrusive. Described as a "large Poincaré system," it belongs to quantum reality. Retroactively, it was the choice of defining observable properties satisfying the requirements of

Hilbert space, properties that referred to a reality where nothing happens, that made probabilities seem relative and the measurement instrument responsible for whatever "happens."

It remains that probabilistic spectral representations do not dictate their reasons the same way that Hamiltonian or Hilbertian representations do. Maxwell's Queen of Heaven ruled over a world that was supposed to "resemble" those rare cases where his equations could be solved: every system described in terms of dynamic equations "had" to have the same type of behavior, the behavior exhibited by the periodicity of two-body motion. And Hilbert space gave this claim the power of an axiom: all Hilbertian quantum systems can be represented in terms of the stationary model of the energy states for an isolated atom. However, the "new spectral representations" assert the *constructive* nature of the solution they embody. For, the ability of a system to define how it should be addressed and represented does not preexist the construction of that representation. For each case, "reasons" must be sought, its "intrinsic concept," as Deleuze and Guattari would say. In other words, with the arrival of "singular" functions, the general power of the = sign disappears. The factish proposed by Prigogine is deprived of the power to subject dynamic systems to a form of "intuition" that recognizes "sameness" beyond diversity, an intuition that is, of course, physical-mathematical rather than perceptual, but just as capable as perceptual intuition of masking the construction of the problem with the evidence of a solution. That is why it is appropriate to speak here of a new type of factish, the *problematic* factish.

Far from being a weakness, this feature of spectral representation (physicists refer to it as being *model dependent*) could be of interest for certain key protagonists in the field of the physical sciences, namely, mathematicians. For them, this feature may be situated within the general problem associated with "rigged" Hilbert spaces, capable of accommodating the nonregular functions that arise with continuous-spectrum systems.

The question such spaces ask is: how, with what can they be rigged? From this point of view, Prigogine's spectral representations provide constraints and meaning. They show that what, from the point of view of mathematics, imposes itself as a "natural frame" can become an effective frame for the construction of an exact description of "large Poincaré systems with persistent interactions." This might be (or might have been) the "small difference" needed by a "great narrative" to accompany the "laws of chaos," enabling them to accede to the status of an event, and creating its before and after.[12] The interest of mathematicians would make a difference where Prigogine's "physical" reasons, rendered inaudible by the hierarchization of physics, were dismissed out of hand.

12

The Passion of the Law

Physical-mathematical invention is not pure. Like any invention, in order to be successful, to have access to the status of innovation, it must take into account the concrete landscape it wishes to modify. The story told here, independently of its outcome, is significant in that the attempted innovation it relates is not presented as a "revolution" that overturns frames of knowledge and imposes questions that have gone unheard. The questions to which it gives meaning were not unheard of. It is simply that they had been judged as *unworthy* of being heard, and treated as such. For that reason, such attempted innovation also, and inseparably, has entailed creating and incorporating in the object itself the means to appeal that judgment and, therefore, the power relationship that has historically allowed it to exist. The arguments put forth by Prigogine in 1962 were associated solely with the experimental relevance of the Boltzmann equation: they didn't succeed. The theorization of collisions and correlations suffered from the handicap of being associated with a field—kinetics—defined merely as phenomenological. The proposal to take into account the obligations of chaos and its sensitivity to initial conditions encountered the split between what "we" are able to know and what a physical system in itself obeys.

If the "laws of chaos" were ever to gain recognition, a decisive ingredient would be the eventual alliance with actors who are free in terms of the hierarchy of physics, mathematicians, who were "amused to learn," as one said to me, that one of their exotic objects had acquired physical relevance.

Moreover, this is already the story told by the mathematical physicists who have agreed to look into "large Poincaré systems": *In the beginning was a mathematical being created to satisfy the requirements of consistency that fall within the purview of mathematics. It vegetated in the respectable dignity of beings who were irreproachable but without consequence, when one day it encountered the problematic terrain where its properties ceased to be exotic and became necessary for the construction of a physical object. A new chapter in the symbiotic relationship between physics and mathematics had been opened.* A chapter where questions of kinetic description, its role as a field of constructive exploration, will become part of the story, part of prehistory, while its domain will be designated a privileged field of application, manifesting the fecundity of the laws of chaos.

If ever such a story were to be told, it would not be "false" but it would feature an event that erases the passions and power relationships involved in innovation, and becomes part of the unbroken history of progress. It would give to mathematicians—and to their freedom with respect to the hierarchy of physics—a role that would appear so "normal" that the potentially decisive nature of their intervention would disappear behind the only true subjects of the story, the properties of the physical-mathematical being and the problematic terrain to which it responds.

I wanted to tell a different history, one that does not disqualify such "subjects," one that is not a contrary version—psychological, social, or Machiavellian—denying the first. My history doesn't deny the "great story," because the eventual possibility of constructing such a story is one of the challenges imposed by history. It does not deny physical-mathematical beings their status as subject, for in the history I have related, they, and they alone, could eventually gain the right to require and obligate. On the other hand, it attempts to immerse this hypothetical great story in the contingent and impassioned history that, in this hypothesis, gave it the power to present itself as necessary.

In the eighteenth century, Diderot believed he could predict a future in which the monstrous "pyramids" erected by mankind's genius, that is, the work of the mathematicians, astronomers, and physicists of his time, would instill not just admiration but pity as well in those who would work to decipher, to conjecture, to follow natural processes in all their diversity. This prophecy is still current. It refers to a possibility that remains open. Concerning the history of mathematical physics itself, in Book III I discussed how the history Prigogine continues to extend has gone through a genuine process of percolation. The historical reciprocal capture that coinvented, on the one hand, a physicist capable of consigning, without scruple, to the field of phenomenology anything that expressed the "arrow of time" and, on the other hand, the reality he addressed—one capable of defining how it should be addressed, that is, of dictating its own reasons—might never have taken place. But it did take place, just as the "capture" of the quantum world by Hilbertian formalism took place. And it is on this succession of historical facts that the antihistorical moral of the eventual "great story" I have presented depends. If the reciprocal capture between "fundamental physics" and "a factish dictating reasons that deny the arrow of time" was to be challenged (which, for many physicists, is far from the case), the challenge would have to adhere to rules that confirm the principle of the game, rules that demand that if a different kind of factish is to be defined, it must still be capable of dictating its reasons—other reasons.

What remains to be discussed is the importance I have given to this kind of challenge as well as the question of the laws and objects of so-called fundamental physics. Does this mean that the question, as such, is a determining ingredient of the ecology of practices? Does this mean, in other words, that I intend to retain for physics the status it held, and continues to hold, whenever the expression "law of nature" is used? In other words, am I claiming that the "laws of chaos" proposed by Prigogine are, or may be, laws capable of engendering the diversity of forms and regimes of activity that we observe, and as such of justifying a regime of peaceful coexistence among the various practices satisfying that diversity? Wouldn't the "irreducible probabilistic representations" then simply be candidates for the position of founder of a new physical authority, one that is more liberal, more open but nonetheless the sole source of legitimacy? The succession of these questions traces a predictable rhetorical slope that I cannot ignore, for those first steps taken toward an ecology of practices may serve as an enticement for following such a slope.

How can this slope be resisted when the narrative I have presented is haunted by a passion for law and the problem of a method of expression that would confirm the authority of so-called fundamental laws over phenomenological description? How can we avoid celebrating a new hierarchization that is now simply more accommodating, no longer contrasting the ideal of perfect knowledge with the ignorance that qualifies all other kinds of knowledge, but still defining the "principles" of what exists? How can we avoid presenting the eventual innovation that would allow physics to no longer deny the arrow of time, to no longer reduce probabilities to a lack of knowledge, as good news, describing a physics that had finally produced the scientific truth of the possible, if not of becoming?

It should be pointed out that the possibility of falling into this trap is included in the requirements Prigogine had to fulfill to achieve his ends. And I have wagered that a way to avoid this trap was through the introduction of these requirements as such. In fact, starting in 1972, by "following" Prigogine and his group's research and trying to understand its implications, I arrived at a *constructivist* conception of the physical object. And I did so without at any time feeling that this conception would diminish the interest of that object, quite the contrary. It is this experience I have tried to share with my readers. My correlative conviction is that physicists could, without in any way denying the passion for truth that characterizes their practice, affirm not the relativity of their truth but the truth of the relative.[1] Although the succession of theoretical factishes I have introduced, beginning with the first state functions expressing the power of the Galilean = sign, are part of a history that might have been different, those factishes are not "soluble" in the contingent nature of that history. The requirement of a reality capable of dictating its own reasons might not satisfy any rational necessity; it is nonetheless expressed by a set of specifications, or obligations, that physicists have found effective ways to satisfy. It is *possible,* therefore, because this would not require that they lose what is crucial to them, that they might agree to distinguish what they have actually managed to construct from the meaning so often given to that construction, namely, the disclosure of the "principles" of reality.

This possibility, to which my personal experience has made me receptive, has nothing to do with an argument that would establish the *plausibility* or probability of a transformation of this type, which would turn the physicist into a vector of peace in the landscape of our practices. Moreover, to hope for this would contradict my project given that such an argument would have as its only subject the question of what physicists' practice could make them capable of. This would be to confer upon physics, physics again and forever, and physicists' possible "goodwill," a key role, while my intent is specifically to bring into existence relationships among practices that are purely immanent, which cannot therefore be judged on the basis of any sort of transcendence. The possibility of such an "ecology of practices" can have as an ingredient the fact that the obstacle presented by the "fundamental laws of physics" has been surmounted, that it no longer functions as a sovereignty claim, demanding an act of surrender from all other practices. However, the obstacle to be overcome cannot be confused with the ecology being invented. This should have as its actors *all* the practices it affects, each on its own behalf, based on the requirements and obligations that singularize it, and all of them, including physics, learning, with the help of the others, to situate the truth of the relative they inhabit and cause to exist.

Gilles Deleuze, quoting the poet Mandelstam, tells us that "memory works not to reproduce the past but to set it aside."[2] Which is exactly what I have tried to do. The past cannot be set aside by simple decree, it requires work, and this work cannot be identified with a kind of critical deconstruction. It means discovering and implementing the way it might allow itself to be set aside without allowing it to be replaced with arrogant new certainties or a contemptuous indifference that substitutes for thought. The challenge of my four explorations of physics has been to attempt to set aside a past in which the physics of so-called fundamental laws was identified with a claim as well as a vocation, the discovery of "reality in itself," in opposition to any "reality for us." But the "present" I have been working for confers no crucial importance upon physicists for giving up such an identification. The fact that they can do this without having to admit that their laws are "merely constructions" with no special relation to reality is what matters for "us." For it enables us to set aside a past in which the only choice appeared to be to accept the authority of those laws or critically deconstruct them.

It is certainly tempting to critically address the fascination exerted by the concept of "law." We could even speak of that fascination, exerted on as well as by the physicist, in terms of the passion of reason. But in this case, if this reason were to be identified with "human reason," the past could not be dismissed.[3] To avoid jumping to this conclusion, I had to show that we have no reason to think that the "reason" in question is human reason in general. This is why I have related the very peculiar demands associated with "reason" in physics to an event: the reciprocal capture that turned the possibility of defining how the past determines the present and the present the future into a central feature of the physicist's vocation.

From this point of view, there is a certain aesthetic satisfaction in the work of Prigogine and his coworkers. The definition of "how" has not been directed back toward indeterminacy, toward "free becoming." It has been transformed into a problem that must be solved for each individual case. In one sense, we could say that qualitative diversity is back, no longer the diversity of phenomena as in the nineteenth century but the diversity of events and actors physics "brought into existence" ever since it decided to look beyond the diversity of those phenomena. Yet, because it is compatible with a critical perspective, this kind of argument cannot be used to "set aside the past." Because it takes advantage of the fact that a qualitatively differentiated reality resists the efforts of physics to tame it, there is a likelihood that such a perspective will initiate a new round of polemics, with physicists protesting that they do not tame reality.

My attempted narrative presentations concerning physical-mathematical invention involve ways of defining the past that, as far as the ecology of practices is concerned, must be set aside. This does not need the consent of physicists. It involves, first and foremost, shifting the kind of interest generally associated with physics, an interest that connects physics with a mythical epic wherein "mankind" learns to recognize the nature and scope of the rights of knowledge, where it confronts the limits and paradoxes of the confrontation between knowing subject and knowable world.

My attempt means claiming, without the slightest need to downplay their interest, that the laws of chaos do not transform

our "vision of the world" and that Prigogine's hypothesis about
a probabilistic determination of the present by the past, and the
future by the present, is interesting only to the extent that it is
not a general idea but a highly singular practical reinvention of
an object that satisfies the theoretical quest of physics. And its
primary interest for "us," that is, from the point of view of an
ecology of practices, is the realization that there is nothing general about the very idea that it would be possible to define how
the past determines the present and the present the future, that
the attempt to implement such a possibility is what defines the
singularity of the physicist's quest.

Leibniz, a mathematical genius, clearly saw the far-reaching
consequences of the power to define the present as "conforming" with the past.[4] His strange definition of monadic reality is
one such consequence, and was confirmed by physics: changes
can be represented in terms of interactions and interacting
entities, and also attributed to entities represented as autonomous.[5] Prigogine's invention of a new type of "spectral representation" that introduces physical-mathematical beings again
evolving autonomously, independent of one another, thus constitutes a new triumph of the affirmation of conformity, of the
identifying power that brings together definition and reason,
description and explanation. But here the assertion of conformity no longer has the power to conceal the fact that it answers
a question because the answer to this question does not possess
the "miraculous" power to erase the practical problem it started
with. It has transformed the "case-by-case" identification of
the approximations needed for physical relevance into a "case-
by-case" construction of the object, thus depriving conformity
of the power to tell the truth "hidden" behind the diversity of
dynamic systems, all of which are defined by the same type of
equation. Conformity is defined together with the effective construction of the solution to those equations. The object defined
by Prigogine satisfies categories that belong to the construction

of the solution of the problem. If the resolution of a dynamic problem continues to be identified with the construction of a "spectral" representation of evolution over time, with the royal road to constructing a factish capable of dictating its own reasons, the definition of the object can no longer silence its practical dimension, the requirements and obligations it satisfies. It incorporates the question of how and at what price the past determines the future in a way such that the answer can be provided only by the "case" the physicist is dealing with.

In other words, the laws of chaos have the interesting singularity of breaking the "intuitive" relationship that, even when equations could not be integrated, allowed a well-defined behavior to be assigned to the system, a behavior that would be "self-deducing," "automatically calculating" its succession of states from moment to moment. It is the *construction* of the solution satisfying intrinsic mathematical obligations that determines the identity of the object described. As we have seen, satisfying the requirement of conformity affirmed by the laws of chaos obligates the physicist to no longer ask about the behavior of individuals but about the behavior of ensembles. And it is these ensembles that allow the question of how the past determines the present to be answered. The laws of chaos say nothing about what determines the reality they describe; they define the extent to which, and on the basis of what definition, such a reality is able to satisfy the requirement that brings description and determination into correspondence. Yes, the quest continues, but this continuation inextricably designates the singularity of the question and the invention that is required in order to satisfy the requirements associated with this question. The "laws of chaos" possess the *truth of the relative.*

The importance I have given to Prigogine's proposition—a present that would be able to set the past aside—is thus an "ecological" one. Whatever the verdict of his colleagues, his proposition entails the possibility that other practices may situate themselves positively with respect to physicists' demands that we recognize that physics defines the "laws of nature." To reply that these laws are laws of physics, not of nature, is not to reduce them to arbitrary constructions. "Nature" is involved, yes, but the way it is involved cannot be dissociated from what physicists are looking for. The laws of chaos are of ecological interest because they positively affirm what quantum mechanics implied: the laws of chaos are not laws that "atoms" or any other quantum or classical being would satisfy, the way Galilean bodies were said to satisfy the law of falling objects. And yet they create a knowledge about reality as such, that is, about the extent to which it is—or is not—capable of satisfying the requirements of conformity that Galilean objects fulfilled. And they do so precisely because they have accepted this requirement as the challenge they had to meet.

What is an atom, a neutrino, or any other individual defined in terms of such a "fundamental law" Why can we get them to "act" in our experimental devices, trap them, delegate them as if they existed in themselves, whereas we can only define them in terms of an ensemble representation? Let there be no mistake, the question should not be construed as an objection. Rather, it is at precisely this point, when the relationship between "fundamental," "objective" definition and the regime of experimental existence becomes *problematic,* that physics can be said to create a knowledge about "reality" as such. Recall the remarkable agreement between the predictions resulting from the Boltzmann equation and the experimental measurements that formed the basis of Prigogine's position: physics could and should account for this agreement, should be worthy of the experimental relevance of Boltzmann's model. But being worthy does not mean explaining the relevance but inventing the means to create consequences for the model, eventually transforming it into a first step toward an objective definition. Becoming worthy of the experimental practices that have made atoms exist, act, become experimental actors and may mean that, but it is only in terms of ensemble properties that we can define them "objectively," as objects capable of giving their own reasons.

Whatever the fate of the "laws of chaos," the problematic character of the relationship between atoms, or any other experimental being, and the physical-mathematical factishes that meet the requirements of objective definition is crucial from the point of view of an ecology of practices. Two distinct practices, with distinct obligations, are involved, making it important to resist confusing "experimental factishes" with "theoretical factishes," especially those theoretical factishes that have, throughout the history of physics, extended the Galilean miracle: the creation of an object capable of dictating the categories of its description.

In fact, it is hard to imagine a history that is similar to our own overall, but in which some analogue of Pasteur's microbes might never have seen the light of day, or an analogue of Perrin's atoms in the sense that they resulted in quantifying Avogadro's number, or a molecule of DNA in the sense that bacteria can be made to bear witness to the particular role it plays. However— and this was the meaning of the metaphor of percolation I used in Book III—the history in which the scope and significance of the "Galilean miracle" have been constructed, a history that has included the Lagrangian event and its consequences, the probabilistic interpretation of thermodynamic entropy, and the triumph of Hilbertian quantum mechanics, is characterized by "improbable" points, which leave the imagination free to invent other possibilities. Possibilities in which Duhem, Bohr, even Engels, might have played rather different roles. Borrowing a metaphor from dynamics, I would say that at those points certain questions entered into immediate resonance, questions that, while not strangers to one another, are generally discussed in distinct terms, in the sense that the actors associated with them generally affirm the need to respect their independence (the metaphor concludes here: there can be no discussion or respect in phase space). In itself, the term "principle" expresses the refusal to distinguish that singularizes this "resonant" physical heritage. Do we deal with principles of "reality," principles of any possible rational understanding, the principles of a vocation that the physicist cannot betray without being deprived of the source of her invention? Bringing about the difference between experimental factishes and the singular theoretical factishes that confirm all such principles, showing that experimental existence and laws require distinct "capacities" from what they address, means describing the respective achievements those factishes constitute as situated in their respective spaces of practice. The experimental neutrino, which has existed since the origin of the universe, does not obey the laws that claim to provide the reasons for that universe.

Physics cannot, in itself, be the vector of peace in the sense that the ecology of practices, as I understand it, gives to the term "peace." A peace that is not tied to any form of surrender. A peace invented by each practice, along with the invention of a way to present itself that is compatible with the existence and interest of other practices, satisfying other requirements and obligations. But a step has been taken toward this possibility of peace, for the amalgamation of the neutrino of the laboratories and the neutrino that appears to bear witness to the vocation of physics to decipher the principles of reality contradicted this possibility. The theoretical-experimental neutrino, claiming the combined authority of its practical definition and its theoretical definition, is incapable of coexisting with other beings who create and are created by other practices. It can only present itself by referring those other beings to this "reality for us" that it transcends and judges. On the other hand, the fact that the neutrino must be expressed in terms of two distinct practical modes makes it possible to conceptualize the singularity of the questions and requirements to which, in each case, it responds, rather than accepting the fact that it bears witness to a "reality without us." The neutrino, by virtue of being expressed twice, brings into existence two distinct and contrasting "us's," none of which may be identified with a general "us" that could be contrasted to a "without us," because each of them is characterized by specific obligations. As such, these distinct "us's" can be connected with the multiplicity of all the other "us's" that are, simultaneously but each in a specific manner, producers of, and produced by, what we call reality. The claims of the neutrino belong to a space of coexistence in which all our creations allude to the unknown in any creation.

This first step in the construction of an ecology of practices can also serve to promote other questions raised by the generalized war of definitions that awaits us whenever we abandon the physics of laws. Not in the form of a model, but to the extent that we recall the way it came to pass. Not by criticism, disenchantment, or derision, which would confront a practice with the need to abandon its hopes and doubts, dreams and fears, but by acknowledging that the hopes and doubts, dreams and fears associated with each practice make such an ecology interesting, one that, although relative to history, culture, or relationships of social power, is capable of affirming the truth of the relative.

BOOK VI

Life and Artifice

THE FACES OF EMERGENCE

13

The Question of Emergence

If there is one question that transcends the field of so-called modern practices, it is the question of life and artifice. Whether technical artifice can produce life or only prepare the conditions for it, while awaiting the breath that will animate material that has been worked upon, whether the made being is faithful to its maker or escapes its grasp, whether it escapes accidentally or by vocation, or because the maker has "pierced" it, has partially broken it to escape the monotony of manufactured products— these are the timeless stories, each of which reprises another, older story, that populate our memories. And within this fundamentally anonymous perspective, it is possible to situate the impact of each new technique of delegation. From the medieval clock to contemporary informatics and genetic engineering, prosthetic devices, the synthesis of organic compounds from inorganic molecules, or metabolic activities reproduced in the test tube, every technical innovation capable of nibbling away at the difference between our ability to have something do something for us and what living things do by themselves arouses the same interest, the same confused passion, fear and pride. Every time a delegated agent acquires a new skill, a new figure of the living is made available for our stories, and a new figure of the risk assumed by those who dare to challenge the order of nature or creation.

However, if there is one problem that is far from anonymous, that immediately brings up the question of the "science wars" with which the ecology of modern practices can today be identified, it is indeed the problem of emergence. For in this case, it is no longer a question of human power confronting the order of nature or creation, but the possibility, for a scientific discipline, of assuming power in a field previously occupied by some other discipline.

Of course, we could claim that the question of emergence has endured throughout the ages. Aristotle's disciples were already arguing about composite bodies endowed with new qualities that arose from the elements that composed them. How could these new qualitative properties be explained? Was the form of composing elements weakened or destroyed by composition, or did it remain untouched so that the properties of the composite would be novel in appearance only? We might be tempted to claim that this same question was being asked in the eighteenth century, when antimechanistic chemists claimed there was a difference between composition, which was their problem, and the simple aggregation of physicists.[1] Except that, at this time, composite bodies and aggregates had distinct spokespersons: the difference between them was now inseparable from the question of the relationship between chemists and the supporters of mechanics. Similarly, when Leibniz pointed out the foolishness of those who dreamed of explaining sensation, perception, and consciousness in terms of the mechanics of inert matter, he seems to have been taking part in a quarrel that continues today with the unfortunately celebrated mind-body problem. Except that now the quarrel is no longer a conceptual one; contemporary "materialist" philosophers no longer claim any status other than that of being spokespersons for those who engage in what would finally be a scientific approach to the brain. And it is on

their behalf that they signal a future in which, from psychology to the social sciences and therapeutic practices, all forms of
knowledge concerning human behavior will be understood in
terms of neuronal interactions.

The question of emergence arises from this polemical context. It was initially forged as a weapon against what would be
called the reductionist bias. But any weapon can be used against
its inventor. The thesis of emergence sounded like a challenge:
you cannot "explain" this emerging totality, as such, as the sum
of the parts in terms of which it is being analyzed. Naturally,
the challenge, once stated, was used to organize an explanatory
counterstrategy. In other words, the theme of emergence transforms the question of the obligations associated with the "emergent" into a field of confrontation. Will or won't this emergent
entail the obligation to "add" something to the operation of the
parts and, if so, does the addition in question entail the obligation to recognize the powerlessness of analytic thought?

In this context, the question of "laws," in the sense that we
speak of the laws of physics, is both very close and very far. It
is very close, in principle, because the reductionist argument is
most often inscribed in a unitary vision of the world, in which
the "parts" presented must, in one way or another, "obey the
same laws" as the matter studied by physicists and chemists. It
is very far, in practice, because no one dreams of requiring those
"parts" to actually bear witness to such obedience.[2] The hierarchy that has already been established among the disciplines
here does its work. If chemical transformations and the ensemble of interactions among molecules are claimed to satisfy the
fundamental laws of physics, any biological mechanism that can
be analyzed in these terms should as well.

On the other hand, reference to technical artifacts is far
more prevalent. As early as the seventeenth century, long before
the "science wars," the clock and the automaton, which had once
celebrated the splendor of divine creation, were used as part

of a philosophical operation that already forged the terms later used by scientists. The clock is a weapon against Aristotelian thought, for which matter is unintelligible as such but requires a form, with which are associated both the existence of individual beings, each of which is endowed with its own end, and the possibility of knowing them. In the case of the clock, matter and finality can be understood separately: consisting of inert parts, and as such subject to the laws of mechanics, it owes its clocklike existence to the genius of the maker, who has subjected those parts to their own ends, who has incorporated them into a coherent mechanism defined by a finality—telling time. From Leibniz to Bergson, some philosophers were able to challenge the relevance of the metaphor of the living organism the clock proposes. But in the context of a "science war," it provides an inestimable advantage. The question of finality designates the stronghold that must be defended or conquered. For a certain time, the "teleology" inherent in the living has served as a standard for so-called vitalist biology. Although "mechanistic" biologists might dissect the living at their leisure, organization toward an end would always be something that was "added" to the dissected parts, which those parts cannot, as such, explain. Jacques Monod's well-known book *Chance and Necessity: An Essay on the Natural Philosophy of Modern Biology* celebrated the fall of the stronghold. The teleological nature of living beings is only apparent, for they cannot be explained in terms of "final causes." However, they are "teleonomic," meaning that it is still on the basis of their finality—self reproduction—that they allow themselves to be described. For it is natural selection—significantly referred to as the "blind watchmaker" by Richard Dawkins—that must account for the singularity of living beings, supply a reason for the characteristic ways a living being has of reproducing, existing, behaving.[3]

I will return later to the blind watchmaker, for he is currently being challenged by new protagonists, on behalf of a new type of artifact whose making corresponds to another kind of practice. At this stage, the example of the "defeat" of vitalist biology in the face of the so-called neo-Darwinist offensive, illustrated by the arguments of Jacques Monod and Richard Dawkins, provides an opportunity to highlight the reasons why the ecology of practices I am trying to conceptualize must confront the question of emergence and, more specifically, the way in which this question has been defined in polemical terms.

To what extent is the question of finality relevant to an understanding of living things? Only biologists not engaged in the polemic with vitalism, such as Stephen J. Gould, were interested in asking the question, which then becomes very complicated and very interesting, requiring fine distinctions and risky hypotheses. Every characteristic can present a different problem, can tell a story that will distinctly interlink heritage and novelty, the coherence of previously stabilized meanings and unforeseen possibilities. We'll return to that. What I want to emphasize here is that understanding the challenge to which the living being exposes the biologist is barred to the vitalist biologist just as it is to the believer in neo-Darwinism. In both cases, the polemical position is expressed by the production of an identity that is substituted for practical requirements and obligations the way a solution is substituted for a problem. What the biologist deals with cannot pose the problem of the relevance of the requirements in terms of which it is addressed, for, in doing so, the possibility of a betrayal, of a passage for the enemy, is liable to be created. As for obligations, these are mobilized by the supreme obligation of having the legitimacy of one's own approach prevail.

This mobilization, like any mobilization for war, introduces slogans, watchwords. Thus, the case favored in reductionist literature, the one that serves as proof and slogan, is the emergence of the molecule of water, with qualitatively new properties, from hydrogen and oxygen atoms. Similarly, the power of the laboratory should gradually dissipate the pseudo-problems posed by the "qualitative" emergences claimed by the adversary. And this "holistic" adversary, partisan of the emergence of a living "whole" irreducible to the sum of its parts, will, on the contrary, link his claims concerning the limits of experimental practice to the fact that real "wholes" are proof that they exist precisely to the extent that their properties can be objects for description but not for experimentation. Thus, even if the scientist can intervene in the development of an embryo, she has to recognize the relative "autonomy" of that development. Intervention can create monsters or kill, but the embryo cannot be redefined in such a way that its development is proven to obey a function whose variables would be manipulated by the experimenter.

Therefore, what the laboratory can do becomes the subject of a polemic. For instance, it is the existence of a new type of laboratory, that of the molecular biologist, that Jacques Monod celebrated when he announced that the "secret" of the teleonomy of living beings had finally been pierced. The laboratory of the molecular biologist has succeeded in turning living beings into reliable witnesses, in subjecting them to the variables the experimenter manipulates. Not living beings in general, however, bacteria and viruses. It is their performances that were articulated in terms of the catalytic, regulatory, or epigenetic functions of proteins, such functions relating to the associative, stereospecific properties of those molecules, which is to say, in the last analysis, to the DNA molecule containing the "genetic information" that determines their synthesis. The partisan of the irreducible emergence of the living organism (the "holist") was betrayed by some of the living organisms he intended to represent. Now, he is asked to specify where, exactly, he claims to break the chain of consequences that runs from the bacterium to the elephant, not to mention humans.

That the invention of new kinds of laboratories and new

laboratory beings can in this way be mobilized for the polemic, that experimental questioning can be referred to as "reductive," is one of the most damaging consequences of the science wars. Where was the reductiveness when Pasteur had his microorganism "act" in a context such that its autonomy had to be recognized? Or when Körner, a student of Kekulé, subjected the three isomers of dibromobenzene to a substitution reaction, replacing hydrogen with an NO_2 radical, the distinct isomers he obtained demonstrating, by their relative proportions, the hexagonal structure of benzene? Or when the artificial DNA molecule synthesized by Nirenberg (UUUUUUU . . .) succeeded, on May 27, 1961, in "causing"—using all the necessary enzymes but "out of the body," in a test tube—the synthesis of a protein, an obviously "stupid" protein, composed of a single type of amino acid?[4] Events of this kind mark the creation of new laboratory beings and the new laboratories that correspond to them.[5] But they do not pose the problem of emergence and do not allow any reduction to occur. They mark the success of an operation of delegation. The delegated being, which bears witness to its existence (Pasteur), to what it acts on (Körner), or its specific responsibility (Nirenberg), brings about new practical possibilities. Similarly, bacteria and the other laboratory beings that molecular biology has managed to turn into "reliable witnesses" were in no way "reduced" to an arbitrary assemblage of molecules. Those beings were targeted by operations of delegation or were themselves delegated, and each of the "properties" that supposedly "explain" them celebrates the singularity of the successful operation, not the generality of the power of explanation.

From this viewpoint, there is no need to try to determine which characteristic would rightfully protect an elephant or a man from a reduction that would have succeeded with bacteria. It is much more interesting to point out how the operations of experimental delegation that have treated bacteria as targets or actors have been made possible. The experimental invention of the bacterium takes full advantage of the fact that the bacterium, unlike the elephant or the man, undergoes no embryonic development because it is "born" adult, whether in a test tube or anywhere else, whereas the elephant or the man need their mother's womb. That is why the question of embryological development is not the "same" question, only more complicated, as the multiplication of bacteria. While bacteria have made it possible the impressive construction of the experimental factish of DNA, with its properties of replication, transcription, translation, and regulation, this factish does indeed possess the truth of the relative. It owes its autonomy to the experimental tests it successfully underwent, and this autonomy is therefore relative to the tests the bacterium is able to experience from its environment without losing the stability of its definition—that is, without dying. That the human embryo or the elephant embryo cannot resist similar tests, that they require a "protected" environment, does not protect them "by right" from future experimental inventions. This difference signifies nothing more than that the question of how they are to be addressed will have to be invented. And if the precedent of the bacterium here had to serve as an argument, it would be to announce the possibility of surprises we are yet unaware of. For, prior to its experimental invention, no one could have foreseen the extraordinary sophistication of the models it would impose, and continues to impose, on the biologist. For biologists, the question of determining "what a bacterium is capable of" is only just beginning.

When DNA becomes a "program," claiming to be the ultimate explanation of all living beings and, at the same time, claiming to give natural selection the role of a (blind) "programmer," the sole (teleonomic) "reason" to which living organisms can respond, it is not the power of the laboratory that is being expressed but the power of the polemic that shaped the question of emergence on the field of confrontation. And along with it the various powers that are not interested in the question of emergence at all but are greatly interested in capturing success-ful operations of delegation and the claims of reduction that may accompany them. This provides a twofold benefit: the power to create new ways of "doing" and the power to silence, in the name of "reduction" to an approach that is "finally scientific," those who would contest the way the problem (for which these new ways of doing supply a solution) is expressed.

The same situation occurred when the Pasteurian micro-organism, vector of transmission of epidemic disease, became the "cause" of that disease, the royal road to a "finally scien-tific" medicine that would reduce illness and healing to "purely biological" processes. This was a typical case of the reciprocal capture of distinct interests. For doctors, reference to this royal road means adopting a position that gives them the power to disqualify charlatans.[6] For the majority of industries related to medical practice, the difference between doctor and charlatan has little interest. But its consequence, the fact that the doctor is made dependent on the network of laboratories providing her with a guarantee of an "anticharlatan" practice, interests them much more. Medicine, like all modern practices, each mobilized by conflict with other practices and all of them against opinion, is vulnerable to, and even demands, all the forms of capture that ratify the validity of its position.

In this joyous context, the fact that the "emergence" of mind in its relation to the "state of the central nervous system" may appear to set the stage for a "summit" between science and phi-losophy is a far cry from expressing a privileged purity. Rather, it is the glaring absence of any operation of delegation susceptible to reciprocal capture that ensures the disinterested character of this "great question." The notion of state haunts the rhetoric of the sciences because it constitutes the master reference for reductionist versions of emergence; but this reference indi-cates that emergence, in this case, is purely and simply defined in terms of confrontation. Confrontational reductionism has

no need of the laboratory, and its relation with operational consequences and possibilities is simply a matter of rhetoric. The
only thing that really matters is that the adversary be disqualified, that he be lined up against the wall.

The "state" is in effect responsible for uniting "anything"
that might be a relevant element of understanding the situation
and for expressing the possibility of organizing that multitude
in such a way that "all" the relevant relations become relations
of reciprocal determination from which one should be able to
deduce a full description of what has become a "system."[7] Reference to the state is typically followed by a challenge, with the
adversary lined up against the wall. If he accepts that "everything" has been accounted for in the definition of the state, will
he appeal in order to avoid the "reduction" to "something else,"
some ingredient whose sole meaning will be to express irreducibility? And what is most curious is that this strategy "works."
It succeeds in trapping some of those it targets. In *The Self and
Its Brain,* John Eccles, wishing to "defend" mind, invents for it
the ability to act through "infinitely weak" energy interactions
with large numbers of neurons "in critical equilibrium."[8] What
splendid freedom it is to "choose" between two evolutions from
some critical point.[9] What an astonishing capability those large
numbers of neurons have that they are able to maintain themselves in "equilibrium" at some critical point in order to allow
the "mind" the responsibility of choice.

Eccles's speculation is representative of the astonishing
intellectual regression provoked by the "science wars," a regression that explains why the mind-body problem is one for philosophers—and scientists who wish to "raise themselves up" by
addressing the "great questions." Eccles's presentation of the
problem is none other than that already found in the old thought
experiment of "Buridan's ass," which is faced with the necessity
of choosing between one of two equally attractive alternatives.
If Buridan's ass doesn't have the ability to create a difference

where there is no preexisting difference, won't it die of hunger in the midst of the two alternatives? asked those who wished to see it assigned the freedom, or will, associated with the ability to decide "without reason." Leibniz had consigned the challenge to the ridicule it deserves. To satisfy the argument, the ass must be represented as a pencil standing on its point. It is not at rest, but "uneasy," any small difference will send it toward one alternative or the other. The "paradox" of Buridan's ass, which, absent free will, will never choose one side rather than the other, implies not the fiction of two equally attractive pastures but that of a plane that would cut the ass, as well as the entire universe, in two with no difference between the two halves. If the "mind" is to make "free" decisions, the "critical equilibrium" of neurons must also imply the entire universe. The universe, at this critical juncture, "waits" for John Eccles to choose between two possible futures—in one universe he will pull out his handkerchief to wipe his nose and in the other he will sniffle.

Today we can anticipate a new quarrel involving systems characterized in terms of determinist chaos. Doesn't the "state" of a chaotic system lend itself to an even more convincing reductive argument? It fulfills all the conditions for reductivism because it is "determinist," that is, supports the claim that all relevant relationships are made available when determining the system's behavior. And because of the erratic character of this behavior, all of the manifestations an adversary might use as indicative of the freedom to choose can be incorporated. This adversary will then have to show his true face (dualist, spiritualist, irrational, believer . . .) because he will have to argue the difference between "true" freedom and behavior that is erratic, unpredictable. The very terms of his argument will allow the reductionist to triumphally conclude that "we have left the domain of scientific rationality." Which means: we are entering the world of opinion, where anything is permitted but nothing counts.

Hollow confrontations, power relationships, claims of constituting a royal road, complaints and accusations against the conquering imperialism of a blind and calculating rationality, visions of the world, and reason—all the confrontations that serve as ecology in the modern sciences converge around the question of emergence. Therefore, it is from this field of battle that we must escape. More specifically, this field must be transformed into a problematic and practical terrain. But in order to do this, the meaning of the term "claim" must first be transformed. Emergence cannot be disentangled from claims about reducibility or irreducibility; therefore, a practical, constructivist sense must be given to the issues covered by that term.

14

The Practices of Emergence

It is not often that I have the opportunity to speak well of the work of philosophers of science. That is why I don't want to miss the chance to point out the parallel between the way I approach the concept of emergence and its proposed definition by J. K. Feibleman. He begins with a conventional definition of emergence, which associates the relation between a whole and its parts to the relation between ends and means. According to the "holist" version of this definition, the genuine "whole" expresses its autonomy over the parts in that it can be seen as its own end and its parts will be used as means to that end, or purpose. Therefore, the "whole" is defined as being *organized* as a function of that purpose. But to this conventional definition, Feibleman adds an element that could change many things: "For an organization at any given level, its mechanism lies at the level below, and its purpose at the level above. This law states that for the analysis of any organization three levels are required: its own, the one below and the one above."[1] In other words, the purpose of an organization is not found in itself but is always seen from the point of view of something else.

As a test, let us apply this three-level definition to a favored case of reductionism, the emergence of the molecule of water. The interest in such a swing toward chemistry resides in the questions it brings to light, in this case, those that associate the respective "identities" of whole and part to the practices that allowed those identities to be defined. For, using the three-level definition, the identity of water is immediately doubled, even within the practices of the chemists who defined it. Water plays two distinct roles: one of its identities corresponds to the chemist's purpose in understanding it as a molecule that will interact with other molecules; the other corresponds to the purpose of understanding it as a solvent, that is, a liquid. Consequently, "water" had to emerge twice: as a molecule composed of "parts" and as a liquid with specific properties, composed of molecules.[2] And, in fact, each of these emergences has three levels.

Let's look at the molecule, while remembering to distinguish the atom from the chemical element. Ever since Mendeleev, the element has been a part of the *chemical* definition of molecules and reactions, but it presents no problem for emergence. The chemical element, like matter in the Aristotelian sense, has no properties that could be used to define it "in itself." Its definition entails the definitions of simple and compound bodies and their reactions. The element does not explain the molecule, it is explained along with it. On the other hand, the atom claims to explain the molecule the way the part explains the whole. It owes its scientific existence to practices of a very different kind, which do not address it as a chemical actor; therefore it can, unlike the element, claim a separable identity. "Emergence" can be reduced to two levels if and only if we adhere to Épinal's image of a chemistry that has been "reduced" to physics. In fact, element and atom came to designate the same being only after a series of complicated negotiations in which data from various practices had been articulated and coadapted.[3] And in this process of negotiation, the "purpose" is found "above," on the level of the practice of negotiation itself. The identity of the molecule has been "organized" as a function of a known purpose—the realization of coadaptation, the use of old properties that have been reinterpreted or new properties that have been ardently sought to show that the molecule can be fully explained by the atoms of which it is composed.

Another demonstration, very similar but this time involving statistical mechanics, would demonstrate the emergence of the "whole" formed by a liquid consisting of a population of molecules. But the problem can be made more complicated. For the physical chemist is not the only one "for whom" water is both molecule and liquid. The same is true for the living body. Molecule and liquid "exist" for cellular metabolism in distinct ways, each of which is defined by distinct purposes. In fact, the "purposes" of liquid water, as cellular metabolism as a whole constructs them, are far more subtle than those that made it a "solvent" long ago. Moreover, it is cellular metabolism that obligated the physical chemist to understand the subtlety of what liquid water can do.[4] We can thus state the problem as follows: from the point of view of cell metabolism, doesn't the "identity" of liquid water also "emerge" as being relative to the purposes metabolism invents?

The same type of problem can arise in the case of "detection." It is not only from uncontrolled anthropomorphism that biologists talk about "detectors" when they describe a metabolic function. In one way or another, living metabolism, as well as the laboratory, implies the construction of devices whose "purpose" seems to correspond to detecting (assigning an identity to) a molecule.[5] The irresistible character of the metaphor must be taken seriously, but not literally. Perhaps, borrowing an idea from Bruno Latour, who borrowed it from Michel Serres, we can make use of the prefix "quasi" to mark both the relatedness and the distinction between biological "practices" and practices of human understanding. A molecular quasi identity emerges from biological quasi detectors—a three-level quasi identity given that it relates to the quasi purposes of detection and to the quasi means constituted by the interactions among atoms used by quasi detectors.[6]

Let's return to the general question of emergence. If, as I have done, we include in Feibleman's definition the "purposes" associated with the practices of understanding, the question assumes a practical and political sense. It signals a way of relating two practices characterized by the fact that one includes in the definition of what it studies a reference to the object of the other in the form of a "purpose," which is to say, it includes the possibility of transforming what it studies into a means of explaining that object. In other words, the question of emergence is never "passively" asked, it is always actively asked. The whole and its parts always refer to a third term, a practice whose purpose is to articulate their relation. Practice or quasi practice: the articulation of relations between neurons and the ways of experience do not interest neurophysiologists alone but had to have been an issue throughout the history of living organisms with brains.

Once the question of emergence arises, whole and parts must be mutually defined, negotiate among themselves what an explanation of one by the others implies. The holist version of emergence denies the possibility of this negotiation because it identifies as a purpose for the "whole" the manifestation of properties that confirm that it cannot be reduced to parts. The reductionist version of emergence transforms the negotiation into unilateral determination because it is interested in the "whole" only to the extent that it promises to explain itself on the basis of its parts. It remains to be seen to what extent, when queried from the point of view of this negotiation, the question of emergence can cease being a battlefield where definitions of "whole" and "part" confront one another, each claiming both autonomy and the power to assign meaning to the other. This possibility, if it is to escape good intentions, can only be confirmed by the appeal of its effects. The explicit recourse to an ecology of practices that my definition of emergence expresses will have to shift the appeal that competing visions of the world always promote.

To assimilate, as I have just done, purposes associated with practices of understanding with those that can be attributed to the living organism is somewhat forced, because the analogy is only partial. One way of making this partial character explicit is to point out the relative indifference of the experimenter (more specifically, the experimenters as a community) to the way in which "new water," redefined as a compound "emerging" from those parts known as hydrogen and oxygen, will redistribute the properties that could be attributed to old water. What is important is the construction of a new story. The experimenters' *appetite* is now directed toward the creation of new devices, new kinds of proofs and tests, far more than on the means to "recover" all of water's former properties. The question as well of finding out how composite water and solid-liquid-gaseous water are to be related is relegated to other research projects.

The experimenters' appetite for the world from which they take what will become the substance of their questions often assumes an aesthetic form. Thus, when Jean Perrin celebrates the "vast host of new worlds" that atomic reality allows physicists (us) to peer into, he also celebrates the defeat of values associated with "reality" by phenomenological physics, a reality defined as regular, predictable, and measured by instruments that assume its homogeneity.[7] The thermodynamic phenomena and their variables, which corresponded to laboratory practices that were quite distinct from those that, at the end of the nineteenth century, brought into existence the discrete world of microscopic events "beyond phenomena," are, of course, said to emerge from that host. But it is this host itself that caused Perrin to speak, which made him a visionary and a poet. In fact, we can go so far as to say that the question of emergence here is asked "backwards," for it is the parts that emerge from the "whole," from the observable phenomenon. Contrary to what has often been claimed, there is nothing reductionist about Perrin, for whom discrete reality is not a "means" of explanation. On the contrary, observable phenomena interest him only to the extent that they are reinvented as a "means" for that discrete reality to be characterized in observable terms.

The appetite of molecular biologists is quite different, but in this case as well the problem of emergence is presented asymmetrically, privileging the means. When they subject bacteria to tests that challenge their survival and their ability to proliferate, these biologists have effectively succeeded in occupying a position from which bacteria appear as being organized for survival and reproduction, and the mechanisms they study then appear as so many means at the service of that purpose. But this position is unique. The role biologists have invented for themselves with regard to bacteria does not constitute a right for the scientist with respect to the living organism. This role reproduces the one that bacteria are liable to confer upon the environment, the challenges they are capable of undergoing without necessarily dying from them.

The "vision" that confers upon DNA the status of a program, because it implies the omnipotence of selection in the role of the blind programmer, assumes and affirms that the uniqueness of bacteria is the truth for *all living organisms*. Every living being "says" the same thing as bacteria, except in a more complicated way, and *must* therefore be able to confer upon its environment the same kind of role. Regardless of the feature studied, its only explanation is found in its selection: in one way or another, it *must* have had a selective value, increasing its bearer's chances of survival and reproduction. In other words, the power of selection, which constitutes the "level above" from which the living organism can be endowed with a purpose, can survive and reproduce, would be limitless, so that no problem of articulation can arise between molecular "means" and the "whole" that constitutes the living organism. That is why, from the point of view of selection, the purpose can be attributed indifferently to genes or to the living organism. As Richard Dawkins has stated, and his witticism is quite to the point here, we could also say that the organism is a means that the gene gave itself so as to ensure its own transmission from generation to generation.

One of the most unexpected aspects of the "revolution" known as molecular biology is to have created the concept of "absolute" emergence, as Jacques Monod called it, satisfying no reason other than that of selection. Like the clock, which owes to the laws of mechanics properties of secondary consequence only, and everything to the intelligence of the watchmaker who made and assembled each piece, the living organism of molecular biology is "compatible" with physical chemistry but owes nothing specific to it. Jacques Monod has never celebrated the prodigious activity of proteins and their interactions, but rather the *cybernetic logic they obey*. In fact, molecular biology, while it celebrates the reduction of life to a gigantic network of catalytic reactions, associations, and intermolecular regulatory activities, also celebrates the triumph of technical artifice over Perrin's teeming matter. It was not without reason that the specific performance to which proteins are susceptible has been compared to microscopic "Maxwell's demons." Just as the demon embodied the rights of the probabilistic interpretation that enabled it to intervene at the level of molecular activity and to impose a form of collective behavior that broke with the rule of irreversibility, the performance of proteins *subjugates* chemical activity, turning it into a biochemical "means" for achieving an "end" that is foreign to it, that relates to a history of selection alone. Selection operates on a field that is always already defined by a logic of subjugation since it operates on the result of unpredictable mutations that primarily express the imperfection of subjugating chemical reactions governing the replication of DNA, the imperfection, therefore, of making those reactions subservient to the logic of conservation for which they are the means.[8]

Neither Jean Perrin, Jacques Monod, Richard Dawkins, or any other spokespersons of the all-powerful genes address emergence as a *problem*, thereby allowing the three "levels" corresponding to the problem to resonate. On the other hand, not far from them we can discern the figure of a true practitioner of the problem of emergence, whose appetite is stimulated by the possibility of emergence as such. This figure is the creator of technical artifacts, of beings who, if they manage to exist, will have overcome challenges that are associated not with the requirements of competent colleagues but with the possibility of reliable performance, endowed with meaning for an essentially heterogeneous collective and related to essentially disparate constraints.

The technical-industrial innovator has nothing to prove, in the sense that proof seeks to differentiate between fiction and fact. Her milieu is fiction. She is not, however, released from all obligations, quite the opposite. Her practice obligates her to start with, if not create, a twofold indeterminacy. An indeterminacy regarding the way in which the being she creates satisfies the constraints of the "level above," the level she addresses, that is, the level whose constraints that being will satisfy in giving them a determinate meaning. And an indeterminacy regarding the way in which that being will distribute the respective values of what it mobilizes from the "level below": what it will define as a "means" and what it will define as a possible source of breakdowns or problems to avoid.

The verb *envisage* is appropriate to this practice and its obligations. To "envisage" a problem does not imply its resolution, at least not initially, but relating the terms in which it has been expressed to the solutions it may authorize. The approach of someone who envisages is oriented, but not unilaterally. It involves answering a question, a possible, but the problem as first formulated is only a hypothesis. Indeed, a "world" is implied, and it will become an integral part of any "solution," which may require that the problem be formulated differently. Of course, we can state that the experimenter also "envisages," but the space her practice delineates has a stable topology. She knows, a priori, what "the world" (that is, her colleagues) asks of her, and she also knows what will identify a well-formulated problem. The obligations of proof, the creation of a reliable witness, satisfying the requirements that put it to the test, supply stable criteria for success. The technical innovator does not know, a priori, how she is obligated nor what she may require. She inhabits a space for which a relevant topology must be drawn, one that subsequently will be deciphered in terms of "means" that are implemented and "needs" that are satisfied.

The question that orients the approach of the innovator (a neutral term that refers to a group) does not fall within the perspective of discovery, and what is constructed has no ambition to see any kind of preexistence recognized, the way a microorganism or DNA might claim it. The delegated agents do not have to explain themselves, their actions do not have to bear witness to the properties of corresponding actors.[9] They can do so, but that is not what is asked of them. Questions and agents respond to one another within the perspective of a new emergence that must define both its *prerequisites,* what it requires of materials, of the processes and agents it is going to mobilize, and the way in which it will be inscribed in the world, the purposes that will identify it.

Here the contrast between the possible and the virtual, the real and the actual, found in Deleuze may again be relevant. As Deleuze noted, the virtual, has the "reality of a task to be fulfilled." It is not just something that is susceptible to actualization, it confronts us with the problem of actualization. My earlier reference to the virtual concerned the question of quantum indeterminacy (see *Cosmopolitics,* Book IV). In that case, the "task to be fulfilled" was reduced to a mutually exclusive choice between a determinate number of measurement possibilities. With the question of the innovator, the virtual and its actualization rupture any relationship of nostalgia or mourning concerning the reality that would resist its "potentialization," the reduction of choice to a selection between already determinate possibilities. The innovator does not address a reality that would be "potentially" defined by categories of knowledge yet to be constructed, to preexisting "potential" actors, lacking nothing but the transition to scientific reality. Actualization is covered by the "and . . . or" of distinct possibilities of emergence, rather than the "either . . . or" of mutually exclusive possibilities of determination through measurement. Correlatively, the irony of Copenhagen is no longer relevant. The "and . . . or" does not impose abandoning a possibility. It brings about a new kind of appetite—appetite for the "field" as speculatively implied by the possibility of emergence, a field where both the emergent's requisites and the finalities that will be attributed to it must be actualized.

Yet, while innovators are practitioners of emergence, their practice does not allow the question of what might be a practice of emergence within the coordinates of science to be resolved. The technical-industrial-social factish to be constructed does not depend on the interest of colleagues, it has no ambition to raise new questions, to gather around itself practitioners who will connect it with other fields and other purposes. The appetite for the field its construction brings about usually has a limited horizon as the success of the factish imposes the (relative) stabilization of the purposes and means it distributes.[10]

However, the appetite for the field characterizes sciences such as geology, evolutionary biology, climatology, meteorology, and ecoethology, as sciences that address situations that cannot, as such, be "purified," reduced to laboratory conditions, that cannot, therefore, be reinvented in such a way that they become (in some cases) capable of supporting a position of judgment.

The scientist "in the field" is always on a specific terrain, never one that can claim to represent all the others.[11] The appetite of scientists in the field in no way resembles that of the experimentalist, and those who study such scientists have to learn to develop an analogous appetite. For the stabilized operations that ensure judgment in the laboratory are also those that create the distance between the competent inhabitant of the laboratory and those who venture forth in this place where they know that their questions will likely be judged idiotic, naive, and incompetent. The relative absence of such stability in the field can expose the one who studies fieldwork to temptations of ironic relativism. Each scientist would define his or her "own" field, all of them being equivalent before the ironic eye of the one who sees nothing other than the one thing that interests him, the power of fiction.[12]

The practice of the innovator spoke of emergence, not science. The practice of scientists in the field does not speak (directly) of emergence, for what is in play is first of all the question of how to "describe" rather than how to "interrelate." As we shall see, however, the practical problems presented by description have a direct connection with the question of a scientific practice that addresses the problem of emergence.

A geologist, a paleontologist, an ethologist does not "stroll" around, contemplating a scenic landscape; they do not explore a place the way a photographer does, in expectation of an event, of the photograph that will be risked. They set themselves up with their equipment and their skill, and these specify their questions and confer their meaning on the rather mundane photographs they come back with.[13] But what the field gives to them is not *the* answer to *the* question that such equipment and such skill refer to, but the description of a case, and nothing guarantees nor can guarantee that that case will serve as a reliable witness capable of creating a trustworthy, and generalizable, relation between question and answer. Also, the answer is not capable of being subsequently stabilized and narrated economically, as is the case after a successful operation of delegation, or any other experiment. The answer provided cannot economize reference to *this* exploration, carried out on *this* field. Nonetheless, we can speak of an answer provided by the field because of the learning such an answer entails, learning that does not result in conclusions but in narration.

Unlike the experimental factish, which, by definition, "explains itself" in the answer to the questions it authorizes, the field induces and nurtures questions, but it does not supply the ability to explain the answer that will be given to them. Of course, the practice that causes it to exist and is addressed to it assumes that the relationships that allow themselves to be deciphered are "conditions" for the answer, but they are insufficient conditions. However, the loss of the determining power of the condition, the fact that it is incapable of providing an explanation, are not negative categories here. For there to be a field, the indeterminacy must be interesting as such, the questions addressed to the field and the relationships it articulates must welcome the possibility of a mutation of their supposed meaning. The needed appetite for such a possibility and the role played by the field, which is liable to lend the narration the quality of an intrigue, constitute a practical difference between the experimental sciences and the field sciences. The latter, as I have characterized them in *The Invention of Modern Science,* construct stories in the sense that the causes they present can no longer claim to have the power to determine how they cause. The question "What can the cause cause?" here assumes an importance that is foreign not only to the cause associated with the Galilean object, which provides the = sign with its power, but also to the causes associated with all the practices of staging and delegation common to experimentation. Operations such as staging or delegating assume stable relationships and roles, which is precisely what the "field" challenges and for which it substitutes the interest of intrigue.

The appearance of scientists endowed with the appetite I have just described is an important ecological fact within the population of contemporary scientific skills, but the meaning this fact may harbor depends on that ecology. For a long time, "Darwinian" science has been presented in a form that enabled it to claim the same power to judge as the laboratory sciences. Natural selection had to be all-powerful so that its representative could claim the power to judge, to explain, even rhetorically, the history of living creatures. That we would arrive at the "just so stories" of sociobiology when speaking of primate or human behavior was, in this sense, entirely predictable. What is much more interesting is that some Darwinian biologists today seem capable of presenting themselves differently. I am referring to *Wonderful Life: The Burgess Shale and the Nature of History,* in which Stephen J. Gould states that the field sciences are now capable of claiming the uniqueness of their practice, of inventing themselves as different without fearing the judgment that would call them inferior.[14]

Moreover, although the new EcoDevo (ecological developmental) biology explores the embryo's development with the full array of sophisticated tools provided by experimental science, it is nonetheless something like a "field science." The field in this case is the amazing "causal choreography" associated with processes of development that had been characterized by both finalists and neo-Darwinists as directed by a cause (the final cause or the program). The characterization of the continuously self-redefining developmental entanglement mobilizes all the words we have to describe encounters that affect the very fate of the encountering terms. From infection or mobilization to hijacking, seduction or reciprocal induction, the common feature of these narratives is that any simple relation between "cause" and effect" is lost without regret.

When the interest I associate with the field sciences is addressed to living organisms customarily judged in terms of purpose, as creatures of natural selection, directed toward an end, what is learned instead is the risk of such judgment, a risk that cannot be overcome, that will recur at each successive step. Such sciences not only speak about the hazard of circumstance, they create interest in the intrigue that binds heterogeneous elements whose meaning is produced in the encounter itself. In doing so, they serve as a decisive ingredient in the problem of emergence. For, the two confrontational positions that destroyed this problem are similarly challenged. Neither the finalist biologist, for whom the ends of organization define the irreducibility of emergence, nor the "reductionist," who accepts his adversaries' purposes as such only to relate them to the power of selection, have any desire to conceptualize the dual indeterminacy of "ends" and "means." The question that now arises concerns practices that would eagerly welcome this dual indeterminacy, practices that would require an alliance with the field sciences in order to construct the problem of articulation between the requisites of emergence and the purpose that will be associated with what emerges.

To approach this question, I want to examine the answers supplied by the experimental sciences and, more specifically, those sciences that, during the past years, have claimed to "renew" the question of emergence: the physical chemistry of nonequilibrium and the study of neoconnectionist networks. I will try to show that, in both cases, a mutation is produced with respect to the domain of origin. The physical-chemical being "far from equilibrium" may cause a divergence between "condition" and "determination" whose coincidence was formerly ensured by the state of thermodynamic equilibrium. The artificial "neoconnectionist" being brings about a divergence between "fabrication" and "mastery," which the watchmaker's artifice celebrated. Such divergences are what the term "self-organization," shared by both domains, reflects. For the scientists who suggested it, the loss of power, that of the ability to determine or master, has given rise to new values, new interests, and, of course, new claims.

Thus, new, practical faces of emergence, resulting from the experimental sciences as well as from the sciences of artifice, arise, which will allow us to explore possibilities of articulation between laboratory creations and field creations. Some may criticize these new faces as masks that conceal a new strategy for conquering the terrain. Indeed, self-organization can be seen as a new "all-terrain" response. In fact, it was the clearly differentiated—enthusiastic or disparaging—but all too often caricatured responses engendered by nonequilibrium physics that forced me to take the first steps toward what I have here referred to as the "ecology of practices."

These reactions also indicate the limits of the "interdisciplinary project" of which *Order out of Chaos: Man's New Dialogue with Nature,* which I coauthored with Ilya Prigogine, was a part. Whenever the question of scientific practices is involved, interdisciplinarity, whether it finds the source of its references in physics or cybernetics, information theory or some "theory of complexity," suffers from the same weakness as the concept of an "idea" (or an ecology of ideas). The idea seeks to "be applied" and is eager to exaggerate any resemblances. It entails no requirements or obligations, and therefore travels freely as some kind of shared currency that would permit an "exchange" or "dialogue" among the sciences but that dissimulates the glaring difference among the use-values it is able to claim in various scientific domains. So, it is not in terms of "interdisciplinary promise" that I conceive of the possible faces of self-organization, but in terms of the test I associated with emergence as a problem: a practice of articulation that brings about and stabilizes abandonment of the position of a judge who has no need of a terrain because he knows ahead of time what that terrain has to say. Whenever there is a question of emergence, indeterminacy must become a part of the meaning of what is constructed in the laboratory.

15

Dissipative Coherence

In the next few pages, I want to return to physics, but not the physics of laws. The physical chemistry of nonequilibrium refers, through "equilibrium," to thermodynamics, a "phenomenological" physics that was said to have been reduced to the terms of the probabilistic interpretation that led to the triumph of the laws of the Queen of Heaven (see *Cosmopolitics,* Book III).

I want first to briefly recall the rather curious structure of so-called equilibrium thermodynamics characterized in Book III. This science stands out in that its object is not energy, or thermodynamic, processes as such but their "rational mimicry": the displacement of equilibrium, where process time is replaced by the progressive manipulation that forces the transition from one equilibrium state to another infinitesimally close. In this way, we arrive at the three-part definition of entropy as a state function. In the ideal case, when the change of state it measures is a reversible displacement between equilibrium states, entropy is defined in terms of the system variables. When this displacement does not fully satisfy the ideal of a transformation that never brings the system at a finite distance from equilibrium, entropy remains a state function, but its definition becomes indeterminate: it includes some "uncompensated heat" that expresses the fact that any deviation from the ideal results in "dissipation." And when the problem is not displacement, ideal or not, from one equilibrium state to another, but an evolution toward equilibrium, only the maximum value of the entropy is defined, corresponding to the equilibrium state. The evolution of an isolated system toward equilibrium "causes" undefined entropy to increase until it reaches its well-defined maximum.

The definition of entropy, and other thermodynamic potentials, thus gives a central role to the concept of an equilibrium state. More specifically, the two concepts define each other: the equilibrium state is defined by the maximum or minimum value of the potential (according to the definition of this potential) and the potential guarantees the stability of this state. Once at equilibrium, the system remains there and any evolution that would spontaneously move the system away would contravene the second law of thermodynamics. For example, in an isolated system, it would correspond to a decrease in entropy. The existence of a thermodynamic potential function thus characterizes a dissipative evolution by its final state, when all dissipation will have vanished. In short, the thermodynamics of equilibrium is by and large characterized by the opposite of the obligations of a field-based approach: its questions revolve around a state that is unique precisely because it has the power to silence all questions, that is, to provide a final reckoning for a process for which it has neither the means nor the need to give an account.

In *Cosmopolitics*, Book V, I introduced Ilya Prigogine as the successor, then as the heir, to Boltzmann. However, when the work that resulted in his 1977 Nobel Prize is being discussed, he should be referred to as a student of Théophile de Donder. Successor and heir are a matter of choice, being a student is primarily a "fact," even if this fact also implies a choice (not every teacher becomes a "master" for her students). De Donder was a mathematical physicist and a correspondent of Einstein. For him, science was something that "embodies the purest image that the sight of Nature can bring to life in the human mind." And when he was required, out of professional duty, to teach thermodynamics, he did not find that purity. So he decided to create it. For that to happen, Clausius's mute, uncompensated heat would have to learn to speak, would have to participate in the harmony of functions and reveal the musical truth of the indistinct noise known as dissipative evolution. And de Donder turned to that field of thermodynamics where dissipation is entirely intrinsic, where the ideal of a reversible transition from state to state is the most obviously artificial—chemistry. For, measurement by means of reversible displacement was able to normalize chemical reactions only by stripping them of their most important characteristics: the spontaneous heat given off or absorbed by every reaction and the reaction rates that qualify them and that kinetics studies.

Dissipation and chemistry. It required the freedom of a mathematical physicist inhabited by the beauty of his science to challenge the hierarchical structure that sanctified the power of that science. De Donder did this in two ways: by asking about irreversibility, which had been deprived of any meaning on the fundamental level, and by attributing to it, as its "topos," as the site where the corresponding problem could be constructed, a chemical activity that had been reduced to the interaction between the atoms of physics. In discussing the growth of entropy, I spoke of an "enigmatic factish" that raises questions it is, as such, incapable of answering (see *Cosmopolitics,* Book III). But the enigma in question cannot be separated from the final decades of the nineteenth-century crisis concerning the values and obligations of physics. For de Donder, who was in the service of harmonious beauty rather than the power to impose requirements, the crisis never existed. The enigma was free to become a question, and that question created an interest in what it designated as the terrain on which it could become a problem:

the dissipative activity of matter.

With de Donder, thermodynamics, a science that is deliberately blind to what it cannot subject to a rational equivalence through which it can articulate its variables, would reorient itself around a new physical-mathematical being that, in itself, said no more than other thermodynamic properties but raised a question where those other beings gave only answers: it is the *production of entropy* that describes the growth of "uncompensated heat" over time. Concerning the production of entropy, thermodynamics as such doesn't say much, except that it is the most general of thermodynamic potentials. Regardless of the conditions that define a system (isolated, constant temperature and pressure, etc.), the production of entropy at equilibrium is, by definition, identically zero. Correlatively, all irreversible evolutions toward equilibrium are, by definition, "entropy producing." The enigma has become a problem: with what kind of variable can this production of entropy be associated?

Chemistry is privileged in that, aside from thermodynamic variables, chemical transformations are characterized by other variables that immediately introduce time: the kinetic variables that refer to equilibrium as the state in which processes continue to occur but with velocities such that their effects are canceled. De Donder would make the production of entropy the setting in which an explicit link could be forged between irreversibility, process, and time. Forging refers to the art of making alloys, which force disparate materials to become one. In this case, kinetics, which evaluates chemical velocities, and thermodynamics, which qualifies chemical reactions on the basis of the equilibrium state in which the production of entropy is canceled, although previously rivals, will now be forced to participate in the definition of the production of entropy.

The production of entropy is the crucible where the link is forged, in the sense that it defines the question asked by de Donder of every chemical reaction: What do you contribute to the production of entropy? Here, the term "contribution" is critical. It seems to refer to the possibility of "judging" a chemical reaction on the basis of a "value," the value of its contribution. Earlier, Clausius had judged the respective values of the conversion of heat into work and the passage of heat from one temperature to another using as a fixed point their equivalence defined by the ideal Carnot cycle. When the system has returned to its initial state, conversion and passage are exactly balanced. But the "value" defined by de Donder (in this case, the contribution of each reaction to the production of entropy is defined by the product, Av, of its thermodynamic potential, the affinity A, and its velocity v) is part of the question, not part of the solution. The overall production of entropy must be positive, yes, but this thermodynamic definition leaves the specific contribution of each chemical reaction to this production *indeterminate*. This allows de Donder's thermodynamic Av, which defines the overall production of entropy for the ensemble of reactions that participate in a chemical transformation as being positive everywhere, except at equilibrium, where it is zero, to take into account kinetic description, which describes this ensemble as a "system" of reactions that are *coupled* to one another. From the point of view of kinetics, each reaction has a velocity that depends on the concentration of its reagents, that is, on the other reactions that contribute to the production or destruction of the reagents in question. And this coupling of reactions can result in the fact that some of them provide a *negative* contribution to the production of entropy.

The production of entropy, therefore, can be used to present a problem—the difference between each separate reaction and the ensemble of coupled reactions. And this problem is open-ended. The thermodynamic condition expressing the second law, the positive production of entropy, is inadequate for determining its solution. Except in one case. At equilibrium, it is thermodynamics's glorious simplicity that triumphs:

the rate of each reaction is separately compensated by the rate of the inverse reaction. However, the production of entropy as de Donder redefined it only makes the significance of this simplicity explicit. At thermodynamic equilibrium, the coupling of dissipative processes—kinetic description—is without consequence. But the clarification, as is the case whenever it is creative, transforms the meaning of what it makes explicit. Equilibrium is no longer the thermodynamic "state," but a particular situation within a landscape that asks to be explored. And if the system is maintained *out of equilibrium*? If, instead of letting it evolve toward the state that has the power to make coupling insignificant, the experimenter forces exchanges between the system and the exterior—for example, through the permanent flow of chemical reagents that prevents it from reaching equilibrium? What might then be the effect of the negative contributions to the production of entropy that makes coupling between processes possible?

This was the problem Prigogine set for himself, and he provided it with a clear purpose from the outset. For Prigogine, a student of de Donder, was preoccupied with the question of emergence as exhibited by biology. And the uniqueness of his position among physicists arises from the fact that he wanted physics to have the ability to address this question. Living organisms do not have to be answerable to physics, it is physics that has to be answerable to the fact that the living organism is actually possible. In other words, Prigogine required that physical-chemical processes become a relevant "terrain" for the question of life. He insisted that "irreversibility"—the production of entropy—be able to tie its fate not with the evolution toward equilibrium but with the processes that, in one way or another, constitute a living organism.

In his doctoral dissertation, Prigogine generalized to all physical-chemical processes the relationship that de Donder had forged between chemical kinetics and the production of entropy. In 1945, he showed that if exchanges with the environment *constrain* the system to remain outside equilibrium but maintain it *close* to equilibrium, the evolution will reach a stationary state (non-zero velocities) determined by the minimum (non-zero) value of the production of entropy compatible with those exchanges. In this way, the power of thermodynamics can be extended to the neighborhood of equilibrium. And in collaboration with the biologist Jean Wiame, Prigogine immediately published an article in which he examined the possible relevant relationships between his theorem and the question of the living organism.[1] In it, he showed how the stability of the stationary state when entropy production is at a minimum may be associated with various properties of the organism.

The limited scope of his theorem was not an obstacle to Prigogine, for he had established a critical finding: the equilibrium state has become a special case (where the production of entropy is zero because the constraints are zero) and stationary states close to equilibrium can be characterized by a certain "order." For instance, spatial differentiation of the concentration of chemicals can appear in a system subject to a continuous temperature differential: thermodiffusion "couples" the thermal diffusion that produces entropy to the chemical "antidiffusion" resulting in negative entropy production, which would be impossible in isolation. Isn't the order associated with the living organism also impossible when that organism is cut off from its exchanges with the environment?

During this time, Erwin Schrödinger published his highly celebrated *What Is Life?*. The contrast between the two "physicalist" approaches to the order that characterizes the living organism is revealing. For Schrödinger, the order of the living organism is "negentropic," characterized by "negative entropy," which implies that this order questions, in one way or another, the principle of entropy increase. The living organism imposes the concept of an order capable of *resisting* dissipation and disorder. And Schrödinger, celebrated as a precursor by molecular biologists, assumed that it was the chromosome (DNA had not yet been discovered) that contained and transmitted the secret of this order, an order that was defined, given that it was a question of struggling against physical irreversibility, by the language of artifice: chromosomes contain both the law and the key to the means for implementing that law, they must explain both the "program" the living organism obeys and the mechanisms that give that program the power to direct the development and operation of the organism. But for Prigogine, the living organism was dissipative. It did not have to maintain itself against entropic disorder, it challenged the simple identification of the increase of entropy with disorder. For, quite obviously, the living organism causes entropy to increase. In order to live, it must feed itself. To extend Schrödinger's concept of negentropy, we could say that for Prigogine it is the processes that produce entropy that we must turn to in seeking the key to those "negative contributions" required by the order characterizing the living organism.

Yet, it was only in 1969 that the now famous term "dissipative structure," expressing the association of order and dissipation, came into use. It celebrates Prigogine's new assurance that he had resolved the contradiction between entropic dissipation and the emergence of order required by living organisms. A crucial element of his work, which extended over more than twenty years and for which he was awarded the Nobel Prize, is the divergence between "condition" and "determination" required by certain couplings of dissipative processes far from equilibrium.

The production of entropy, which is minimal near equilibrium, has the power of thermodynamic potentials. Based on the conditions it determines—the intensity of the flux that imposes a fixed distance from equilibrium—its minimum value can be used to determine the stationary state and ensure its stability.

And it is this thermodynamic power that Prigogine initially attempted to extend far from equilibrium, until his work, conducted in collaboration with Paul Glansdorff, underwent a *practical* mutation. For the interest in far-from-equilibrium states came to be associated with the fact that they escape the power expressed by the possibility of defining a potential, or a state function: identifying the end point of dissipative evolution and ensuring the stability of that final state. The central question instead became that of stability or instability, and it is the coupling of processes that the answer to this question depends on. In other words, process coupling no longer serves only to define the production of entropy but replaces the production of entropy as the focus of the definition of the *activity regime* toward which the system will evolve. In short, far from equilibrium, the states that prolong the stationary state near equilibrium ("thermodynamic branch") may become unstable. "Dissipative structure" is the name given by Prigogine to the new activity regimes, which owe their stability to interprocess coupling.

The remainder of this history is outside the scope of this book. But I want to discuss two aspects that help clarify the practical novelty of Prigogine's approach. One is the association—the source of several fads and misunderstandings—between dissipative structures and "order through fluctuation."[2] The other is the term I introduced above, "activity regime."

Order through fluctuation expresses, in an immediate way, the emotive charge of the event for a specialist in thermodynamics: the loss of power for the second law, which, through a potential function, ensured the stability of the state, that is, the regression of the inevitable and unceasing "fluctuations" (the state being described in terms of macroscopic values, or means). The fads and misunderstandings arose because the concept of fluctuation was associated with a "cause" or "responsibility": chance fluctuations would be responsible for a "choice" or else would be creative. Rather, the expression "order through fluctuation" indicates that the *practice* of physicists has changed. If they are no longer able to require that a potential function be defined, they are obligated to address the problem of possible instability. They are obligated to "test" an activity regime deduced from its equations in order to determine if that regime will be restored when subjected to a perturbation or, on the contrary, if the perturbation will increase. The perturbation introduced by the test expresses the question imposed by an activity regime that must be conceived as intrinsically fluctuating, once the insignificance of those fluctuations can no longer be guaranteed. That is why, in the case of instability, the physicist can describe a fluctuation as what will be amplified and the new, stable activity regime as a "giant fluctuation" stabilized by irreversible processes. But "chance," here, has no value independent of the (nonlinear) coupling that creates the landscape of possibles and the question: which will be realized?[3]

To speak of an "activity regime" as I have done expresses the fact that it is no longer possible to speak of a "state," for the definition of a state always follows from a power relationship, the satisfaction of the requirement that the definition of a system's identity means the ability to define its state(s). But this power relationship also disappeared along with the disappearance of thermodynamic potential. Whatever controls the external variables (pressure, temperature, reagent flow) no longer controls the system. Not only does the second law of thermodynamics no longer guarantee that uncontrolled local fluctuations will regress without consequence, the *very identity of the system* can be transformed. A factor that is insignificant at equilibrium, such as the existence of the gravitational field, can play a crucial role, that is, it can make distinct activity regimes possible. This "sensitivity" of the activity regime when far from equilibrium to factors that are insignificant at equilibrium transforms the nature of the questions asked. For now it is the activity regime of the system that determines its relevant definition, what this definition must take into account, what the system can become "sensitive" to. Therefore, to study an activity regime is also to study the stability or instability of a definition this regime might justify under certain circumstances but might cause to be modified in others. Correlatively, the notion of "constraint" assumes a meaning very different from "limit." Relationships with the "exterior" constrain the system to remain far from equilibrium, but the "limit conditions" do not provide the ability to determine (as was the case near equilibrium) what, from among the various possibles, will eventually be realized. Although limits are usually associated with meaningful imposition, far-from-equilibrium constraints are given their meaning by the activity regime they make possible. And this meaning will be determined by the *production* of the solution to the problem posed by the constraint. In any event, the constraint will be a condition, but it will lose the ability to determine what it might be a condition for.

Although physical chemists have not abandoned the concept of a system as such, that concept no longer corresponds to the power relationship resulting from the ability to deduce possible behaviors from the definition of the system. They preserve the concept because they can. What they address has been prepared in the laboratory, and they know what the definition of a system at equilibrium allows them to treat as negligible. That is why they can trace the landscape in which stable and unstable activity regimes can be distinguished, and the bifurcations that indicate a transformation of distribution between stable and unstable. But this ability to define the landscape of possibles as preexisting the realization of a given possible is now strictly correlated to the power given to the physicist by preparation in the lab. The concept of an activity regime as such entails a distinction between the "abstract" problem, expressed in terms of constraints, and the concrete solution produced by the effective coupling of processes in space and time. It is this distinction that has been, biographically, my pathway to the Deleuzian distinction between actualization and realization, or the virtual and the possible.

Under far-from-equilibrium conditions, the scientist can no longer "require" but the activity regime can "obligate." It is this new configuration of requirements and obligations that is referred to by the term "self-organization." The term was, significantly, borrowed by Prigogine from the tradition of prewar "antireductionist" embryologists (notably Paul Weiss), who noted the ability of the embryo to determine for itself what would be a "cause" and what would be insignificant. It tells us that physicists who adopt it now consider themselves to be directly confronting the problem of emergence: far-from-equilibrium physics is able to "comprehend" the arguments of embryologists against the reductions that had, in one way or another, joined forces with the defining power of a state. However, the term "comprehend" has two distinct meanings—to include and to understand—and only the distinction between the two can create the difference between a physics that claims to "explain" emergence and a physics that becomes a partner in a practice of negotiation through which emergence can be constructed as a problem.

Chemical clocks that exhibit periodic behavior, Bénard cells that imply coherent collective movement at the macroscopic level in a crowd of innumerable molecules—far-from-equilibrium physics results in the creation of new experimental factishes that signal the emergence of activity regimes that break with the general ideas associated with microscopic disorder. Here, laboratory and theory work in tandem, constructing new descriptions about emerging "order" and natural processes that can be included within the new framework. However, this inclusion concerns beings who "depend" on a "constraining" environment, but, unlike living organisms, *have not created a milieu for themselves.* It is true that far from equilibrium, coupling, or the interrelation of processes that produce it "create" a being whose behavior cannot be identified as the effect of the constraints imposed by its environment. But the being in question appears or disappears, depending on circumstances, without a fuss. Additionally, the coupling between processes is silent about differences that may matter if a "purpose" comes to be involved; for instance, the difference between two possible activity regimes, one chaotic and the other periodic. Physical-chemical self-organization is "factual" in nature, and the term "organization" here does not correspond to any kind of "purpose" that might articulate, for anyone but the scientist, risks, values, and challenges.

Yet, those same beings are also factishes of a very different kind, one that does not bear witness. For a long time I have searched for an adjective that would reflect this difference and, finally, it is the word *promising* that seems most appropriate, because it binds ordinarily disparate semantic uses. What is "promising" often refers to a self-interested approach, whereas a promise is often associated with a commitment. The "technical" innovator I have presented as a practitioner of invention most often assumes, as her point of departure, a "promising" possible, only to discover that there is a great distance between what she thought was promised and the actualization of that promise. In fact, what is "promising" promises nothing in particular to anyone in particular. Unlike the promise, it has no recipient but stimulates, on the part of whoever allows himself or herself to be captured, the appetite (quite unlike that of the experimenter) I have associated with the verb *envisage*. Those who envisage on the basis of a "promising" problem or possibility know they are obligated by a world, even if they don't know how that world obligates them. They know that actualization of what they envisage as promising implies a creation of meaning about which they cannot freely decide. Unlike the experimenter, they negotiate with a world they must encounter as partly indeterminate, susceptible to new relationships of meaning, but which they can encounter as such only if they first recognize it as material for obligation.

The promising factish, therefore, creates interest in the "terrain" where a promise and its recipient will be actualized. But should we take seriously the intense association between the promise and a face-to-face situation, when looking into the face of the one who promises? Yes, possibly, providing that "faciality" is understood in the asignifying, asubjective sense given by Deleuze and Guattari. The face is not that of the "other who promises," the foundation of intersubjective relationships, it is an "abstract" machine, inductive of deterritorialization, and it is as such that it can, in certain circumstances, become a condition of the signifier and the subject. Face as "voice carriers" are not, for Deleuze and Guattari, an anthropological universal: "These are very specific assemblages of power that impose signifiance and subjectification."[4] Similarly, here, there is a very specific appetite that the "promising factish" stimulates when it induces a new kind of relationship between the laboratory and the world. The question of what a being "far-from-equilibrium" is capable of suddenly refers to the price paid for the power of the laboratory. For, it is "outside" where the answers become more interesting, where the power of the laboratory is replaced by the possibility of "reading" histories, of "following" the part played by an activity regime whose intelligibility has been produced by laboratory practices within those histories.

If physical-chemical self-organization is a fact, it also raises questions that the "fact" is unable to answer but for which it is available as something "promising." A self-organized activity regime could indeed be what the emergence of living organisms requires. What is requisite and not what explains—this distinction reflects the difference between *understand* and *include,* both of which the verb *comprehend* refers to. The requisite refers to what a problem needs, without which it could not be presented. It implies that the scientist situates herself with respect to the living organism as she would when confronting a problem, rather than a "fact," for example, the fact constituted by the ensemble of far-from-equilibrium activity regimes in a physical-chemical system. If the possibility of such regimes designates what is requisite, this means that they invite us to wonder about the meaning *they might have assumed* within a story they do not explain. They "promise" in the sense that they can be used to identify an issue: what must be "narrated" is the way an activity regime far-from-equilibrium eventually came to play a *role*, that is to say, assume a meaning for something other than the scientist. In this case, and from the point of view of this role, not all far-from-equilibrium regimes are equivalent. A story could be told in which the stabilization of one regime rather than another—but also, possibly, mutated assemblages, transformations of coupling—could assume a meaning the laboratory cannot provide, the meaning of events affecting a "body," that is, events capable of an evaluation that implies and initiates a distribution between "interior" and "exterior," between "function" and "milieu." The requisite creates an appetite for situations in which the "fact" would become an issue *for something other than the scientist.* It creates an appetite for the problem of emergence, for stories in which the question of the "value" of the possibles it allows would be invented.[5]

It is this new appetite, drawing the physicist "outside the laboratory," that the original title of *Order out of Chaos, The New Alliance,* expressed. The physicist evoked by the "new alliance" would no longer be interested solely in the "world" she has learned to judge in the laboratory, it is the diversity of "cases" she desires. She must then form working alliances with the diversity of knowledge practices liable to identify couplings, arrangements, and coherent collective behaviors whose meaning would "emerge" from asignifying local activities. But here too the question of an ecology of practices is relevant. Physicists may indeed limit themselves to proposing *models*, whose value is relative to the pertinence of the knowledge they will capture and rearrange. But with them the authority of physics moves forward, for it alone, situated at the summit of the hierarchy of the sciences, is authorized to determine what has a right to truly exist, and what must be categorized as an illusion.

Order out of Chaos was inhabited by the hope of a new coherence in our knowledge, one that would heal the deep rifts created by physics's denial of becoming. It called for a dialogue among the sciences, united by the open question of becoming, a question none of them could appropriate.[6] What this well-intentioned offer neglected is that those sciences were also modern, and haunted by the power of disqualification and conquest, more eager for an alliance with the generalizing power of the "new model" revealed by physics than for the risks, questions, and challenges the physicist's "desire" might arouse. At best, in practices such as biochemistry or the study of "social insects," which already dealt with the contrast between multiply coupled activities and coherent overall behavior, connections were made without too much fanfare. I'll return to this later. But at worst, and with considerable commotion, new paradigms (with short life spans) were announced, turning dissipative structures or order through fluctuation into one of those all-purpose concepts that seems to proliferate wherever the will to science takes the place of practice. The "promising factishes" invented by physics can become ingredients of a practice of emergence only if the world they encourage us to investigate is populated with knowledge that is not awaiting the surplus legitimacy and power that physics-inspired "model making" would confer upon it. Such knowledge would need to be capable of obligating the physicist to take an interest in what the model must capture in order to actualize the promise and its recipient.

16

Artifice and Life

The problem of emergence in the context of the sciences of the artificial presents itself quite differently. Here, the question is not at all that of the possible "emergence" of meaning "for" a world. The artifact always has meaning. It can always be understood in terms of a logic that relates means to ends. If we consider the two distinct fields, "artificial intelligence" and "artificial life," that, in the past thirty years, have claimed to "explain" intelligence or life, the common trait that characterizes them (and expresses their shared connection to John von Neumann's work on computers) is the radical distinction between "information processing," which must be understood in logical terms (computation), and the material "implementation" of such processing. Here, the concept of an artifact should no longer evoke the image of a clock, nor that of a robot laboriously assembled. What human art intends to reproduce is the "form" that controls matter, that is, which can be conceived independently of the matter it will control. Computer beings are not actualized; they are indifferently realized for a given physical medium. This physical medium can be a source of breakdowns or crashes, but not differentiation. Here, the artifact is staged in a way that is foreign to the natural sciences: "mind" controls matter. The ideal is submission not to laws but to a project.

I want here to turn my attention to the field of "artificial life." The promoters of this field associate its ambitions with what they define as the failure of "artificial intelligence." It is not possible to construct a "brain" capable not only of reasoning but, especially, of learning to explore a milieu and extract from it the ingredients of "adapted" behavior, unless we have first endowed that brain with a computing "body" capable of encountering that milieu, of moving, falling, touching, and taking into account the consequences of its own actions. Which is to say, unless we have raised the question of the "evolution" either of a population of such bodies or the behaviors characterizing that body. We must first construct a body one can call living, one that is capable of learning, before we can construct a being we could call thinking.

It is quite remarkable that the physical-chemical self-organization I have presented and the "artificial" self-organization I am about to present converge from opposite horizons toward the question of the "body," a being endowed with a topology that creates a substantive difference between "interior" and "exterior," to which corresponds a differentiation between two types of "variables."[1] The body forces a distinction among variables that refer, to return to Feibleman, to the "level below"—variables that, if they belong to a body, no longer characterize physical-chemical interactions but relationships that have a meaning "for" the body—and those that refer to the "level above," which correspond to the milieu that exists for the body and for which it exists, a milieu of welcome or catastrophic encounters, a milieu in which not everything has the same value from the point of view of the risky wager that has produced a given body.

The two horizons are indeed opposed. The "promising fac-tishes" of physical chemistry pose the problem of the emergence of an "exterior" that serves as a milieu. In contrast, if there are to be "promising factishes" created by artificial life, the definition of a "milieu" would not pose a problem. On the contrary, it is the value "for the exterior" that generally fully defines the artifact.

Here, the problem will be one of the emergence of variables that can be called "internal," and which must not be defined from the viewpoint of an "external" finality. In other words, the term "self-organization," shared by both fields, does not have the same meaning. In physical chemistry, the "autonomous" character alluded to by the prefix "self-" is something acquired, but the possibility of speaking of "organization" has yet to occur. In the case of "artificial life," organization is something acquired, but the possibility of characterizing it as autonomous is in question.

The field currently known as "artificial life" may have claimed that the ambitions of its predecessor—the emergence of intelligence—were very premature, but its own ambitions are not much more modest. Chris Langton, a leader in the field, wrote the first manifesto for the inaugural conference in Los Alamos in September 1987. Every word was carefully weighed and every visionary accent carefully deliberated: "Artificial life is the study of artificial systems that exhibit behavior characteristic of natural living systems. It is the quest to explain life in any of its possible manifestations, without restriction to the particular examples that have evolved on earth. This includes biological and chemical experiments, computer simulations, and purely theoretical endeavors. Processes occurring on molecular, social and evolutionary scales are subject to investigation. The ultimate goal is to extract the logical form of living systems. Microelectronic technology and genetic engineering will soon give us the capability to create new life forms *in silico* as well as *in vitro*. This capacity will present humanity with the most far-reaching technical, theoretical, and ethical challenges it has ever confronted. The time seems appropriate for a gathering of those involved in attempts to simulate or synthesize aspects of living systems."[2]

Some twenty years later, it cannot be said that "artificial life" has kept the prophetic promises of its promoter. Artificial life was a gamble because its federative ambition depended on its ability to mobilize the various fields enumerated by Langton.

In order for the bet to pay off, computer scientists or robot manufacturers, for example, would have to agree to identify their products as part of this field, to refer to it, to situate themselves within the perspective it promotes. And to do that, they would have to see some benefit in it. Their products, when situated within the framework of emergence, of the manufacture of life, of a contribution to its logical identification, would have to become more interesting than if they were situated within the more traditional framework of technological innovation. However, this was not quite the case. It is quite probable that these products will confront "humanity," as Langton writes, with a number of far-reaching challenges. Yet, how can we fail to recognize, in the way in which he presents those challenges, a mobilization, in the furtherance of scientific ambition, of the ancient figure of "man" defying the order of creation—man's confrontation with the product of a knowledge that would finally fulfill his ultimate goal, the definition and reproduction of "life" as such, independent of the contingency of his earthly origins? Knowledge products do create, and will, of course, continue to create, challenges, although more dispersed, arising from the labyrinth of technological innovations that capture and reinvent for their own use what Langton wishes to mobilize.

However, "artificial life" does not simply satisfy a mobilizing rhetoric. Something has happened, a "factish" has been invented and recognized that has created the possibility for a relative mutation of what we understand by an artifact. While the scope of this event will not be what Langton hoped for, nonetheless, it may enable emergence to partly escape the traditional frame of the "human artifact/living organism" analogy. But to address this issue, it is best to abandon Langton, who, from his computer's keyboard, wished to be the creator of "worlds" populated by quasi-living creatures.[3] It is not a matter of judging the scientific value of a work but of turning away from the slightly premature questions it has inspired: "What of man's view of himself? He now takes pride in his uniqueness. How will he adjust to being just an example of the generic class 'intelligent creature'? On the other hand, the concept of 'God' may take as much a beating as the notion of 'man.' After all, He is special now because He created us. If we create another race of beings, then are we not ourselves, in some similar sense, gods?"[4] We need to abandon the apparent grandiosity of such claims for they conform to a mythic mold that has been reduced to cliché. I now want to turn to the man without whom I would never have investigated the question of artificial life, Stuart Kauffman.

For Kauffman as well, God is not far away. This is how he recounts the passion that has driven him ever since he began trying to understand life: "I've always wanted the order one finds in the world not to be particular, peculiar, odd or contrived—I want it to be, in the mathematician's sense, generic. Typical. Natural. Fundamental. Inevitable. Godlike. That's it. It's God's heart, not his twiddling fingers, that I've always in some sense wanted to see."[5]

So, Kauffman wants to "see" order as godlike and not "become god" as Langton did. Also, he doesn't want to see "God's twiddling fingers," which are the required intermediary in the clock metaphor of creation. According to the metaphor, we cannot identify with God in terms of his ends, which for a believer are impenetrable; but we can recognize the work of his "fingers," the arrangement he has imposed upon matter. Particular, odd, contrived—these are the adjectives that describe the genius of the designer, his freedom of creation. They bear witness to the power of the mind that conceives the project, a power that is all the more evident because it imposes upon matter a way of being that is foreign to it. In contrast, the words *typical, natural,* express a mathematical requirement: order should be "generic." Consequently, the relation of ends and means becomes misleading. Not all performances are of equal value. Anyone who wishes to understand the obligations associated with the order of living organisms must reject the triumph of someone who succeeds in getting his artifact to do what he wanted it to do. By way of affirming a value that refers to the type of order that would be needed to characterize "artificial life," Kauffman relates a requirement and an obligation that question the possibility of referring to the creator's project as bearing exclusive responsibility for the artifact's creation. As for the value he affirms, we still need to examine the two terms used enigmatically by Kauffman to characterize it: "generic" and "heart."

In mathematics, the term "generic" designates a behavior that is not only "robust" in the sense of being relatively stable compared to perturbations or the imprecision of initial conditions.[6] The property of genericness implies that behavior is also qualitatively stable, in terms of the details of the *relations, connections,* and *interactions* that bring it into existence. We can say, trivially, that evolution toward equilibrium is a form of generic behavior for physical-chemical systems because it may be characterized by the diminishing significance of interprocess coupling. But the term can only be used for equilibrium retroactively, following the discovery of much more unexpected kinds of generic behavior.

Kauffman himself participated in the early history of the field. In 1965, as a young student already excited by the themes of complexity and self-organization (in the tradition of "second-order cybernetics" associated with the names of W. Ross Ashby and Heinz von Foerster), he assembled a rather unusual network of Boolean automata.[7] Kauffman's automata are logical artifacts; the term "Boolean" refers to the functions the different automata obey. Each of them "calculates," using one of sixteen Boolean relations, an output value (0 or 1) based on its input values. The fact that they are networked means that each of them, in synchronization, will (if it outputs 1), or will not (if it outputs 0), send a signal to those automata with whom its "output" is connected, based on the signals it has received from other automata in the previous step. Until then, the performance of networks of Boolean automata had been predetermined. But the young Kauffman connected a hundred automata "randomly" and found that the collective behavior of the resulting network was one of unexpected simplicity, given the ensemble of possible a priori "states." Moreover, this behavior was robust: up to a certain point, it resisted changes in its connections until it "shifted" into another, different behavior (the landscape of states is characterized by "attractor basins").

Kauffman's model was the origin for the field of "neoconnectionism," an explosion of new technological tools and mathematical theories that allowed researchers to "understand" what had initially been discovered. Along with the "cellular automata" for which Conway's Game of Life was the prototype, it served to bolster the belief that "artificial life" was not mere rhetoric. It ushered in a new model of the artifact that satisfied, as is frequently remarked, a bottom-up rather than a top-down approach.[8] The artifact's creator no longer needs to be represented as a designer endowed with twiddling fingers that enable him to carry out his project, to impose downward what he has conceived topward. The creator "profits" from a new form of causality we can call "coupling causality," which is neither linear nor circular as in cybernetics. It is the fact of coupling that is important, not the type of interaction (physical, chemical, logical, electronic) or the purpose for which they are arranged. The creator is interested in behavior that is already qualified, *already endowed* with a relatively robust landscape of possibles "emerging" from that coupling.

If the generic properties exhibited by the Boolean network make it a "promising factish," "God's heart" should singularize the new interest these properties arouse on the part of someone who addresses a "randomly connected" network, the new, practical relationship between the artifact and its maker. For, the "neoconnectionist maker" is not only looking to map the stable behaviors of the network. He wants to modify, to model those behaviors in such a way that on the map of possibles, "bottom" assumes the meaning for "top." The most typical example of such a relationship is the one in which the network acts as an "agent" for the recognition of shapes.

The example of shape recognition is interesting in that it refers to an apparently simple performance—something we do without even thinking—but which had always been difficult for artificial intelligence to get right. For example, what is a "B"? Yes, it is possible to formulate criteria for identifying the shape "B." But those criteria must satisfy a rather formidable requirement, they must allow for the recognition of an indefinite multiplicity of Bs, one more "poorly written" than the other, some of which resemble "D," others resembling "8," and others even resembling "A." This is why it is crucial that neoconnectionist behavior be robust. The fact that the relationship between an initial distribution of the values, o or 1, of automata and the resulting stable behavior is resistant to modification of the initial configuration "promises" that if that relationship could be constructed as a "recognition" of the configuration in question, that recognition would be *indulgent* by definition, robust with respect to variations. No longer is it a question of the production of criteria that make explicit how the recognized shape is to be specified but of the "learning process" that will make the difference between a welcome indulgence and one that is unwanted. It is a question of establishing an optimal coincidence between the attractor basin for all initial configurations leading to the same behavior and the ensemble of all initial configurations that, *for us,* are "Bs." In this case, learning involves a modification (based on a process that is fundamentally random but automatically controlled) of the connections or weighting of connections among automata until the network adopts the same behavior for everything that *we* recognize as "B," and adopts other behaviors for everything that is not "B" for us.

A "random" network can learn, but it's important to understand that it doesn't learn all alone and, of course, it has no knowledge of what it learns. The learning process involves two elements and cannot be reduced either to a design, no matter how tentative or negotiated, or to spontaneous evolution, no matter how controlled. The maker proposes but the network disposes, in the sense that, given the maker's proposition—the initial configuration that was imposed—the network evolves toward a form of stable behavior that belongs only to itself, which the maker may acknowledge but about which he harbors no ambition of predicting. For the maker, such behavior, regardless of what it is, will be the answer, the translation, emerging from the networked ensemble, of what was initially proposed, and it is based on that response that learning will begin. For all the initial configurations that the maker judges or wants to be similar, the translation must remain the same, and for other propositions, which he judges or wants to be different, the translation must be different. No matter how approximately we write them, we recognize that twenty-six distinct letters compose our words. The network must be able to distinguish them. Leaving aside the technical aspects of the algorithms used to modify the network so "learning" can take place, the important point to remember is that we are dealing with an *inter*action in the strong sense.

"The network is capable of learning!" "It is an artificial *neuronal* network, the first appearance of the absent body of artificial intelligence, that has just been invented." Such statements are not the laborious conclusions of specialists, but they clarify the premises of their interest, the conviction these networks brought about almost immediately. Namely, that the network's operation is a vector of meaning and, yet, incapable of justifying the meaning that "emerges" from that operation, creates the topology of a "body." The "internal coupling" whose robustness can be used to make the transition from the ensemble of

interactions to the meaning of that ensemble "for" its operation is distinct from any relationship to a milieu, for there is no
"milieu"; a completely artificial environment determines the
initial configuration of the network. The invention of learning
practices creates a "body" by exposing what I have called the
causality of coupling—the causality that singularizes the network—to another, heterogeneous "causality" that couples the
network to operators that will set about teaching it to actualize
their own objectives.

With the appearance of the neoconnectionist artifact, every
speculative argument concerning our mysterious ability to
"recognize" things without being able to specify the criteria of
resemblance, which has engaged philosophers from Plato to
Wittgenstein, has been captured. There is no need to have an
"idea" of a table to be able to state "It's a table." The recognized
object "emerges" as a collective response, in the here and now,
without a model or a localizable memory. More specifically,
here "self-organization" causes a "quasi object" to emerge for a
"quasi subject," which should not be confused with the network
as such. The network itself is inseparable from the "quasi purpose" it fulfills, but the meaning of that "quasi purpose" relates
to the one for whom emergence occurs.

It is this "emergence" of a body through coupling between
the network and the maker that can, I believe, give to the term
used by Kauffman, "God's heart," an interesting interpretation,
even if it's not the one he intended. Whatever he intended, he
used a charged analogy that contrasts the heart not so much to
the fingers but to the rational mind of the designer who causes
his fingers to move on the basis of his project. Such analo-
gies always reveal much more than what their user may have
intended. Judith Schlanger, in her marvelous *Penser la bouche
pleine*, used the example of an Egyptologist who "demarcates"
his object, the Egypt of Egyptologists, by disqualifying all the
other "fictive" Egypts.[9] Nevertheless, all of them, she claims, are

included—the Egypt of myths, the Egypt of films and novels, the Egypt of dreams. They all coexist alongside the demarcated Egypt that disqualifies them within a dense milieu that makes interesting the demarcation that apparently excludes them. And it is this density, this muffled and stubborn "cultural memory," that allows us to understand the interesting innovation. The demarcation, if it were to create a vacuum, would be stable, attached to its evidence. It isn't, because whenever it produces new consequences, these are liable to create resonance in the dense milieu that feeds it, a milieu that, for the speaker as well as for the hearer, becomes the vibrating matter of a new actualization.

In our case, the "heart," in contrast to a "reason" capable of accounting for its operations, does indeed mobilize a dense cultural memory, in which the capacity that identifies reason continues to hesitate between recognition of unique legitimacy and indictment for arrogant pretense that bars access to a different order of truth. But it is not as one of philosophy's "great themes," first with Pascal and then in psychotherapy, that the problem of the heart finds the means to insist.[10] On the contrary, what is innovative is the way in which the problem is liable to be reorganized around one of its components. What Kauffman's "God's heart" expresses is that the consequences "promised" by the factish concern the way in which the "psychosocial" identity of the makers of artifacts will be demarcated.

Andrew Pickering has compared the development of a new, classical detector, one that uses a physical or chemical process to identify an entity or process that is also chemical or physical, to a kind of two-step dance.[11] The scientist adjusts the machine, then withdraws and allows it to operate. He observes what it "does," in this case, what it detects, and interprets the reasons for what he judges to be its defects. He then goes back to work and readjusts the machine, and continues to do so until the machine detects what it is supposed to detect. There is certainly an interaction, but once the machine is stabilized, the scientist has learned a great deal and can tell the story quite differently. Now, the machine assumes a passive role, the action is entirely redefined in terms of what the scientist "did not know" at the start, problems he hadn't noticed, distinctions he hadn't thought to make. The maker, when involved in the "dance," may indeed have experienced extraordinary things, becoming a detector, confronting a world a distinctive feature of which he seeks to capture. But his psychosocial identity incorporates the way in which the story will achieve its conclusion, with the final separation between himself, on one side, the world and the machine, also physicochemical, on the other. The world bears witness through the machine, the machine's operation is explained by the world. The same is not true in the case of neoconnectionist networks, however. Here, the site of the "dance" is a coupled causality that will never be disentangled. The maker will never know how his device operates. And the device doesn't detect in the ordinary sense. Its purpose is not to become the witness of distinctions that could be said to belong to the world and need only be recognized. It must produce conventional distinctions, those to which the maker attributes value, among the resolutely confusing shapes proposed by "the world." The culmination of the process is not the separation of the maker, on one side, the machine/world, on the other, but the maker/machine, on one side, whose values are mutually adjusted, and the world, on the other, always as confusing and bound to remain so. In fact, success for the maker occurs when "his" machine has succeeded in recognizing the "B" that he had so carefully mangled when writing! And when the network finally, spontaneously "recognizes" what has been put before it, its operation can never be compared to a fragment of "nature" that may well have been selected, staged, and purified, but should still obey the same "reasons" as nature. The maker's judgments have passed into the machine, the only "reason" for its operation being the conventions it has "learned" to obey.[12]

The learning network is not a hybrid comparable to the clock, for example. It does not necessitate a historical and constructivist reading that struggles against the triumphal syntax wherein we distinguish the laws of mechanics, on the one hand, from the design associated with the human project, on the other. It is as a hybrid, exhibiting the processes of stabilization and negotiation from which it originated, that the device is presented. And the maker is someone who has caused it to emerge as a hybrid, to the extent that a part of himself has "passed" into the machine and has bound with the properties of genericness inherent in the machine to form a composite that no one is supposed to ever be able to separate. What God has united . . .

When scientists talk about God, they are often talking about themselves. The God of Einstein, a mathematician, occupies the site Einstein hoped to construct. The demon-god of Laplace knows the world as Laplace, the astronomer, thought he was capable of knowing the planetary system (which he believed to be stable). Maxwell's demon sorts particles that the physical chemist cannot at the macroscopic level. Langton's God plays on the world's keyboard. Kauffman's God has a heart, which refers, I believe, to the *inter*action and hybrid world of reciprocal capture that is productive of meaning. A world in which the "factish" made promising by its generic properties explains nothing as such, but implies and assumes a maker who interacts and evaluates, and whose values are "passed on" to the world, becoming, in the strong sense, an integral part of that world, inseparable from it, an ingredient of an order that nonetheless remains "typical," "generic," and, as such, impenetrable, "even to God."

Learning the alphabet is a poor example, however, because the maker's values cannot be affected by the process. It is not impossible that the new psychosocial maker of these new artifacts will one day refer to open-ended learning, where the maker's "values" would be partly generated by the answers of his device ("this just gave me an idea . . ."). In this case, construction should be told in the form of a story: a story in which the demarcation between the maker and the machine continuously transforms itself; a story in which the maker's identity—what he seeks, the possibles he intends to actualize—would "emerge" along with the behaviors of his device; a story in which the roles would remain radically asymmetrical but would no longer put before us the owner of a project and the device that is supposed to realize it. A constructivist story.

Identifying the possible creation of a new psychosocial type of maker is a form of speculation. But the possibility of such a speculative stance is part of the resonance effects resulting, within the dense cultural milieu that entangles the themes of fabrication, autonomy, emergence, and the link between creator and creature, from the redistribution of agencies that may be associated with a new type of artifact. In our story, the creation of the clock that ideally satisfies, autonomously and solely on the basis of the laws of mechanics, the intentions of the clockmaker, has had effects of which we are the heirs. Theology has been able to emancipate itself only by turning God into an absent God. Biology is still an heir, and has given natural selection the figure of the clockmaker, or, more accurately, according to Richard Dawkins's expression, the blind watchmaker, adjusting, permutating, modifying the mechanisms of a population of "clocks" that, in the most highly diverse ways, tell the only time that "counts" for the clockmaker, the rate of transmission of genes over succeeding generations. And it is to biology that Stuart Kauffman turned in attempting to read in it the consequences of a possible "marriage between self-organization and selection."[13] For Kauffman, the blind watchmaker must "marry" the generic properties of coupled causalities. Through its metaphors, language acknowledges the dense milieu in which such references are distinguished: let no man put asunder.

Yet it is also, whenever living organisms are involved, at the point where the clockmaker goes blind, where the figure of God and the maker must both disappear, that the question arises of determining how the "factish" of a coupling causality can become an ingredient for the problem of emergence and evolution.

For Stuart Kauffman, the issue is that of "theoretical biology," but the notion of theory is profoundly ambiguous in this case. If it were to function as it does in the theoretical-experimental sciences, it would imply the construction of a power to judge that should minimize the reliance on history and turn the terrain into a theater of proof, much like a laboratory. But redefined in terms of the practices of negotiation I associate with the problem of emergence, it can signify an approach to what biological evolution requires as we are able to puzzle it out. In this case, the mutation imposed on the notion of "theory" by the theoretical biology with which Kauffman nourishes his dream would imply a mutation of the "theorist." Whether this mutation is clandestine or mutilated, whether mutant theoretical practice is persuaded to claim it resembles what it disagrees with, the way Darwinian practice was persuaded to claim it retains the power to judge, or whether it has the freedom to assert itself, is an issue for the ecology of practices.

The research that, for Kauffman, ushers in the new field he calls "theoretical" is collected in his massive *The Origins of Order,* which can be considered the leading work of contemporary "theoretical biology.[14] But the book will be unreadable for anyone who expects theory to provide the miracle of an approach that comprehends diversity within the luminous affirmation of a principle to which it is subject. The book contains a series of studies of formal situations, which introduce relationships judged to be typical of biology, but in a highly simplified manner, through the use of *toy models* (here toy signifies both that the model is something to play with, rather than one that claims to provide a faithful representation, and that we *can play,* that it can be manipulated). The behavior "emerging" from the model is compared to observable biological data that, when compared to the behavior of the toy, become interesting, capable of providing, in certain situations, information in the language of the model.

The common feature of all of Kaufmann's toy models is that they accept the hypotheses of Darwinian evolution but, contrary to neo-Darwinism, do not assume that selection is all-powerful. They present the effects of a hypothetical selective pressure on the exploration of a landscape of possibles, where the very point is that everything is precisely not possible because exploration, from mutation to mutation, has as its subject the transformations to which beings characterized by internal coupling (for example, "interconnected genomic networks") are susceptible.[15] It is the network itself rather than any given trait that is characterized by a coefficient of "aptitude," and the characteristic connection rate of the network measures the number of genes on which the meaning (in terms of aptitude) of a mutation affecting a gene depends. In other words, Kauffman's models are not based on any new biological hypothesis. They are limited to *taking seriously* what every biologist knows: the correspondence between a trait (more or less adapted) and a gene is in no way representative of the living organism. Whereas the neo-Darwinian evolutionary biologist generally tends to minimize the complications resulting from this minor problem, Kauffman's models propose making it "the problem," primarily by studying the effects of selective pressure as relative to the type of being to which it applies.

A single general hypothesis finally falls out of Kauffman's exploration and it is upon this hypothesis that his desire for a theory is concentrated. If selection favors the ability to differentiate, if it "encourages" the network to explore a spectrum of diversified "activities," selective pressure should cause emerging behaviors—and, therefore, the connection rate characterizing the coupling from which they emerge as well—to evolve toward the *edge of chaos*. Here, perfect order is behavior that is completely predictable and robust. Kauffman refers to it as "frozen." The system is locked into one and only one mode of operation. In contrast, perfect chaos is compared to a fluctuating, erratic liquid, in which any alteration of an element can trigger a cascade of consequences throughout the network. When order dominates, the freeze percolates throughout the network, but it leaves behind isolated, unfrozen, pools. In the predominantly chaotic regime, however, it is the liquid regions, fluctuating chaotically, that percolate, leaving frozen islands here and there. The edge of chaos thus corresponds to a generic behavior that preserves the "best" of both worlds: the possibility of cascading innovation and relatively stable modes of operation resistant to chance.[16]

"If it proves true that selection tunes genomic systems to the edge of chaos, then evolution is persistently exploring networks constrained to this fascinating ensemble of dynamical systems."[17] In other words, selective pressure does not confer differentiated "adaptive" values only on those beings that emerge from coupling, but also on the coupling itself, as requisites for an evolution capable of providing its fecundity to the "marriage between self-organization and selection."

Kauffman's "toy models" obviously do not constitute a theory in the sense that the multiplicity of forms of marriage might find their respective contracts referred to a single institution that would define the truth of marriage as such aside from any anecdotal differences. Quite the contrary, it is the apparent generality of "selective pressure" as a vector of evolution, the possibility of assigning to it the responsibility of evolution independently of what it bears upon in *each* case, that is annulled, whereas a generality of a different kind is offered in its stead: a hypothesis like that concerning systems "balanced" at the edge of chaos can orient questions, not answer them. It can bring into existence, as a problem, the emergence of "adaptive values" that cause life and artifice to converge. Broad statements such as "natural selection must have . . ." are replaced by the indeterminacy of "we do not know a priori." We do not know how to formulate the question of "value" in general terms, in that it may refer to a particular trait or to generic properties of interconnected ensembles, such as those that characterize the "edge of chaos." And, in this last case, we do not know which coupling situation is the subject: the genomic network, specific ontogenesis, the dynamic of inter-specific coevolution? This is what needs to be *conceptualized*.

But the "promising factishes" of Kauffman's toy models are vulnerable, as is self-organization far from equilibrium, to the theoretical ambition that refers to itself, now and always, as the power to economize the terrain. This vulnerability is primarily expressed by the possibility of grand considerations that appear to communicate scientific practice and wisdom. And in this case, it is a "stoic" wisdom that celebrates a universe that "awaits" us, in the sense that we are the expression of chance, yes, but also an expression of the generic order promoted by theory, a fraternal universe because coupling is everywhere, but a dangerous one because of cascading consequences. "Our smallest moves may trigger small or vast changes in the world we make and remake together. Trilobites have come and gone; Tyrannosaurus has come and gone. Each tried; each strode uphill; each did its evolutionary best. Consider that 99.9 percent of all species have come and gone. Be careful. Your own best footstep may unleash the very cascade that carries you away, and neither you nor anyone else can predict which grain will unleash the tiny or the cataclysmic alteration. Be careful, but keep on walking; you have no choice. Be as wise as you can, yet have the wisdom to admit your global ignorance. We all do the best we can, only to bring forth the conditions of our ultimate extinction, making way for new forms of life and ways to be."[18]

We could say that, in this case, Kauffman, as he did when he spoke of "God's heart," thinks with his mouth full. But there is a difference, and it is crucial for the ecology of practices. In the latter case, what is being expressed is what, for Kauffman, understanding life demands. In the present case, stoic wisdom *includes* an ensemble in which everything—from paleontological data to the historical, technological, and political dynamics that "identify" us—bears witness in one way alone, that of an allegory of exploration exposed to selective pressure (we all do the best we can) and the price of that exploration (the unforeseeable catastrophe). Here too, questioning the obligations of a practice of emergence entails questioning the kind of appetite this practice induces for the "terrain." Is the factish's "promise" the submission of the terrain to a theoretical-ethical-speculative generalization or does it create an appetite for the terrain, where the indeterminate promise to which it gives meaning might be actualized.

17

The Art of Models

It would be a misunderstanding to confuse an appetite for the terrain with the creation of "good" science, respectful of beings and participating in the secret harmonies of Being. If the practices that bring about the terrain-as-problem evoke a precedent, it is not one of utopian reconciliation, where knowledge would break any connection to power. Rather, it is the problem of another form of power, analogous to the kind of power that, according to François Jullien, Imperial Chinese civilization favors, as evidenced by the omnipresence of the word *chi*.[1]

Chi is a word with as many meanings as our term "energy." It refers to a dynamic configuration associated with nature as well as with art and calligraphy, the composition of poetry, government, and warfare. The use of the word in Chinese thought contradicts any possibility of contrasting *phusis* and *technē*, spontaneity and manipulation, submission and action, conformity and efficiency, whether these refer to human government or the grand cosmic design. *Chi* implies the disposition of things, of characters, of intrigue, of political or military power relationships. And it refers equally to the arrangement that produces their respective propensities and to the intervention that will, without force, noise, or, apparently, effort, take advantage of this arrangement and lead the situation, as if by its own dynamics, to the desired issue. A part of *chi,* therefore, is the art of relying on *chi* for some advantage, the art of manipulation and enticement. The art of the great warrior is letting his enemies kill one another or betray their agreement while he remains invisible, so that the enemy army grows demoralized to the extent that the final battle is no more than a formality. Here, reason does not triumph over force, it weds force, it becomes force, and does not respond to any criteria other than those of efficient manipulation.

The art of *chi* despises violence, not because it would contradict a moral ideal but because it is not effective, because it indicates failure by opposing the propensity of things rather than confirming that propensity by taking advantage of it. Nor is it eager to discover a truth beyond dispositions and mechanisms, or seek confrontation or harrowing dilemmas. But it would be especially stupid, because this art escapes our excesses and closes the perspectives in whose name we have committed great crimes, to see in it the position of wisdom we are said to have betrayed. On the other hand, it is worthwhile pointing out that the practical mutation that could transform the dual identity of the artifact and its maker, as well as the question of the "marriage" between biological selection and self-organization, find their most apt metaphors in the art of conforming to the propensity of things.

No doubt the Chinese would have understood Kauffman's statement that "Evolution is not just 'chance caught on the wing.' It is not just a tinkering of the ad hoc, of bricolage, of contraption. It is emergent order honored and honed by selection."[2] But they would have certainly understood it without the slightest sentimentality. "Honoring" and "honing" have nothing to do here with moral respect; it is a question of using another's force to bend him to our own purposes. This may be characterized as (and all such characterizations are pejorative for

us) "manipulation," "suggestion," "seduction," "appropriation," "instrumentalization." The interesting point is that we
are accustomed to using these pejorative terms whenever they
refer to relations between human beings. However, they are now
presented as metaphors for a new type of relationship between
phusis and *technē*. The psychosocial image of the technician has,
until now, emphasized a practice conceived as submitting an
ideally inert material to a purely human project. And it established the figure of free choice and will as the problematic point
of contrast between the "emergence" of assemblages that were
respectively natural and human. The "technician of *chi*" has not
renounced his will in order to make room for the democratic
or revolutionary utopia of a "self-organized" nature that produces order, beauty, and truth through the free spontaneity of
its self-creation. He is "without principles," no longer respects
the master word used to organize the hierarchy between knowledge and application: "Understand the principles nature obeys
in order to bend her to our purposes." It is enough that he can
make nature *bend,* follow her folds, marry them so he will be
able to create others.

It is interesting to approach from this point of view the mutation the term "universal" underwent within the problematic of
"self-organization." The law of gravity is said to be universal in
the sense that any mass, no matter where it is in the universe, is
supposed to obey it. However, the "promising factishes" of the
physical chemistry of systems far from equilibrium and network dynamics also allow one to speak of universality.[3] The very
beautiful word *attractor* accurately expresses what this notion of
the universal entails, the type of necessity with which it communicates. This necessity is always relative to a mathematical
or logical model, a hypothetical schema of relations expressed
by the model. Furthermore, when we deal with situations that
make evolutionary sense, the model aims less at representing the situation than at relating it to a problem. The universal

defined by the model cannot claim to be that to which the situation is subject. It only claims to be relevant for an understanding of that situation. Although the model introduces a robust attractor, characterized by generic properties that apply regardless of the circumstances, it designates a situation one of whose ingredients may have been the *question* of the universal that has, literally, captured it, infected it with these generic properties. Various situations may be "judged" according to the terms of the universal into whose grasp they have fallen. However, they are not necessarily defined by the categories of this judgment because they are capable, in return, of defining it in their own terms.

The problem of emergence may be approached through the art of models. The identification of a universal is no more the answer to this problem than a propensity is an answer for the "technician of *chi.*" Such a universal is characterized by the insistence of a question for which an answer may eventually emerge. The necessity with which it communicates implies that, *if the model is relevant,* the modeled situation, in one way or another, *must have* taken it into account and assigned a meaning to it. Does this situation express it immediately? Do the generic properties serve as an opportunity? Has an activity regime acquired its meaning and purpose because of them? Has it succeeded in becoming a requisite for other activity regimes for which the model would then provide an ingredient? Or does an aspect of the situation that the model failed to take into account become interesting and intelligible precisely because it allows the situation to *avoid* being captured by the universal? The universal is a question, a proposition. As for the intelligibility being constructed, it is related to the way in which the situation has disposed of that proposition. The necessity—if the model is relevant—arises from the fact that, in one way or another, determining "how," the way in which the proposition has been disposed of, must have taken place.

At this point, the model severs its connections with the theoretical-experimental practices that have made it a weaker substitute for theory, a representation that is not supposed to resist the challenges that a theory must overcome. A model, as it functions in the theoretical-experimental sciences, has a domain of validity that is carefully delimited, for, through its definitions, it employs simplifying expedients whose scope is explicitly relative to this domain. On the other hand, anyone who speaks of "theory" assumes the risk of claiming that the theory must remain a reliable guide, even when used outside the practical domain for which it was constructed. Once it is a question of the "field sciences," however, the model is no longer defined in contrast with a theory. The model is no longer defined by its simplifications or by ad hoc hypotheses. It no longer belongs to a practice designed to "prove," because the validity of a given proof would, in any event, be valid only for a given situation. Rather, it is a question of producing a problematic tension between what the model requires and what the field discloses. By identifying its requisites, a model makes a wager and assumes a risk: what it requires of reality should be necessary and sufficient for making intelligible what has been learned in the field.

We can compare this use of the model with what Gould defines as "Darwinian discovery." "We define evolution, using Darwin's phrase, as 'descent with modification' from prior living things. . . . We have made this discovery by recognizing what can be answered and what must be left alone."[4] Darwinian evolution requires the prior existence of living things. All of the reasoning it employs presupposes this. It gambles, therefore, that biological evolution, in putting forth its own problem, has no need of a solution to the question of the origin of life. In other words, it positively denies a hypothesis like that of "vital force," which would be *simultaneously* responsible for life's origin and its history.[5] What has been "discovered," in the sense that the model actively implies the reality it proposes, is the possibility of using a "disconnect," the possibility of separating the question of life's origins from what happens once living things exist. The model of evolution cannot investigate the origins of life, for it requires selection, which assumes the presence of living beings; it requires the specific relationship that every living thing invents with its milieu, its congeners, most often its predators and, in some cases, its prey.[6]

Whenever it's a question of evolutionary models associated with the field sciences, realist ambition—what the model requires of reality and the obligations entailed by the model's claims to relevance—relies on *requisites,* on what the model takes the risk of treating as securely given in order to proceed. This ambition is not trivial. Most models in the social sciences and economy fail to satisfy this requirement. Equations are written expressing the consequences of rules, norms, laws, or conventions which, the model claims, "explain" the evolution of social or economic situations. But these rules, norms, laws, and conventions vary over time, and the model would only make sense if this variation were noticeably slower than the evolution the model is supposed to explain. Which, in general, is not the case. If the time scales are comparable, the model is worthless. This was Norbert Wiener's objection to the hope of Margaret Mead and Gregory Bateson, who urged him to focus on the social and economic sciences and make them fully scientific disciplines that would finally contribute to solving the urgent problems facing society.[7]

To overcome Wiener's objection, a model must assert the risks associated with it, the power relationship that characterizes the situation if the model is to be relevant. Only the situation can authorize the modeler to separate what the model will define as variables and constants, or forget certain aspects of the situation in order to highlight others. The dimension of the situation that is responsible for the satisfaction of the model's requisites can be forgotten to the extent that (as is the case for

the history that has given life its "origin") it does not, or no longer, intervenes in the terms of the problem. However, the question of determining *how* the problem will be formulated is part
of what the model must explore.

I want to turn now to models that specifically concern the
problem of emergence. Unlike a model that might be called
"scenographic," because it tests the consistency between the
history it can be used to predict and the history of the "field,"
whose terms, witnesses, and indices are identifiable within the
modeled situation, the model of emergence attempts to articulate a hypothetical emergence with requisites that are associated with other practices, that is, requisites whose meaning is
initially relatively indeterminate with respect to the question of
emergence for which they are, hypothetically, a possibly necessary but always insufficient condition.

It is here that we again encounter the question of "universals" associated, primarily, with self-organization. These
universals are part of a strategy that relates emergence with requisites. They are relative to the construction of the model from
the perspective of mathematical practice: the model in question
belongs to a class characterized by a generic property, a "promising" property in that it is impossible to "escape" it other than
by radically transforming the model. Once recognized, a universal of this type creates a terrain for the question of emergence,
for it defines one of the issues that "must have" polarized the
situation. If the model is relevant, if its requisites are legitimate,
what emerges had to have "confronted the problem" and been
determined by determining the meaning that would be attributed to it. The universal gives the situation the significance of a
critique.

But the role of mathematics in the question of emergence
doesn't end there. It can also "shift" the issues associated with
a scenographic model toward a problematic of emergence. I will
give three distinct examples of such shifts, three typical cases of

what singularizes the questions of emergence when compared to theoretical-experimental questions: in each case, as the modeler learns to formulate a problem, she discovers that this problem has been (partly) formulated before.

Take the problem of eco-ethological models that make use of a predator and its prey. The initial scenographic model, designed to account for situations where statistical series are found to exist because humans, as predators, have been interested in the frequency of capture over long periods of time, is the so-called Lotka-Volterra model, which makes use of predator–prey interactions. The model typically results in a form of periodic behavior. Predators eat abundantly and reproduce easily, but at the expense of their prey, whose numbers decline. Consequently, hunger and famine occur and the number of predators decreases, which benefits their prey, whose numbers increase. This allows the predator population to increase again, and so on. This first example, however, is simply a starting point toward the general case that introduces competition among predators. We can then ask about the evolution of populations coupled by their shared dependence on a set of resources. However, the empirical relevance of the model of interspecies competition encounters limits that have nothing to do with the complicated details of such coupling. In fact, field studies lead to a change in the nature of the model. Rather than being a scenographic model of coupling to which competing populations are subject, it becomes a description of the coupling that some species *manage to escape.* Seasonal changes in reproduction, the choice of resources, the amount of food needed at different times of the year—all these "details," which the model "smoothed," can become interesting to the extent that they *counteract* the effects of interspecies competition. The relevance of the model changes. It is no longer tied to coordinating its predictions with empirical data but to identifying specific behaviors that falsify those predictions.[8] Moving from the question of solving equations to the problem

introduced by those equations, modeling has allowed scenographic practice to "rise" to the "how" of emergence. This rise
is expressed by the correlative appearance, for the modeler, of a
quasi subject, the populations of competing predators, responding to a quasi object that is none other than the very object of her
modeling: the "universal" problem of interspecies competition
for predator populations.

Therefore, the modeler *should not trust* her model, not
because the model might be wrong or irrelevant but because she
does not know, a priori, *how* it is relevant. The Lotka-Volterra
model apparently designates an "object," but it must be used
with *tact* in order to expose the possibility that a "quasi subject"
might have appropriated the problem corresponding to the
model. The question of knowing "how to describe" is no longer
one that concerns the scientist alone. Correlatively, the nature
and scope of "objective" definition are transformed. Objectivity is beside the point. Interspecies competition is a problem for
specific groups, but it does not allow a solution to be deduced;
it raises the question of finding out *how,* with what ingredients,
using what expedients, a solution has "emerged."

Tact is a quality most often exercised among humans, but it
points to a much more general problem—that of a relationship
created with a being for whom a problem is assumed to exist,
a problem that can be identified, or so it is believed, although
how the problem presents itself to this being is unknown. Tact,
therefore, expresses an obligation that limits the power of whoever is situated by her knowledge of the other's problem. She
"knows," accepts, and desires a relationship that incorporates
the *open* question of the "how" and "tactfully" respects the fact
that time is needed for the answer to this question to "emerge"
for the concerned being. Teachers who lack tact do not feel this
obligation, and most often those who are tactful fail to capture the identity of the "how" that has been invented during
the course of the relationship. The goal of the modeler—and it

is in this sense that her tact is part of a scientific practice—is to define the way in which the situation she models answers the model's question. "Tact" then comes to imply a transformation of requirements and obligations compared to those that govern experimental procedure. We could even say that it is no longer the scientist alone who imposes requirements. Of course, the scientist must require that what she addresses has a stable existence in terms of the relationship that is established. Wherever the conditions of a field science are found, the features studied must be robust with respect to the type of intervention that allows them to be studied.[9] But the field also allows itself to be characterized in terms of its own requirements. The relevance of the scientist's problem depends on the fact that this problem has actually required, long before the model that makes it explicit, an answer that gives it meaning. Correlatively, the field "obligates" the scientist to recognize its "preexistence," to recognize that she, the scientist, will only encounter it by acknowledging that preexistence.

This same quality of tact is at the center of my second example: biochemical modeling. Take the behavior of the amoeba *Dictyostelium discoideum* in the presence of cyclical AMP. Cyclical AMP, a creature of biochemistry laboratories, intervenes in the intracellular behavior of amoebas and in their intraspecies relationships. The rhythmic production of cyclical AMP in the milieu serves as a "signal" for the population, that is, it modifies the intracellular behavior of "receptor" amoebas.[10] The data of biochemical analysis culminate finally in a "scenographic" model of nine interconnected equations with nine variables. Can the model be used to explain the behavior of the amoeba in terms of the molecular interactions it introduces? In one sense, fortunately, yes, as this behavior is not that of the amoeba itself but a partial description, one that has already been worked and reworked to allow the question to be asked. But the interesting point is that the work the successful explanation has obligated the modeler to perform can become the starting point for a new question that uniquely designates the behavior being explained as the specific behavior of a living being.

The "system" of nine (nonlinear) equations taken as such defines a literally "unmanageable" system that may generate extremely diverse behaviors, even though it is supposed to explain "what the amoeba is capable of," "what it does," that is to say, behavior that is stable and reproducible. Consequently, the practice of the modeler cannot be reduced to one of simple confrontation between the model's predictions and described behavior. The modeler doesn't require that the amoebas verify her equations, she is obligated by the amoebas to recognize that not all the possibles defined by the equations are valid for them, that some are excluded and others privileged. The amoebas, therefore, obligate the modeler to pose the problem of her model, for it is now a question of understanding how they themselves, in one way or another, "manage" the diversity that the equations define as unmanageable. Can the modeler reduce the number of equations, distinguish, for example, which are "slow" and can be decoupled from the others? In this case, she will have to "trace back," through the values of the parameters that must be selected in order to support the appropriate behavior, to what the model now allows her to identify: an ensemble of biochemical "quasi choices," which have intervened in the very invention of *Dictyostelium discoideum*.

The modeler's practice, the detailed negotiation with the parameter values, the calculation of their consequences, in a sense closely follows the problem of selective evolution as it is made explicit by the model. Selective evolution then corresponds to a figure closely allied to tact. The model builder's initial equations form the matrix of a "luxuriance" of possible temporal behaviors and imply that a mutation that modifies a reaction rate, or introduces, eliminates, or alters a coupling, may have uncontrollable, and usually catastrophic, consequences for the amoeba. Selection no longer has much to do with the figure of the watchmaker, blind or not. The selective history of the bio-chemical mechanism of the amoeba's behavior has much greater need of the precautionary prudence of an apprentice pickpocket working on a mannequin covered with bells. Tact, the clever negotiation to obtain one thing rather than another, more often one thing rather than the unavoidable other, correlates the prob-lems the modeler faces when confronting her equations and the problems selective history (from which the role conferred on cyclical AMP by the amoebas emerged) had to resolve.

The practice of modeling in biology is often the work of researchers who take inspiration from economic models, but the problem with economy is that it radically lacks tact. Its appetite for theorems, used primarily to determine optimal conditions, takes the place of relevance. Why not have the model hypoth-esize, for instance, that unemployed workers "disappear" from the market if that is a condition for a theorem?[11] The economist requires, with a unilateral brutality that is the opposite of tact, that the modeled situation give her the right to publish a theo-rem. When employed in biology, this lack of tact immediately conspires with the omnipotence that neo-Darwinian theorizers give to selection. What emerges must optimally satisfy a given adaptive value, and the existence of the optimum allows evolu-tion to proceed from theorem to theorem. On the other hand, "modeling the field" can, as we shall see, enable us to counteract the theorem-based inspiration of the economist and "return" to the problem that singularizes a behavioral trait.

Take the typical behavior of ants in search of food.[12] The uniqueness of this behavior is its intelligibility on the group level. Although individual behavior may appear somewhat erratic, the behavior of a group of ants is a key example of effi-ciency, and it seems to deserve an explanation maximizing some adaptive value. If we assume an optimum, we can always con-struct it, but if we don't, other questions arise. Not to assume an optimum means understanding the efficiency, not deducing it. The emergence of collective behavior has to be "followed" according to the way in which interactions among ants "modulate" (but do not determine) individual behavior. And the collective behavior that "emerges" from such interactions turns out to be remarkably efficient indeed, capable of preferring a large source of food over others, or systematically exploring a milieu, similar to a projector revolving around a nest. The erratic and nonrobotic (programmed) behavior of the individual ant becomes, in this type of model, an essential component of group efficiency. It requires that the individual be somewhat "entrepreneurial" in order for the group to "explore" the opportunities in its milieu. But a more general concept also arises, which changes the stakes when studying collective behavior. Not only does a given population of ants in a given environment select the food sources that "count," but the way in which it selects them suggests a hypothetical "tracing back" to the problem of that multitude of species we call "ants." The interactions among ants are such that a small quantitative change in a parameter (which may correspond to a random genetic variation) qualitatively transforms the way in which the method of seeking out and selecting resources operates. "Ants" in the generic, multispecies, sense could then coincide with the invention of a relation between individuals and the group, which is the "matrix for significant variants." The relationships that allow the transition from the individual to the group would not only belong to a species, they would (partially) identify that species with a "choice" made on the basis of a genetic matrix of "hypotheses," subject to selection in each different environment, a genuine "machine" that is no longer adapted but adaptive. Here too, the question of emergence appears with the acceptance of the problem, with the possibility not of "reducing" one level to another, but of introducing a quasi-practice of inter-"level" articulation.[13]

The three examples above—interspecies competition,

amoebas, and ants—apply to different aspects of biology, but they have one thing in common: they describe how we read a way of functioning that is stable and capable, to a certain extent, of reproducing itself from generation to generation. The modeler wagers that since it is robust, it must be understood as having invented the means to be robust, and the relevance of her activity as a modeler depends on this wager. If we look at the way Deleuze and Guattari define the concept of a body, by relating it to "informational coordinates of separate, unconnected systems," we can say that the wager is that the situation is "embodied," and as such defines the emergence of a disjuncture between internal and external variables in relation to the milieu, which has nothing to do with the distinction between internal variables and the limit conditions of physical-chemical systems.[14] To define a system by its limit conditions does not imply tact, and the principle of exploration to which this definition corresponds is one of variation (to increase pressure, temperature, the intensity of the temperature gradient, or the imposed relationship of chemical concentrations). To define an "organism" does not imply tact either, if the organism refers to a body judged in terms of a relation between ends and means, where every organ fulfills a function through the harmonious division of responsibilities and tasks. Addressing a "body" imposes this specific art I have called "tact." The model must explore the disjunction as such, approach it from two sides at once. It must negotiate the relevant internal variables with respect to observable external behavior, but also approach that external behavior from the point of view of the milieu it defines for by itself and for itself, that is, identify the selection and values of the variables it requires from the milieu in which it emerged "as a body," identify how, from its point of view, all milieus are not equal.

The body, in the customary sense, is certainly composed of a multitude of bodies in the sense I have introduced above. But it is not at all certain that it functions as a body in the same sense.

In other words, it is not at all certain that a practice whose ideal is the convergence between the requisites of the model and the requisites of the body "itself" retains its relevance in the case where it addresses a living being whose experience includes the feeling that it "has a body." Whenever it is a question of the human body, in particular, and its marvelous or terrifying ability to allow itself to be "modeled" by cultural practices, the question of determining what a model wishing to address "the body" should address becomes critical. To speak of "modeling" cultural practices is itself significant. The "model," in the sense in which it refers to a scientific practice, can no longer be dissociated from other "modeling" practices. The human body is always that of a being belonging to a given family, a given group, a given culture, and this belonging also implies the way in which the body is "fabricated," the way in which it is "understood," and how the requisites of its "normality" are identified. And at this point the power relationship "within the modeled situation," which the scientific practice of modeling requires and benefits from, disappears.[15] Trap, temptation, and curse, the question that arises is less one of the disappearance of this power relationship than of the derisive ease with which it is obtained. The human "collaborates" with the project of elucidating what it requires and, in some cases, even what it is subject to. We are in the process of preparing to explore the transition to the limit, where the relationship between construction and definition will again change its nature.

18

Transition to the Limit

In physics, approaching a limit imposes a number of precautions whenever several variables simultaneously tend toward the infinite or toward zero at the limit. To avoid any confusion, these variables must be individually managed. Physicists must take the risk of emphasizing a single variable in order to construct reasons why the description of the problem that gave them their meaning (for example, what is a gas?) loses its relevance at the critical point, even though they know that they are changing their meaning collectively. Similarly, I must try to "slow down" the loss of relevance to which the transition to the limit corresponds. In this case, that means trying to remain for as long as possible within the framework of my initial question, that of scientific practices in which the scientist can risk requiring, so as to identify where and why this requirement changes meaning. But this is only a first step, for the question of the limit returns. What the initial question assumed was a pathway to the limit, but not *the* pathway, the one that would coincide with the general definition of that limit. On the contrary, for it is from that limit that one can attempt to turn the pathway itself into a problem.[1]

Unlike the situations studied in physics, the "limit" here

does not constitute a given problem, imposed by a change in the properties to be interpreted. More specifically, if properties do change, the way such change is characterized involves a critical question about the very issue of characterization. To make this question and the commitment it demands perceptible is to make perceptible the "critical point" at the limit. Here, therefore, the existence of the limit belongs to the "present" of whoever effects, but first experiences, the transition to the limit. It defines this present relative to the perplexity, the "perplication," of the questions and distinctions that the limit has stripped of their tranquil differentiations.[2] The critical questioning of knowledge does not have the generality of critical thought, which always silently assumes the ability to judge on behalf of what is not questioned. It is part of the risks the present obligates us to take.

In *Cosmopolitics*, Book I, "The Science Wars," I described the problem that, for me, requires a transition to the limit, namely, the "modernist" practices I took the responsibility of characterizing as constitutively polemical. For, in order to present themselves as scientific, they *need* to disqualify the opinions, the beliefs, of others, the nonmodern practices of which some claim to serve as rational substitutes. Identification of the problem and the question it raises situate me because they express the conviction I have tried to implement until now with respect to other scientific practices. The way in which those other scientific practices create their questions and their risks satisfies requirements and obligations whose singularity instantiates a difference with what precedes or surrounds them, a difference that has no need to be reinforced through polemics and disqualification. The same is not true of "modernist" practices, whose claims postulate that the one who asks questions, because she is a scientist, which is to say rational, which is to say modern, escapes the illusions, traditions, and cultural assumptions that, *on the contrary,* define those she is dealing with. Modernity here is an integral part of the definition of science in the sense that it gives the right to invoke a stable difference, a difference that allows one to judge and claim kinship with the power relationship whose invention the experimental laboratory celebrates.

The critical questioning I associate with the transition to the limit refers to this commitment to dissociate modern science and modernist science. The critical point signals the appearance of modernist practices, discussed in "The Curse of Tolerance" (Book VII), where I clarify the "cosmopolitical" question that gives its name to this series of books. The domains I'll address are those where the definition of a scientific practice can no longer benefit from a stable difference between the scientist's practice and what she interrogates. And it is the heteroclite ensemble of practices, modern or otherwise, and the beings, factishes and fetishes, to which they refer and which are ingredients of their existence, whose modes of coexistence will then be (begin to be) examined. But before risking this approach, in which perplexity would have to construct the practical obligations that satisfy the perplication of questions and distinctions, we must slow down, examine situations where the problems that will trigger the transition to the limit are already present, but where modeling is not yet a caricature and still has a chance to express what the being the model describes requires of the milieu with which it has been coinvented.

It would be worthwhile to take as an example those studies in experimental psychology that attempt to penetrate the mystery of an activity such as reading, which has some interesting characteristics: the laborious manner in which it is learned, the way it breaks down under the effect of neurophysiological disturbances (recognizing letters but not words, words but not sentences), and the fact that the reader, once she has "emerged," "reads the way she breathes," that is, cannot prevent herself from identifying a word but, on the contrary, must make an effort to identify individual letters. This is a very interesting example of an external device that is liable to literally "pass into" the human and, therefore, seems to promise a stable definition in the face of life's contingencies. But the example is too complicated to slow us down reliably. The number of young humans who will never be "one" with the alphabet even though they are supposed to have "learned to read" is too high not to suspect that the question of "emergence," here apprenticeship, cannot eliminate everything that "knowing how to read" allows us to ignore. Possibly, among the ingredients of apprenticeship are the multiple components that, in another form, belong to the art and experience of reading (about which "knowing how to read" is equally silent), namely, those that enter into the effective encounter with a particular text. The "true" reader is one who may well be able to read "in general," but for whom the encounter with a text has nothing general about it. The slowdown turns out to be impossible, for even assuming that what "emerges" could really be modeled, what this descriptive, scenographic model would benefit from, that is, the irrepressible nature of "knowing how to read," turns out to be an obstacle to the possibility of tracing the description back to the question of emergence. The model would reproduce the final emergence but provide no clue for the many questions that cluster around what "learning to read" requires.

On the other hand, there is another episode, one that is truly generic in human life. It is the one that leads infants, in one way or another, to transition from the mode of existence of a young mammal, not fundamentally different based on appearances from a newborn primate, to that of a young human engaged in language learning and the relationships their specific identity presupposes but that must "be produced" for each of us individually. This episode is so fascinating that it has brought about the equally fascinating but relatively indecipherable series of experiments attempting to get primates to "talk." And it is the subject of an indefinite number of speculations and variants, where science, myth, and religion freely intersect. But most extraordinary is that, in the face of such divergent interests, one way or another small children continue to successfully manage this transformation, in any event, the vast majority of them.

It seems, then, that we are dealing with an extremely robust "history" whose success is tied to the very invention that defines what it is to be human, a history made to be repeated and that could, *in this sense and to this extent,* be defined as an extrauterine extension of human ontogenesis. It is as if the infant had its own requisites, as if it were capable of a power relationship with an environment that, barring any dramatic circumstances, enables it to learn and become. And yet, we also know that, at the same time, another kind of history is beginning, inseparable from the first. In fact, when the infant manages to stand and take its first steps, and even earlier, it is indeed possible that this history has already begun. But in the case of learning to speak, I feel I can take it for granted that the situation is clear: the infant does not learn to speak in general. Together with words, there is an indefinite ensemble, implicit and explicit, of ways of being, of entering into relationships, of interpreting and anticipating, that is created or stabilized. The two-year-old child is no longer a small, generic being; it is the child of a family, a culture, a tradition. It would seem, then, that the requisites of the newborn do not communicate solely with the notion of a necessary but insufficient condition but with that of a necessary and *necessarily* insufficient condition. Which is to say that they incorporate in their very definition ingredients that must be determined by what is no longer a "milieu."

In any event, this is the reading Daniel Stern proposes in *The Interpersonal World of the Infant.*[3] Psychoanalysts quickly recognized that Stern's book made use of a disquieting approach, one likely to classify as "professional legend" the version of the myth of paradise lost and original sin that were the basis of their own categories. The psychoanalysts' infant would have to "fantasize," it would have to experience the original illusion of a fusion, and the adult, barring nearly irreparable damage, would have to renounce that initial experience of well-being.[4] It seems to me that what Stern is suggesting is a new kind of model, which introduces the requisites of the infant but also assigns a crucial role to the unique nature of its interaction with adults. For, according to Stern, the manner in which adults "respond" to the infant's "behavior" poses the same question as apprenticeship itself, which involves both "repetition" and "acculturation." It would incorporate both their cultural, familial, and personal interpretation of what those behaviors signify and what those same behaviors lead them to feel and do irrepressibly, that is to say, robustly. Correlatively, "emergence" would occur through asymmetrical capture over time. Through its behavior, the child suggests a response from adults, who in turn suggest to the child a new way of being, and the process repeats.

In this case, we can speak of emergence as a *productive and functional misunderstanding,* whose terms change continuously but irreducibly entangle human genericness and cultural-familial specificities, producing a child who has become capable of experiencing itself and others as endowed with continuity, historical materiality, and intentions, but who experiences them in a way that integrates fundamentally heterogeneous ingredients, rhythms and refrains in adult value judgments, whether implicit or explicit, concerning affects, legitimate or illegitimate, expressible or inexpressible. These ingredients can arise from a mode of action that may or may not be deliberate, and they may be consistent or contradictory among themselves. They coexist in distinct ways, each of them understood in what Félix Guattari recognized, in his own terminology, as "incorporeal universes" and "multiple, dislocated, and entangled existential territories."[5]

Several paths are possible with such a model. One that

should obviously be avoided is using the model in the predictable operation of normalizing the description and confusing the
"successful" relationship between parents and infants in our
own culture with what the human infant requires in general.

Another path leads to a consideration of the relationship
between apprenticeship and misunderstanding. *Misunderstanding* is a loaded word, but here it has no Freudian–Lacanian connotation implying the impossible fulfillment of desire,
or the always failed relationship, or the painful lack at the core
of any illusion of belonging. This type of dramatization is very
interesting from the professional point of view of the psychoanalyst, who effects a decentering and creates a highly specific
power relationship that stabilizes the therapeutic process in a
unique and radically unilateral way.[6] I will attempt to follow how
the concept of a model changes meaning without the operation
suddenly having dramatic or disparaging consequences. Stern's
description "models" the young human, but here there can be
no question of condemning a given form of alienation but of
approaching practices that "introduce" a human into a world it
can inhabit only if it learns to comply with the requirements of
what it will encounter there. After all, even, and especially, in a
highly formalized science like mathematics, it is through misunderstanding that definitions and rules are held to be self-suf-
ficient, operating in such a way that compliance, understanding,
and application go strictly hand in hand.

Mathematics, which in Greek meant "that which is readily transmissible," in this sense constitutes a very interesting
example of a "misunderstanding." Even when a mathematical
definition is transmitted for the billionth time, what we call
"comprehension" remains an event, the production of a "before"
and an "after." It is only "after," once we have understood, that the
normative words through which this knowledge is transmitted
assume their *effective* meaning, which transforms them into references, instruments, and constraints for exploring, reasoning,

and constructing. It is only "after" that the words retroactively appear sufficient to define the knowledge that is transmitted through them.[7] Between the "after," where the teacher cannot but dwell, regardless of her good intentions, and the "before," where those she addresses are positioned, transmission implies a genuine practice of misunderstanding.[8] We can even speak of a "categorical" misunderstanding to the extent that, contrary to other kinds of learning (walking, riding a bike, driving a car, juggling, mountain climbing), mathematics is unique in that it confronts the one who learns it with explicit formulations that comprise both conventional rules and normative injunctions. However, as in other kinds of learning, it is a matter of "embodiment," of rules and injunctions "passing into" the body. Whereas the set of definitions and rules appears to introduce a purely "spiritual" operation, the pure product of abstract formulation, reasoning, and proof, what must be produced when one "gets it" are ways of perceiving and being affected in a functional, nearly automatic, way. To be able to recognize "$(a-b)^2$" in an algebraic text and automatically adopt the mental gestures and practices appropriate to the problem expresses the success of the corresponding modeling operation.

In such a case, misunderstanding is not another way of expressing the question of human existence, the failure of language, which never lets us say what we "desire," or the tension between the never satisfied quest for truth and the risk of cynical abandonment. It does not designate the kind of staging that confers the power to recognize sameness throughout each step of a psychoanalysis or the "phenomenology of the spirit." This kind of misunderstanding could, however, communicate with the concept of *transduction* created by Gilbert Simondon in *L'Individu et sa genèse physico-biologique*.[9] Transduction does not refer to the human, to language, or the search for genuine rapport, but to the problem of individuation, through which an individual characterized by discreet relationships with its milieu is produced. In fact, Simondon used the physical-chemical phase transition and the concept of a critical point as an experimental field for creating his concept. But anyone who might claim to draw from transduction the power to recognize that the production of an infant with an individuated relation to language responds to the "same" problem as the genesis of a crystal would be misusing the concept. Comparing the crystal with the infant has meaning only because the first step in the process of transduction is *not* to define the process of individuation but to learn to *resist* the way in which the problem has generally been presented. Transduction applies both to the crystal and to the human to the extent that neither the terms that enable us to explain the individuated crystal (interatomic forces, a configuration that corresponds to the minimum potential energy resulting from those forces) nor the terms that can be used to explain the human (genetic programming or social, cultural, economic, or symbolic structures) allow us to describe the process of individuation.

In all cases, what must be resisted is the temptation to explain the genesis of the individual from *previously individuated* conditions, the way the mold would explain the statue or hypothetical statements a solved problem.[10] Atoms, genes, and structures make the individual the simple realization of the possible they define, which is to say they miss the process of individuation.

Simondon also tried to provide a generic description of the process of individuation through transduction. "This is," he wrote, "the physical, biological, mental, and social operation by which an activity is propagated gradually within a field, basing that propagation on a structuring of the field enacted from place to place."[11] The crucial point is that this operation always implies communication, but first as a problematic tension, between two *scales of reality*, one "greater" than the future individual, the other "smaller." And it is this "primordial heterogeneity" that will be retranslated, once individuation takes place, into two rival explanations, each of which confers upon one of the scales of reality that are able to communicate with one another through transduction the power to obscure the process of communication, that is, the power to explain the individual.[12] In this sense, the first difference between the human and the crystal is that genetics and macrostructure are rival explanations, while the beauty of the perfect crystal relies on its ability to effect a harmonious convergence of two rivals: the forces of interaction between atoms and the energy equilibrium between the crystal and its environment.

Wherever it is relevant, transduction attempts to bring about a form of thought that is capable of resisting the temptation to choose between rival principles of explanation, a temptation Simondon qualifies as "hylomorphism": the Aristotelian duality between form and matter. For Simondon, this duality has served as the matrix of every position that has been adopted since then. Some of these base explanation on a "form" that imposes itself on matter thought to be available, others on "matter" conceived as being capable of causing form to emerge. Is it "symbolic order" or the norms of mathematics that are transmitted unaltered that "inform" an available mind, or is it the "matter" of the operation, a form of generic competence of the human psyche, that is responsible for the possibility of learning? It is as a vector of resistance that does not limit itself to celebrating the "failure" of these alternatives but creates a new appetite and riskier obligations[13] that transduction might assist in constructing the problem presented by the relationship between apprenticeship and "misunderstanding."[14] The asymmetrical capture correlated in time that, for Stern, "models" the infant would then be a primary example of the communication of two "scales of reality" whose heterogeneity is their primordial given.

This would be a good place to slow down, for it is not enough simply to have good intentions. Transduction, because it enables us to simultaneously contemplate crystallization and human modes of individuation and individualization, is *speculative,* and the intrusion of speculation is part of the transition to the limit that I am attempting to initiate.[15] There would be no more unfortunate confusion than to treat this intrusion as a victory, as the conquest of a point of view that considers the distinction between construction for a scientific purpose and speculation to be pointless. Nothing could be worse than to view Simondon's ideas as the basis for a scientific approach to "emergence." The inherent challenge of speculative thought is the creation of concepts that allow us to speak, simultaneously and at the same time, of what our habits oppose (for example, crystallization and thought), but this creation is an experiment in which our habits are both ingredient and target. It does not seek the discovery of a point of view that would guarantee the right to unify what we oppose and to establish a judgment concerning the "proper" way to answer questions that produce hesitation, perplexity, or expectation.[16] The practical effect that singularizes speculative thought is to contradict the temptation of a judgment that recognizes and anticipates. This thought straddles abysses, but the "same" that it constructs, the "anticipation" it feeds must accept the constraint of "accommodating" no one, of not confirming any particular practical requisite, not justifying any power relationship. That is why this thought is fundamentally descriptive, and the possibility of drawing normative consequences from it, regardless of the register of the norm, indicates either its failure or the (mis)appropriation of its use.[17]

Transductive thought produces the effects proper to speculative thought to the extent that we cannot, without possible contradiction, make use of it without also introducing at the same time the "transductive" nature of its use as soon as it becomes a part of practice. Practices of apprenticeship may take their inspiration from various forms of hylomorphic thought, with an emphasis either on the "form" to be transmitted or on "matter," as is the case with constructivism, or from Simondon's critique of hylomorphism; all are distinct examples of transduction, and the latter cannot claim any superiority over the others by virtue of its conceptual reference. It should be added to the others, along with its own requirements and obligations. And its own risk of failure. In other words, transductive thought provides no benefits with respect to the strictly *empirical* problem found in these examples. It provides no guarantee. Its role is to create words that might stabilize thought capable of resisting the slogans and legitimations through which the risks associated with a practice become rights (of reason, progress, objectivity) for which the practitioner is merely the representative.

Speculative reference to transduction thus puts at risk the power of models that claim to authorize an economy of perplexity. Experimental factishes can, through a constitutive vocation, claim to "explain" the world, and it is possible to assert that the world "explains itself" through them. Reference to transduction reminds us that, here, explanation, made possible by the coming into existence of each factish, primarily celebrates the primordial heterogeneity between the requirements of the scientist and the world that is supposed to satisfy them. But the reference to transduction can also help recognize and celebrate the occasions when the scientist, temporarily putting aside any professional plausibility, searches for the words to express the question that the experience of what she is involved in invincibly imposes.

So, when I tried to put into words the expression that appeared on Kauffman's lips, "God's heart," I created the figure of an interaction involving a "Maker" whose values pass into the world. I made use of a figure that expresses transduction, which can be used reciprocally to assert that the maker's "values" do not explain what is made, even though they "explain themselves" through the making process. But in doing so, I borrowed the words that Stephen Jay Gould dared employ at the end of an article in which he tore apart the "just so" stories of sociobiology. The biological theory we need, he wrote, should replace the questionable charm of such stories with the profound joy arising from an understanding of evolution as integration: "the world outside passing through a boundary . . . into organic vitality within."[18] Here, Gould doesn't claim to be a "vitalist," but he effects, with a joy that accepts perplexity rather than the pretense that must deny it, a transition to the limit. Gould also uses the word *integration* "with his mouth full." It is in terms of "integrative insight" that Gould, in the same book, evokes the way in which Barbara McClintock allowed herself to be invaded by the apparently disparate multiplicity of "data" produced by maize in an attempt to understand that data. And Gould compares this integration with the experience of Dorothy Sayers's detective hero, Lord Peter Wimsey: "He no longer needed to reason about it, or even to think about it. He knew it."[19] And when Gould talks about his own experience, he writes, "And so my work has been integrative; that's what I'm best at doing. I do figure out Dorothy Sayers's mysteries because Peter Wimsey is constructed as that kind of thinker. If you read *Whose Body?*, her first novel, I'm sure that Dorothy Sayers had a theory of thought and that she wrote those novels to counter the Sherlock Holmes tradition that thought was simply deductive and logical."[20]

Here, Gould describes in the same terms, with his mouth full, a theory of life that should help biologists do their work, a theory of thinking, and equally his own experience as biologist and writer when the outside, the scattered elements and bizarre connections of a situation, move inside and contract into a living unity—"he knows." It is here that speculative thought can assume its "ecological" scope, bring into existence the perplexing joy of this convergence, and give it the means to produce its own divergence, one that would prevent it from becoming a pretense, a skeleton key that would open all doors and would be confirmed in all cases: the birth of the kind of all-purpose response produced by a transition to the limit that is brought about unnoticed, accompanied by the exaltation that the feeling of truth provides. Convergence mustn't be avoided, it should be celebrated, but in suspense, held in its problematic space, "countereffectuated" and not precipitated into triumphal solution.[21]

The divergence to be recognized, and which indicates an approach to the limit, affects all the terms required by scientific practice. Consider the term "confirm." Even Popper had to admit that scientists are right to seek experimental confirmation when their theories are bold and fragile. Confirmation, he proposed, is not a proof but the nourishment the fragile creature requires. But here confirmation will always be experienced as proof or as an argument that can be put forth to support a proposition. It will always express the power relationship between the one who asks the question and the one who answers it. Adults who encourage the young child to take its first steps also "confirm" its attempts, and this confirmation is itself likely to be as vital for the child as "fact" is for the bold proposition. However, it will never serve as a proof or an argument. What about the distinct and entangled modes of confirmation negotiated by teachers and family for children in school? What about the analysand whose dreams "confirm" the interpretation of her analyst? What about the experiences that, it is said, confirm "faith"? To follow and map such divergences, it is necessary to countereffectuate the proposed convergences and deliberately ascend the slope, resist being carried down by the power of resemblance.[22]

I have associated the art of modeling emergence with tact, but tact is no longer a secure thread. Even when it is associated with human relationships (doctor/patient, adult/adolescent), it always refers to a power relationship that is able to control itself and can create the space the other should come to fill in its own way. The modeler, the doctor, the adult all propose and know that it is up to the other to dispose.[23] Yet, it is the very meaning of propositions that is affected by the transition to the limit, that is, the meaning of the confirmation we expect from the other.

Suspending triumphal confirmation—through the device or disposition implementing the proposition—forces the history of our satisfactions to ebb, and it is the "our" that begins to blink, that causes what those satisfactions have identified to diverge. We have benefited, and will continue to benefit, from all power relationships, from every stratification that may allow for a stabilization of the difference between the question asked and the answer that confirms it. We can find out how a "body" defines its milieu. But what it means to "live" or "die" does not follow from the thread of our definitions. Experiencing this marks the critical moment when constructivism escapes, as event, from the stories in which the practices that allow us to claim to know what we know are constructed. The moment when the question of the impersonal nature of the infinitive insists through "us," when "knowing" begins to resonate with its opposites.

In "The Science Wars" (Book I), I limited the scope of an ecology of modern practices to the question of determining if new psychosocial types could be generated, new "we's" not defined by polemics and hierarchies. Resolving the question of the ecology of practices through the speculative becoming of practitioners would be a trivial solution. Rather, the question is one of asking which "type" of practitioner would not have a phobic relationship—"but if we introduce this type of problem, we can no longer work"—at the moment of reflux, when their categories are confused. That is why it's important to acknowledge that speculation is not part of a fascinating "beyond" but already inhabits those moments of confused joy when the scientist thinks with her mouth full. That is why the way physicists have learned to define gas and liquid in terms of a transition to the limit is also interesting. For this transition does not require criticizing the gaseous state and the liquid state, but integrates into their definition the question of the critical point at which the distinction between those two states is, in fact, at issue.

Practitioners familiar with those "critical points" at which

they think with their mouth full would probably be less interested in hollow generalizations, reflexive recursiveness, or
other irresolvable paradoxes. But the question doesn't end
there, doesn't affect only "us," our knowledge and its relationships. I'd like to return, one last time, to the starting point of
the transition to the limit I have attempted, the description that
Daniel Stern provided for the "emergence" of the infant.

What is unique about this emergence is that it concerns a
becoming that is of interest to us all, and by all I mean all cultures, all traditions, *modern and nonmodern*. Stern's description may, initially, challenge hylomorphic models of all stripes
that modern researchers have proposed for giving the infant the
ability to establish their hypotheses. That is why, for example,
the Sternian baby challenges the Freudian baby, which is made
to establish that what will follow its emergence will confirm the
power of "matter," the universality of the unconscious conflicts
of psychoanalysis, as well as the behaviorist baby, which celebrates the power of an exterior "form" to inform matter and the
availability of matter to form. It challenges the Lacanian baby as
well, which it prevents from dramatizing the misunderstanding,
the discordance between "interior" and "exterior." But, distinct
from the risk of its normative becoming, to which I've alluded
in passing, the Sternian baby presents another, more insidious
danger. It is capable of allowing us to claim that we have now
understood how "the others were not mistaken." Those others
are the "nonmoderns," who, for example, believe that the newborn is a stranger from another world, who speaks another language, a stranger whose identity must be discovered so it can be
named, and who must be welcomed and humanized.[24] Couldn't
we see in this a marvelous illustration of Stern's description, a
wonderful confirmation of the definition he proposes? Thanks
to Stern, we "now know" that the way in which we "welcome"
the newborn, the way in which we conceive of, anticipate, and
interpret its behavior, is a vital ingredient of its becoming. Isn't

it wonderful that others have, without having read Stern, created the words and references that inhabit and guide parents in
this process?

But that is the danger, for this "wonder" is liable to be celebrated as follows: *"Thanks for confirming the progress of our
knowledge, the validity of our new definitions. Thanks, and forgive
us for understanding you better than you understand yourselves, for
having constructed in your place the meaning of what you are doing.
In order to protect you, we will avoid telling you that we understood
what your beliefs 'really meant,' what they enacted without realizing
it. Your ancestors and your fetishes no longer surprise us, neither do
they disgust us. We have taken from them what we needed to learn,
and they confirm that our descriptions are right. They will serve as an
argument against our backward colleagues."*

Suspending the confirmation, safeguarding the moment
when the impersonal—"to speak to a child" or "to come into the
world"—vibrates, are essential here. Not in order to avoid the
unavoidable, the feeling that "we have understood," but in order
to stand back and experience it in such a way that the suspension
of its confirmation is incorporated in its occurrence. For the tri-
umphant confirmation I have presented above qualifies us. If we
yield to this triumph, we will trample, with the best intentions
in the world and with the additional satisfaction of remembering our own arrogance, the inappropriable space "where angels
fear to tread."

BOOK VII

The Curse
of Tolerance

19

The Curse of Tolerance

Nothing is easier for modern man than tolerance. How could it be otherwise? How could we not be tolerant? I am not referring here to "others," to those in whom we encourage tolerance. I'm speaking of "we," and this "we" does not refer to a concrete group to which one may or may not belong, but to all recipients of the message of modernity. It is a message that, as a "master word," is instantly applied as soon as we hear, understand, and accept that "we" are not like others, those we define in terms of beliefs we are proud, but possibly also pained, to no longer share.[1] Our era is no longer one of crusades. The master word is occasionally paired with a nostalgic vibration. Tolerant is he, or she, who measures how painfully we pay for the loss of the illusions, the certitudes, we attribute to those who we think "believe." Therefore, happy are those whose confidence has remained intact. They dwell where we, moderns, cannot return to other than as caricatures, sects, and despots.

But nostalgia and tolerance toward others who are lucky enough to "believe" barely hides our immense pride. We are "adults," we are capable of confronting a world stripped of its guarantees and enchantments. This tired refrain is well known. As if by chance, it serves as a favorite theme among modern specialists of what Freud called the "three impossible professions": education (the transmission of knowledge), government, and healing. These three professions have in common the appropriation of practical concerns that all human societies share. And the second thing they have in common, introduced by Freud when he characterized them as "impossible" and assured his followers of his "sincere compassion," seems to be, in our so-called modern societies, that their practitioners consider themselves the privileged heirs of the loss of illusions that defines modernity.[2] As if their primary obligation were to affirm the inevitable nature of the destruction their practices inherited, as if their first requirement were to obtain recognition for and propagate the heroism of the impasses and aporias their practices explore in detail and celebrate using modernist slogans.

To the extent that what Freud called a "profession" is duplicated in our societies, the question of tolerance, which should be "cursed," is twofold. It refers to a group of knowledge practices that are defined by their ambition to be recognized as "scientific" in the modern sense of the term. Here, I am referring to what are generally known as the "social sciences," which cover a spectrum that ranges from psychoanalysis, psychology, and medicine to various forms of sociology, teaching, anthropology, and many other fields that serve as references to various forms of social assistance and intervention. But the term also refers to practices that, in one way or another, serve as caretakers for our sociocultural "ecology," practices that address, for better or worse, the relationships that may be produced among heterogeneous groups. Naturally, the groups are intertwined, but their intertwining can, depending on the situation, appear differently. Anthropology, for example, whether it wants to or not, produces a relationship, but one that is most frequently fundamentally asymmetric. It provides "us" with a knowledge of other groups, but the relationship from which that knowledge derives does not appear in the foreground, or it appears only in the service of the science that produces it. Conversely, "social workers," whenever they need to mediate, refer to knowledge they did not produce but that is supposed to prepare them to deal with situations in which an individual, a family, or a group has a problematic relationship with its environment. They are not supposed to "generate" knowledge themselves, but merely to produce modifications in relationships, enabling negotiation whenever there is a threat of confrontation or repression. As such, there is no reason to criticize the distinction between knowledge-producing practices and mediation-producing practices. Their obligations are partly distinct. But only partly. To address such distinctions, the question of what those obligations might have in common should first be asked, a question I associate with the question of tolerance.

Throughout this book, I have approached scientific practices from two points of view: requirements that must be satisfied by whatever it is they are dealing with and the obligations they acknowledge and that apply to the way they progress. Now, I am confronted by practices that present the problem of tolerance because, in one way or another, explicitly or implicitly, they assume, between "we" and the "others," a difference in kind as expressed by the possibility "we" claim of judging "others" in terms of beliefs without ever encountering them. From this point on, the leading question will be that of obligations, because the constraint I have referred to as a "requirement" is, at this point, no longer a reliable guide.

Until now, the practice that required was put at risk by its requiring. The values that bring the experimental laboratory into existence are risk-based, and the obligations of the experimenter correspond to that risk. It is not just a question of finding out whether or not the account of the phenomenon presented, staged, in the laboratory confirms the arguments of the one who produced it. It is a question of differentiating between two meanings of the term "artifact." In any case, an "experimental fact" is an artifact, a fact of human art, capable, unlike facts in the customary sense, of authorizing an interpretation and silencing rival interpretations.[3] But what succeeded in giving it this capability must resist the accusation of being responsible for it. The experimental account must not be an artifact in the sense that the artifact now disqualifies the experimental claim, defines the experimenter's procedure as responsible for an account that has been extorted, "fabricated," incapable of supporting the challenges demonstrating that it can be reliably attributed to the phenomenon being questioned. The obligations of the experimenter correspond to the definition of what Bruno Latour called an "experimental factish."[4] Certainly, the fact is "fabricated," it can even mobilize an impressive crowd of technical devices, each more sophisticated than the other, but its fabrication is directed at the "invention-discovery" of a being that can claim to exist autonomously, independently of the practices that enable us to "prove" that existence. Pasteur's microorganism, once it entered scientific existence, also became capable of claiming to have existed before humans and to have been the vector of epidemics, even though humans read supernatural intentions into the scourge that assailed them.

With the question of emergence, the relation between requirements and obligations is already complicated.[5] In order to free emergence from the polemics and rivalry that have made it one of the focal points of the "science wars," I have chosen to reexamine it through the question of its specific requirements, which are forgotten once answers are turned into military flags. Determining if, and to what extent, "that which" emerges can be "explained" based on "that from which" emergence is produced is not a matter for experimental observation or theory but relates to a practice of articulation focused on the difference between "that which" and "that from which," that is, between two distinct approaches corresponding to distinct practices.

At this point, new types of "factishes" appeared, which I have referred to as "promising." These factishes are promising in the sense that "based on" what they demonstrate, questions may be asked that they are unable to answer as such. Learning if, how, and in what formulation the promise has been kept must be produced "elsewhere." The answer must come from the encounter with the histories, intrigues, reasons, and relations that the scientists who interact, in their own field, with beings said to be "emergent" learn to decipher and narrate. Requirements and obligations then relate to the practice of articulation. This practice requires, at its risk and peril, that what is deciphered in the field can, at least partially, be defined as a response to the promise-question. How this response is expressed is inde-terminate—one must address the field—but that there will be a response, in one form or another, that is, that the question will be relevant, is the risk that supplies a framework to the encoun-ter in the field. Correlatively, I have associated the obligations this practice of encounter brings about with the values of "tact." Whoever is endowed with "tact" knows, or thinks she knows, what the other's problem is, but she also knows that this knowl-edge will be worthless if it is delivered to the other. Therefore, she also knows how to create the space in which the other will be able to determine, in his own time and in his own way, how this problem will be formulated and the meaning it will assume.

Between the "experimental factish" and the "promising factish" the relationship is pharmacological in nature in the sense that the pharmakon is unstable—cure or poison—just as the sophist, physician or seducer, is unstable (see *Cosmopolitics*, Book I, "The Science Wars"). As soon as the "promise" claims to be endowed with any ability to determine how it should be kept, that is, as soon as the promising factish claims a power and autonomy that assimilate it to the experimental factish, the practice of articulation tips the values of tact toward those of proof. The field is no longer a site of problematic encounters but a theater of operations of conquest that are appropriately said to be "reductive." Molecular biologists have succeeded in encountering bacteria in such a way that the promises expressed by the DNA molecule were able to be kept by an articulation between "genetic information" and the metabolic performance of bacteria. But this unique success wavers once "that which is true" for bacteria is said to be true for all living organisms.

We now come to the problem of practices whose power of judgment is said to depend on differentiating between the scientific, rational, or objective approach to a situation and the beliefs, customs, habits, illusions, and so on that define the actors in that situation. In saying that the question of requirements is no longer a reliable guide, I mean that the requirements associated with both types of factish, experimental and promising, maintain a pharmacological relation with those practices. The satisfaction of the requirements of a science is always associated with power—we can require because we insist that what we address has the power to confirm the legitimacy of our questions. But that which is addressed by both the "social sciences" and the practices of intervention that are authorized by those sciences push this "we can" into the indeterminable. For the "power of science," the power of whoever it is who is said to "do science," here becomes an ingredient that is inseparable from scientific practice. And this happens in two ways. It is on behalf of science that the scientist feels she has the right to ask others some very strange questions, to subject them to highly unusual situations, to describe them in a way that, in any other context, would be judged highly uncivilized. And the power that precedes and accompanies this is also an integral part of the attitude, the availability, the submission of those the scientist interrogates.

I am not referring here to the inseparability of powers, the possible connivance, intentional or not, between the ethnologist and the colonizer, between the social worker and the police, between the teacher and the work of social selection performed by the school, and so on. Rather, I am trying to limit myself to the practice of producing knowledge. Not because I feel that these problems can be generally separated but because they must be distinguished if we are to avoid jumping to conclusions, the amalgam that precipitates every problem at once. The power of the experimental factish, which is capable of resisting challenges that question its autonomy, does not deny the multiplicity of the other ("technopolitical") powers that set it in motion. But it makes it possible to create distinctions that avoid confusing, in the same accusation, Watson and Crick's joy at discovering the double helix of DNA and the mobilization (also overseen by Watson) of biology and medicine around the challenge of mapping the human genome. But questioning requirements whose satisfaction would create an intrinsic bond between the social sciences and power, identifying the pharmacological ease with which terms such as "obey," "understand," "interpret," "predict," "control," and "verify" change their meaning when they involve humans interested (in one way or another) in the knowledge that can be produced about themselves, is not my intention. It is only the starting point for extending into this field the question of scientific practices, the values they bring about, and the risks that identify them.

Therefore, it is on the basis of the question of the specific risk that identifies the social sciences—the threatening possibility that, even with the best intentions in the world, the effort to create knowledge in their case might turn into the abuse of power—that I have chosen to construct the question of obligations that might singularize those practices. By acknowledging the very specific link that unites "science" and "power" once what it is that science addresses has to be defined as "sensitive to power," those obligations will bring into existence, as a value, the ability to escape the trap represented by that link.[6]

To give a sense of the tenor of these new obligations, I want to offer a quote from William Blake:

He who mocks the infant's faith
Shall be mock'd in age and death.
He who shall teach the child to doubt
The rotting grave shall ne'er get out.
He who respects the infant's faith
Triumphs over hell and death.[7]

The words are those of a curse. And to escape the curse, it is not enough to tolerate "the infant's faith," for the "respect" that triumphs over hell and death has nothing to do with abstinence, tact, or the protective kindness that always culminate in self-betrayal, and betray as well the secret derision they conceal. "A robin red breast in a cage / Puts all heaven in a rage."[8] Cursed is he who frees the bird to please the rebellious child.

Therefore, I have chosen to introduce the practical question presented by the scientific enterprise whenever it addresses beings capable of wondering about the one who investigates them "What does she want of me?" by treating tolerance as a *curse,* and by tying the question of the obligations the social sciences might bring into existence and that might bring those sciences into existence to the risk of that curse.

The "curse of tolerance" evokes the highly nonneutral nature of my questions. It is not a question of weighing the pros and cons, but of bringing into existence a risk, that is, to treat as an "abuse of power" what some social sciences may consider to be success. A curse upon anyone who thinks they are free to redefine, in their own terms, the way in which the "other" inhabits this world, even when they are willing to tolerate them, even when they regret their own lost innocence. For innocence disqualifies the other, the one who does not yet know, the one who has not yet endured the "great divide" that forced us to recognize that a bird is not only a bird, and that the heavens are indifferent to our constructions.[9]

The curse against "tolerance" is not based on a concern for justice. With regard to those "others," it is not a question of defending them, as if the practices that we disqualify needed us to do justice to them. It is a question, now and always, of our own practices, of obligations capable of stabilizing the irreducibly pharmacological, unstable nature of the reference to the general advancement of science as a duty we must fulfill, a reference that is here liable to change into poison risks that are legitimate whenever laboratory or field practices are involved.

The practices I have analyzed until now are also unstable, but I have tried to stabilize them using their own resources, their own passion, without creating obligations and requirements that differ from those they create for themselves and that create them. That is why I accepted the fact that what they addressed was, thereby, "put at the service of science," judged in terms of requirements whose satisfaction confirms practice. The choice of the "curse of tolerance" as the operator of challenge indicates that, from this point forward, being "put at the service of science" has changed meaning, has become something that might serve as an obstacle to the invention of the sciences. However, I don't want to waste my time "denouncing" practices that cannot resist this challenge. The production of obligations is a creative act. To create the curse of tolerance as an obligation doesn't mean cursing whatever would betray this obligation. It means constructing a question. How can escaping tolerance communicate with the obligation of what I have called a practice? Isn't it, by definition, a unique adventure, a becoming that precisely embodies an escape from the requirements and obligations of a science? This is the specific question that defines this book and will lead me to the heart of the speculative question of an ecology of practices.

20

The Curse as Test

To turn the curse of tolerance into an operator guiding the construction of what would pass its test is to gamble with what scientific practices may, or might, be capable of. It is a question of prolonging them, rather than bringing them to a sudden stop, which would be relatively easy. All that would be required to do this would be to establish, between the ensemble of "objective" scientific practices and the question of knowing what humankind can know about itself, the supposedly impassable barrier that separates object from subject. But we mustn't fool ourselves. Such a separation is extremely costly whenever it becomes foundational, that is, when it is no longer limited to serving as a reference for statements intended to "protect" the subject, but is promoted as a starting point for "another kind of knowledge" that would be addressed to the human being "as a subject." The practitioner of this other kind of knowledge is asked to participate in a form of asceticism, for it is a question of purifying the relationship to knowledge, of eliminating everything that, in the practitioner as well as what she addresses, might become material for objectivation.

This is true especially of phenomenology, whose ambition has been expressed by Merleau-Ponty: "The phenomenological world is not pure being, but the sense which is revealed where the paths of my various experiences intersect, and also where my own and other people's intersect and engage each other like gears. It is thus inseparable from subjectivity and intersubjectivity, which find their unity when I either take up my past experiences in those of the present, or other people's in my own."[1] If the phenomenological world "is not the bringing to explicit expression of a pre-existing being, but the laying down of being," everything must play out, now and forever, as if for the first time, within an action-passion that is both dialog and infinite meditation.[2] There is no room here for any sort of "economy," obligation approaches infinitude through the twofold task of "revealing the mystery of the world and of reason."[3] Phenomenology, Merleau-Ponty rightly concludes, "rests on itself, or rather, provides its own foundation."[4] Which means, baldly stated, that it can no longer be a question of practice in the sense evoked in "The Science Wars," which I associated with the question of the *psychosocial type.* Anyone engaged in the phenomenological quest is not constrained by requirements that can be satisfied and obligations that can be met. And they cannot be a researcher, in the sense that the researcher always addresses other researchers within the collective practice of research. The problem is not resolved but suppressed, because the response rests entirely on the inexpressible quality of someone who has to be able to claim to "reveal" meaning, whereas everyone else is constructing meanings. Escaping the curse of tolerance in this way is precisely what I wish to avoid.

The curse of tolerance thus refers to the speculative possibility of a psychosocial type of researcher, not the sublime figure of the sage, the phenomenologist, or the mystic. A speculative possibility does not simply fall from the sky of ideas. Speculation originates in unique situations, which exhibit the possibility of an approach by the very fact that they have already undertaken it. That is why, unlike the other books in *Cosmopolitics,* this one

will engage me with other authors—principally Bruno Latour
and Tobie Nathan—without whom it could not have been written. For, here, it is a question of "thinking with one's allies,"
not in the sense that, like Monsieur Jourdain, they would have
done unawares what I describe in their place explicitly, but in
the sense that what they have done, the risks they have accepted,
the issues they have made perceptible, have been obligations
for me. And they have resulted in the creation of the obligation I
refer to as the "curse of tolerance."

However, before introducing my allies, I would like to present a "quasi ally," an author who "could have" accompanied me
all the way but who, in fact, will enable me to evoke another connotation of "curse." As an exclamation, the word is often used
when we find ourselves trapped *once again* by what we thought
we had escaped. It is something that might be uttered by someone who thinks she has escaped from a carnival maze only to run
into a wall that bars the passage and sends her back to where
she came from. It is this dimension of the "curse of tolerance,"
the way in which we discover that we are again in a position to
be "tolerant," just when we thought we had overcome the challenge, that I want to explore in the work of Georges Devereux,
the founder of a field designed to help the modern researcher
encounter the nonmodern "other"—ethnopsychiatry.

Devereux titled one of his most important books *From Anxiety
to Method in the Behavioral Sciences.* He used the theme of anxiety
to address the "pharmacological" question of the methods that
what he globally referred to as the "behavioral sciences" borrowed from the other sciences.[5] For Devereux, anxiety is not the
near-universal affect known to Freudians and method is not the
repression of something that cannot be satisfied. For Devereux,
anxiety and method raise the question of what is required of the
specialist of the behavioral sciences when her method demands,
implicitly or explicitly, that she forget the difference between
pouring a drop of acid on a lump of dead flesh or on a living

organism. "Minute differences apart, much the same chemical reaction occurs in both instances. In both cases the flesh reacts to the acid chemically, but . . . the living organism 'knows'—*which is a form of behavior*—while the excised flesh *does not 'know,'* and, therefore, does not *behave.*"[6] Similarly, for Devereux anxiety and method raise the question of what the obligations of the behavioral sciences might be if their method didn't protect them by "a similar *downgrading* of the observed. What sound behavioral science calls for is not an (actually or fictionally) decerebrated rat, but a recerebrated behavioral scientist."[7] A wide-reaching program, you might say, with little exaggeration, for the "cortex" in this case does not refer to an intelligence that behavioral specialists would be deprived of. The problem is not one of the distressing inferiority that singularizes the practitioners of those sciences. It is on the basis of the anxiety to which their practice exposes those practitioners that the double mutilation against which Devereux struggled has to be read: that of the animal or the human endowed with "behavior," that is, capable of observing the observer, and the questioner.

The problem raised by Devereux intersects my own, which is that of a definition of science that entails "mutilating" what it takes to be its object, even if it proclaims the necessity of respecting it after all. And in both cases, what is problematized is less a moral question than a change in the nature of practices that produce "facts" and "proofs" whenever they are directed at beings that are not indifferent to the way they are treated. In those scientific practices where requirements might serve as a reliable guide, the fact that those requirements provide a stable reference for the scientist with respect to what she questions would express the stable difference between the two meanings of the term "test." The scientist subjects what she questions to various tests in order to verify that what she is dealing with is capable of yielding a proof, but those tests have nothing to do with the tests undergone by beings able to experience them as "tests." It is precisely when we are dealing with the specific problem presented by what Devereux calls "behavior" that this "nothing to do with" changes meaning and becomes a defense against anxiety. The possibility that scientists operate without their own cortex arises specifically whenever they maintain the postulate that there is a difference in kind between their tests, which are meant to provide proof, and a testing experience inflicted on powerless beings. This denial entails what Devereux characterizes as a mutilation of thought.

Devereux's work provides a fecund example of the approach I have tried to introduce with the concept of an ecology of practices. In fact, his work is the product of a practitioner trained in physics who was shocked by the poverty of what are known as the "social sciences," including ethology. But unlike many others, Devereux avoided a two-part trap. He refused to reason in hierarchical terms; that is, he refused to believe (as do many physicists who turn to biology or the social sciences) that a good dose of "real science" would save those poor, retrograde sciences. And he deliberately turned his back on the temptation to respond to the question of behavior with an effusion of good feeling that would ensure that the scientist remembers to respect the "subject" before her and interact with it while respecting a freedom that too easily excludes the rat suffering in the name of science.

For Devereux, all knowledge practices share a common trait: in one way or another, they culminate in a cognitive utterance: "And this is what I perceive." But this common trait is not a reductive instrument, which could be used to establish resemblance. It is a common thread when formulating relative differences and practical contrasts among the sciences. The choice of this common thread relates to quantum mechanics, which Devereux was quite familiar with, and which constituted a decisive step on his path toward a *constructivist* approach to knowledge. If measurement implies a "demarcation" between what will become a phenomenon, with observable attributes, and what will be defined as an instrument, whose definition expresses the practical question of the observer, it corresponds to the *creation* of data in the sense that such data have nothing to do with raw facts. The creation of data constitutes a twofold actualization—for the one who asks the question and for whatever the question is about—and both entail risk. The expression "And this is what I perceive" is in itself the fulfillment of an operation through which "this" and "I" are distinguished, thereby creating a boundary that, for Devereux, is nothing other than the "ego."[8]

The ego defined by Devereux has no attribute independent of the way in which the demarcation is produced. When the demarcation is practical, transmitted by practitioners, stabilized by the "method" and the instruments of that practice, this ego presents the question of what I have referred to as the practitioner's "psychosocial type," defined in terms of requirements and obligations. But this psychosocial type is here defined in a way that will allow Devereux to emphasize the uniqueness of the behavioral sciences, and what makes practitioners so readily choose to "deprive" themselves of their cortex. For Devereux, the behavioral sciences are unique in that they expose themselves fully to the risks that characterize any practice where measurement is involved. Theoretical variables always codetermine instrument and object. But in the case of the drop of acid poured not on a lump of flesh but on a living organism, those variables should make fully explicit and put at risk the theory of the observer about what the organism in question—rat or human—endures: not only what it is made to suffer but the fact that it "knows" that it is made to suffer. And when the behavioral sciences concern humans, if we are to avoid the *self-destruction* of the theory they make explicit, those variables must be capable of explaining the behavior of the observer herself.[9]

To characterize that observer, Devereux once again invokes

quantum mechanics. The intervention that creates data and explanation is always a perturbation of the observed, the demarcation between what observes and what is observed also being the demarcation between what perturbs and what is perturbed. This is well expressed by the concept of delegation because what the delegated agent "does" expresses the scientist's questions. But once again this general thesis assumes its fullest expression in the behavioral sciences. For, theoretical variables, if they are to be capable not only of identifying the studied, and therefore perturbed, behavior, but also that of the perturbator, obligate the scientist to make explicit just how she is herself perturbed by the other's behavior, which her perturbing activity has caused. The perturbation of the other is in this case strictly inseparable from the perturbation produced by the other, and identifying the perturbation experienced by the practitioner must, therefore, become an integral part of practice. How the practitioner "knows," or refuses to "know," how she is perturbed by the other affects how she "perceives" and "interprets" the behavior of that other. That is why the question always arises about what the observer "can bear knowing" about herself. The question of anxiety.

Among the behavioral sciences, Devereux emphasizes psychoanalysis, of which he gives a "constructivist" version. For Devereux, psychoanalysis, to the extent that it supposedly recognizes that its technique *creates* the phenomenon it subsequently explains, serves as the most accomplished example of what a behavioral science should be. The analyst cannot be satisfied with "This is what I perceive" as a translation of a stable demarcation between herself and the other: "Any analyst who believes that he perceives directly his patient's unconscious, rather than his own, is deluding himself."[10] In fact, it is always "the analysis of *his own* disturbance (fantasy) which the analyst then communicates to the patient, calling it an analysis of the *patient's* fantasy."[11] And analysts' practice of the analysis of their own countertransference, of the anxiety that has led them to demarcate themselves in one way rather than another, is the only practice that does allow them to speak of analytic "data."

Psychoanalysis in Devereux's rather particular sense provides a model for the "science" of behavior, and that model seems to offer a way to escape the "curse of tolerance." There is no one less tolerant than Freud, according to Devereux. He created a technique that proscribes nostalgia and complacency, and which expects both analyst and patient to deliberately subject themselves to the challenge of putting at risk their demarcations, how they perceive and assign to others roles and intentions, being finally forced to accept that this expresses their own unconscious fantasies. Similarly, psychoanalysis would then prevent the ethnologist and the sociologist from using the various methodological defenses that have allowed them to distance themselves from the anxiogenic object (such as "cultural relativism"—the experiences of others have nothing to do with my own) or to deny that distance (expressed as the participatory observation: I disturb nothing nor am I disturbed by anything).

And yet it is with respect to this "constructivist" psychoanalysis, which creates what it observes, that Devereux took a step that forces me to exclaim "Curse!" Whereas the sorry, monotonous mutilated makers of puppets that mime the autonomy of the scientific factish provoke no more than a sigh, the exclamation marks an experience of betrayal that demands elucidation, that confers upon the curse the power to create a problem. What happened? In Devereux's case, what happened is that he tried to retain for psychoanalysis, even in the constructivist version he had proposed, the privilege that Freud had attributed to it. Not only would it be the "only *psychology* whose exclusive and characteristic objective is the study of what is human in man," it would have the means to achieve that end. And in doing so, Devereux turns to a definition of ethnopsychiatry that exposes him to the "curse of tolerance."[12]

It is not a question of weighing the clinical merits of what Devereux called psychoanalysis or questioning his fidelity to Freud in defining the unconscious as an inseparable operator in the production of both analyst and patient ("only a 'preparation' can 'analyze' another 'preparation'").[13] Rather, it is a question of learning how Devereux maintains the privilege he confers upon psychoanalysis: not only the privilege of creating practitioners who are aware that it is the analysis of "their own perturbation" that they communicate to the patient, but also the privilege of allowing those practitioners to claim that this makes them capable of serving as reliable witnesses to "what is human in man." For, if Devereux were right, the ethnopsychiatrist, a hybrid who has accepted the dual challenge of field ethnography and the analysis of her countertransference, would correspond to the psychosocial type of researcher capable of stating "I understand" anywhere on the planet, without being herself (dis)qualified by the claim.[14] At that point, the ecology of *scientific* practices could close in on itself. The modern practices of knowledge would be, based on their own definition and using their own resources, able to understand "nonmoderns" not differently *but better* than they do themselves.

And here, of course, Devereux must introduce a "requirement" that weighs on what the analyst addresses, namely, the "human psyche." For the analyst's unconscious to be a reliable witness of the patient's unconscious, in order for it to faithfully express the patient's message without adding or subtracting anything, the analysis of countertransference must not be a "preparation" but a true purification. For Devereux, any deformation of the message expresses the existence of a "reserved, frozen and inaccessible segment of the analyst's own unconscious, which cannot echo and translate the patient's fantasy."[15] In other words, the psyche itself is defined in terms that warrant the possibility of "dissolving" what is frozen and reveal a preexisting possibility of separation between "what is human in man" and everything else, which has to do with the family, history, and culture. For Devereux, psychoanalysis can achieve its "universal"—all-terrain—objective "because one's unconscious is a relatively undifferentiated function or portion of the psyche and can therefore resemble that of another individual more than can one's highly differentiated consciousness."[16]

This distinction between differentiated and undifferentiated, almost casually introduced into the text, is fundamental to the ability to define "what is human in man." More specifically, it is fundamental to the ability to supply a definition that, far from being an integral part of psychoanalytic technique, can claim to provide a foundation for that technique. The fact that psychoanalytic technique obligates the analyst to believe that her own unconscious "resembles" that of her patient and that the analytic scene "prepares" the patient for confirming that resemblance would not in itself present a problem. But Devereux intends to give the presumed resemblance produced by analysis the authority of an experimental fact. He wants us to recognize that what the patient confirms is capable of satisfying the experimental requirement of differentiation between the two meanings of the term "artifact." It is a "technical fact" depending on a "preparation," but this preparation also serves as a reliable witness to the human mind as such. Occasionally, Devereux assumes a prophetic tone when stating that "We share a common humanity!" But, if it were a prophetic statement, it should entail obligations, and those obligations would affect both the one who made the assertion and his or her community. The possibility that a determinate practice benefits from privileged access to "what is human in man" entails no such obligations. Rather, it affirms a power relationship between the one who speaks on behalf of that practice and those affected by its requirement. It is not the prophet but the founder of psychoanalysis as a *science* of behavior who *requires* that the "psyche" the psychoanalyst addresses validate an approach that transcends cultural differences.

When Devereux states that his "conception of normality . . . is not relativistic and culture bound, but absolutistic and culturally neutral, i.e., psychoanalytic," any appearance of symmetry between psyche and culture suddenly disappears, a symmetry that should have followed from his reference to complementarity as defined in quantum theory.[17] The existence of complementary relations between the "psychic" data associated with psychoanalytic practice and the "cultural" data produced by ethnology and sociology should mean that both are operational practices, unable as such to determine "what is human in man." However, the definition of the unconscious as "psychically undifferentiated" suppresses the symmetry between "psychic" and "cultural." The analyst, if "well prepared," is not obligated to actualize, through the "perturbation" caused by the patient, the awareness that what her technique addresses as purely psychic "fantasies" may be so defined relative to this technique only, relative to the fact that this technique operates by excluding from consideration cultural or social questions as such. The analyst's "preparation" rather allows her to reproduce, for her own benefit, the power relationship between the scientist and her object—and to recover the asymmetry between "modern" scientists, free to seek and find the appropriate culturally neutral point of view, and everything (here everyone) else defined as determined.

By what miracle did Devereux's ethnopsychiatrists acquire this amazing ability? But, what price did others have to pay for this miracle, which defined ethnopsychiatrists as mobile and inventive—as scientific—while they are defined by knowledge, irrespective of whether they share that knowledge or not. We must, then, conclude that for Devereux ethnopsychiatry remained "white man's" science, a two-faced science: not only does the ethnopsychiatric therapist-researcher remain a psychoanalyst even though she knows that, from the point of view of the sick native Indian she's treating, she is a healer, but she also claims that her analytic practice enables her to understand what a healer means to the Indian. She understands the other, but the other cannot understand her.[18] Therefore, the miracle in question is unfortunately nothing but the retranslation of the tolerance moderns demonstrate so easily in the presence of the nonmodern whenever they claim to represent science. Here, the possibility of such a retranslation does not strictly follow Devereux's premises. In fact, it occurs at the point those premises are abandoned, specifically, whenever Devereux "forgets" to follow them as far as they can go. It occurs when Devereux yields to the seduction of convergence between two claims: the prophetic-humanist claim that all humans have something in common and the professional claim that any human psyche equally satisfies a requirement that privileges psychoanalysis.

This may help us understand why the question of requirements, far from being a reliable guide, has actually become pharmacological. And when it actually becomes a poison, all the questions raised by Devereux begin to shift, primarily the meaning and scope of the notion of anxiety. For anxiety, which served as the common thread in Devereux's ecological approach, enabling him to understand why modern scientists so easily stop being creative when they address behavior, then becomes a poison. Anxiety is an operational concept for psychoanalysis; it is what the analysis of countertransference should elucidate. As a result, the behavioral scientist's anxiety does not so much concern the question of "the others"; it is formulated in such a way that it is reduced to a question "without an object," a question that introduces no "other" aside from the researcher-analyst herself. The question raised by scientific practices has thus been appropriated. Both its definition and its solution belong to psychoanalysis. The operation is truly one of invagination: the others, whom ethnopsychiatry knows how to make exist as humans, now find themselves inside, "at the service of science."

Now, heaven's rage can be expressed differently. Cursed be the adult who expresses as "anxiety," to be addressed through analysis, the effect of the child's outrage upon her.

21

Anxiety and Fright

When Tobie Nathan, Devereux's intellectual heir, reinvented the field of ethnopsychiatry, he also challenged the postulate of the "universal human" implied by analytic theory and the reference to anxiety as an all-purpose anthropological tool. In doing so, he shifted the issues of ethnopsychiatry. Devereux's anxiety allowed him to follow the way the landscape of scientific practice was highlighted by the contrasting question of the "perturbations" that express what scientists require from what they investigate, the tests to which they subject what they question. Nathan has no ambition to unify the landscape of modern scientific knowledge. He is interested in constructing an approach that prevents researchers from disqualifying nonmodern psychotherapeutic practices under the pretext that they cannot be defined as "scientific techniques." It is not analytic technique or anxiety as such that he is challenging, but their claim to benefit from a special relationship to the human "psyche," beyond cultural differences. That is why Nathan's challenge will lead us to ask about the ecology of our knowledge, the way in which we require of all techniques—ours and others'—that they be evaluated in terms of a "universal" reference that selects those that are rational and explains their efficacy.

Although anxiety is an integral part of Freudian technique, outside that technique it does not have, Nathan notes, any claim to a universality that would give the analysis of countertransference access to "what is human in man." However, Nathan did encounter another concept that is important in traditional therapeutic technique and thought, the concept of "fright." Yet— and this is what I find interesting—Nathan does not claim that "fright" is the "true" universal, which could be used to establish a science of "what is human in man." For him, it is part of an "interactive etiological system that brings about multiple psychotherapeutic procedures, which are also always interactive."[1] In short, it belongs to a *technical* definition of both illness and healing.

For Nathan, Freudian psychoanalysis emphasized anxiety over fright because the latter "brings with it an infinite array of external causes, whereas anxiety, more austere, purer, justified a sophisticated technique in which thought is everything, similar to the neuroses whose treatment psychoanalysis would devote itself to."[2] Consequently, the theoretical crux of the distinction between anxiety and fright would be "primarily clinical, even technical. To acknowledge fright as a central psychic affect would have led to the conclusion that some 'other' has intruded, has influenced or modified us, possibly even caused our metamorphosis. This also would have meant admitting that the psychoanalyst, through his presence alone—or rather through the psychoanalytic setting—had triggered immense fright leading to repeated transference events. . . . For, the essential fright is that the truth of what I perceive, of what I feel, of what I think, resides in an other. But if truth is found only in that other, if I have no way of knowing how the other is frightened as well, then the entire effort is merely a new form of fright, and not therapy."[3]

When seen from the point of view of the fright caused by the analyst, psychoanalysis is not disqualified. For we are confronted by the question of the obligations of the "frightening" therapist, including those of the Freudian therapist, even if in Freudian therapy the notion of fright is not explicitly introduced. More specifically, the Freudian therapist would be obligated to dissociate her practice from its scientific claims and assume responsibility for managing a technique a unique active dimension of which is that it does not make explicit the fright it works with. This is a practical responsibility from which the analysis of the countertransference will not release the therapist.

Starting from the question of the therapist's obligations, Nathan was also able to contrast processes of divination—reading coffee grinds, for example—associated with therapeutic practices with diagnostic procedures, such as the Rorschach test, which is part of a purportedly scientific practice: "In fact, [reading] coffee grinds provides a considerable methodological advantage over the Rorschach test. Naturally, it is also a kind of projective test, but one undertaken by the clinician—who is obviously, in this case, a *seer*—rather than the subject. And since it is administered in the presence of the subject or an object representing the subject (a photograph or an article of clothing), the reading of the coffee grinds can only provide information about the state of the relationship between clinician and subject, not the hidden hypothetical nature of the so-called subject. So, the reading of coffee grinds would, strictly speaking, be a technical procedure intended to force the clinician to speak only about the interaction that he has established with a person, and consequently to produce usable clinical material."[4]

Divination must be described in terms of obligation: the obligation of the clinician-seer to "produce" "material" under conditions in which she cannot yet "possess"—in the customary sense, as well as the much more disquieting sense the term may connote, in this quite similar to the pharmakon—knowledge about the patient. It is also the obligation to expose oneself, to become the site of a dramaturgy that brings other universes into existence. For Nathan, relating illness to other universes is itself a component of a technical approach, for it causes interest in the patient to shift and to mutate: "Whenever you are a patient and someone does something for you, the entire group benefits from this and is conscious of this fact."[5] The meaning of the message associated with the illness now concerns the entire group. The patient has become a "point of intersection" between the everyday world and the "other universe" from which the message originated.

The obligations involved in the process of divination have the abstract character I associate with constraints, and it is the *production* of abstraction in the strong sense of the term, exclusive of truth or fiction, that seems to be implied by the process (see *Cosmopolitics,* Book I, "The Science Wars"). Lucien Hountpakin, discussing the Yoruba *babalawo,* or "father of the secret," writes that the "active word" he gives his patient "acts as a specific asset, a word unlike any other. The *babalawo* is a true artist. He has to gather elements scattered throughout the culture, he has to cover them with 'common words,' which are immediately accessible, and slip it all into a formula that resists time and space so that, even ten years later, the patient can still use that word to enrich his existence."[6] To resist time and space and produce a renewed relevance in circumstances that have yet to be determined, few philosophers have dared offer such a beautiful definition of the construction we call abstraction.

I want to point out a crucial element in Nathan's writing on "nonmodern" therapeutic practices: it is as if it were imperative for these therapeutic techniques to confer upon the illness a significant value based on a strategy that creates obligations where modern practices have requirements.[7] The multiple "supernatural" universes that such practices imply, confirm, and bring into existence do not satisfy the requirements associated with modern proof, but they are therapeutically relevant. They refer to the therapeutic process as one that unravels hidden intentions and causalities through which a patient can be "produced," capable of interpreting what is happening to her and, if need be, of joining a "real group" to which her trouble means she belongs and which creates specific obligations for her.

There is nothing romantic about such "supernatures" [*surnatures*]. They are not a spiritual supplement that can be used to reenchant a silent world, a source of heightened inspiration, a fabled return to a lost innocence, a reconciliation with a nature in which birds warble the words of gods. The gods, djinns, and ancestors of the therapists described by Nathan do not speak in the transparent immediacy of a lost paradise; their message requires a technician, technical and constructive thought. They belong to the technician's practice because this practice is directed not toward the satisfaction of requirements but, *like his patients, their relatives, their family, and at the same time as them,* toward the decoding of obligations. Seen from the modern viewpoint, these multiple "supernatures" can only be understood in terms of beliefs that must be tolerated. Their therapeutic efficacy may even be recognized by modern therapists, the way Plato's heirs regretfully recognized the art of sophists capable of influencing a crowd by their persuasive speech. But it is precisely at this point that the possibility of an escape from our own judgments arises. It is not a question of knowing whether Tobie Nathan's interpretation is the "correct" one, if it is true, if it "corresponds" to what healers do in practice.[8] The fact is that this interpretation, which is inseparable from an ethnopsychiatric practice initially created to meet the psychotherapeutic problem presented by populations of immigrants in France, forces us to consider what a real "technique of influence" means, when distinguished from the "effects" of influence, or suggestion, that we are all supposedly able to produce at all times, under all circumstances, whether we want to or not.

The work of Tobie Nathan and his colleagues allows us to appreciate the challenge of "fetishes" and practices we disqualify as "fetishistic." Nathan does not offer us access to such practices "in themselves," in their authentic truth. Rather, what he presents are such practices characterized in a way that forces us to think. And primarily he forces us to think about the contrast between the ideal fabrication of the analyst, in the sense that Devereux has effectively taken from Freud—the "purification" through the analysis of countertransference of what blocks the analyst's ability to listen—and the "fabrication" of the nonmodern therapist.

That the psychoanalyst and not the medical psychiatrist plays a role in this contrast indicates a common trait: in both cases the therapists are supposed to "have been ill" and to have followed a path that makes them capable of practice. But this common trait is limited, especially given that nonmodern paths of initiation are not only multiple but also largely secret. Therefore, what forces us to think is not any particular element of the process, whether it consists of objects, words, songs, gestures, or rhythms. It is not a question of conquering knowledge that the process hides from us, or discovering the way in which it orients or guides practitioners. Being forced to think has nothing to do with feeling that we have the right to make explicit what the other might know "implicitly." On the other hand, what might force us to think is the contrast between the analytic purification whose ideal is so much a part of our traditions of truth, both ascetic and scientific, and the contrived multiplicity proposed by traditions we refer to as "fetishistic." The possibility of suggestion, of influence, is for us a key, a quasi-anthropological characteristic, and psychoanalysts are obsessed with the question of knowing how not to influence, how not to "deform" the message addressed by the patient's unconscious to that of the practitioner. In contrast, nonmodern technique turns "influence" into a sophisticated question. The psychoanalyst follows an apparently modern path, contrasting "influence" and "reason," and in fact the "reasons" of the unconscious do not have any other definition: they are what take precedence to the extent that the analytic scene has been purified of all the influences that might have been presented as reasons. It appears that nonmodern practitioners, however, must furnish both their thought and their practical environment with any number of appliances if they are to become capable of "fabricating" thought.

How can we characterize an "influence" that is "produced," actively composed of heterogeneous elements—supernatural intentions, objects, words, gestures, family groups, various materials, natural and cosmic, ancestors, the attacks of sorcerers, obligations betrayed or ignored? Can we use words such as *belief, suggestion, effectiveness of the symbolic,* which we employ to refer to "others" and to that part of ourselves that, as Pascal wrote, "reason does not know"? Or, on the contrary, are we obligated to recognize that such expressions are so many "master words," which can be used to pass judgment without having to encounter or experience, which can be used to avoid turning the practices of others into witnesses of a problem that is liable to "frighten" us, that is liable to call into question our own modern requirements?

The contrast is likely to call into question what our master words define as the norm. If we accept that what nonmodern therapeutic practices do not "lack" is proof, the imperative that modern psychotherapeutic explanation must refer to, that would prove their adequacy assumes a different meaning, creates another, unexpected, dimension of the contrast. It is true that the Freudian ambition to merge therapy and research, that is, to enable therapy to "prove" its own adequacy, has always fallen short whenever it confronts those familiar with experimental proof. But let us assume the correctness of Freud's ambition: the successful transformation of unconscious conflicts, resistance, and repression into "factishes" endowed with an autonomous existence that forces every researcher involved with the psyche to take them into account. Such success would not have been

produced along "with the patient" but elsewhere, among competent colleagues.[9] More generally, whenever a patient meets a
modern therapist, she implicitly deals with this group of competent colleagues who maintain the rational adequacy of technique.
For Nathan, the imperative of proof requires that all "modern"
therapies fabricate illnesses defined by symptoms that must be
"welded to the individual." The individual then finds herself
isolated and categorized, which is to say, subject to the knowledge of the therapist and his colleagues. "The psychopathologi-
cal categories that psychiatrists—as well as psychoanalysts and
psychotherapists—use to classify their patients are never based
on *real* groups. Have you ever heard of groups of people suffering from obsessive compulsion . . . gathered together in the
same place, acknowledging that they all share—who knows?—a
common ancestor, before subjecting themselves to the same
therapeutic ritual? . . . Of course not. Because psychopathologi-
cal categories are concepts that separate, only 'regrouping' individuals statistically."[10]

What produces this contrast, which calls into question the
requirements associated with proof? Somewhat unexpectedly,
I feel it leads us to that showcase of our rationality, the laboratory. Until now, the laboratory has been described primarily
as a place where the requirements that our "modern" factishes
must satisfy, where the invention and discovery of how a being,
an electron, a molecule of DNA, or a Pasteurian microorganism
satisfies the requirements that define its mode of experimental
existence are celebrated. The laboratory, however, is also a place
saturated with obligations. How does a physicist today know that
electrons exist? If we question her critically, we may come to the
conclusion that she "believes" in the electron, and the critical
inquirer will then be able to gloat over such "fetishistic" belief.
But if we abandon the crude dichotomy between knowing and
believing, we are forced to conclude that physicists today *are*
nothing without the electron (or molecular biologists without

DNA), for the majority of devices that populate their laboratory assume its existence. If the field is one in which we can state that the present is child of the past, in which the practitioner would be disarmed, powerless, if not authorized by the past, the field in question is that of experimental practice. If there is a practice where thought must not be expressed solely in terms of requirements but also obligation, where the beings to whom practitioners refer have as their primary function to force them to think, to participate in the production of thought, it is experimental practice. And if there is a place where the beings created have the means to obligate those who create them because they must appeal to such beings to confer meaning on the technical apparatus without which there would be no thought, it is indeed the laboratory.

The laboratory is the place of proof, and its possibility depends on the fact that the questioned being satisfies requirements presupposed by experimental proof. If the electron is an experimental being, if its existence has been proven, it is not only because it has managed to satisfy the requirements and proofs that were specifically addressed to it, but primarily because it satisfied the general condition on which the very *possibility* of proof depends: the electron is indifferent to the proofs we subject it to. A rat "knows" when it is in pain, while the many roles assumed by the electron in the laboratory and in innumerable technical appliances seem to confirm that it "knows" nothing about itself in Devereux's sense of the term. This general condition means that if the questioned being, whether a human or a rat, cannot be considered indifferent to what it is forced to undergo, the distinction between the two meanings of an artifact, which must be made in experimental laboratories, is pointless. The "fact" about the rat in Skinner's cage will always be dependent on the cage in which it was produced. It cannot be used to construct an autonomous factish that both exists "outside the laboratory" and explains what is happening to it in the laboratory.

From the point of view of the requirements of proof, the idea of "behavior" is disruptive. But the question of obligations can be used to transform the opposition between a "real laboratory" and a "pseudo-laboratory" (such as the Skinnerian laboratory) into a contrast. What characterizes such pseudo-laboratories then becomes the fact that not only the scientists who work there but also their "objects" "serve science."

Even those who force rats to run around mazes know just how much the relationship with the rat "counts," as if the rat had to "feel obligated" to carry out the requested performance. To what extent did the "torturers" in Milgram's experiment agree to administer increasingly painful electric shocks to their victims simply because they felt obligated by being of service to science? And didn't the reference to being in the service of science and the legitimacy of scientific experimentation also allow the experimenter to require that his subjects reveal to science, but also to themselves, their "potential" as torturers. The "behavioral laboratory" is not only the place where both meanings of the term "artifact" can no longer be distinguished; it is also where the need to distinguish between them is expressed by the confused and terrifying submission to obligations that produce the behavior of the experimenter as well as that of her subject.[11] In other words, it cannot be the place where our "modern factishes" are produced because *all* agents there are defined in terms of their service to science.

The practice described-constructed by Nathan, and the "supernatures" it incorporates, are not unique because those supernatures are vectors of obligation. So is science in the pseudo-laboratory. It is unique because those obligations force both therapist and patient to think, as well as the entire group affected by therapy, and because they are recognized and cultivated as such. In other words, although we can speak here of a "laboratory of influence," the important difference is that, unlike our "behavioral laboratories," such a laboratory does not create an obligation that incorporates a clandestine and uncontrollable ingredient, does not maintain the pretense, in the name of science, of a well-defined differentiation between the one who knows the questions to be asked and the one who has to answer them. If we must continue to speak of belief, our own requirements turn against us. They identify us as "believers," as those who believe they are released from the obligations that would qualify others.

How can we reconnect with the obligations of our own practices, when the requirements they claim to extend appear to turn us irresistibly into believers? That this question can be asked expresses the fact that the possibility of escaping the curse of tolerance is no longer a matter of having a bad conscience, but a practical question. Tobie Nathan, within the framework of the ethnopsychiatry that he renewed, suggests a practice, and perhaps a key. The co-therapists brought together by the process undergo a strange trial: they must participate in, or themselves engage in, the construction of "active" words and objects required by the "nonmodern" tradition to which the patient (originally, an immigrant) in their care belongs, whether the patient wishes to ignore it and adopt European categories or not. Here, analysis of the countertransference ceases to be a panacea, a "universal" remedy: when undergoing this trial, it is no longer possible to associate the anxiety experienced by the co-therapists with general, undifferentiated reason, such as the fantasy of a "magical," "all-powerful" relationship. Rather, the process obligates them to confront what we call "beliefs." A co-therapist is cursed who seeks to maintain a judgmental relationship with respect to those "magical" gestures and words, accepting them only as conditions of a relationship she must establish for the "good" of the other. For, in so doing, she fails to acknowledge the frightening power of intrusion associated with the supernature to which those gestures and words refer.

The "fright," which here replaces anxiety, does not imply

conversion.[12] The ethnotherapist is not frightened herself, does not attribute to the active object the power to act on her the way it acts on those belonging to the group that produces it. It does not express an obligation to "belong" to a given group but, rather, the obligation to recognize, in a way that is not a form of reflection, one's own belonging. The "perturbation" Devereux spoke of no longer refers to a psychic "core" common to all humans but a cultural core, one that produced modern humans in such a way that they recognize and deplore "fetishes" wherever they find their manifestations. As for the curse, because it is related to a practice, it loses all its prophetic character and is relative only to the creation of values that engage the practice of ethnotherapy. Anyone who feels free to "mimic" "magical gestures," all the while remaining aloof, is as misguided on the ethnotherapeutic scene as the experimenter who feels free to "cheat" in the laboratory or the field scientist who feels free to attribute to the causes she identifies the power to transform narrative into deduction. For reasons that concern her but are unimportant for practice, she is simply unable to participate in the production of thought this practice assumes.

So, we end up with certain ingredients of the definition of a new psychosocial "type" of researcher, a researcher whose practical field of belonging actively prevents maintaining an opposition between modern knowledge and "traditional belief." For Nathan, this researcher alone would be capable of sharing in the realization of Devereux's grand project—the creation of a finally scientific psychopathology. This psychopathology wouldn't try to require a definition of "illness" that justifies the privileges of a specific healing practice, one that satisfies the same categories as the definition. If every psychotherapeutic practice "creates" the illness it heals, that is, if it incorporates it into an operational procedure, where words, actions, meanings "produce" it so they can act on it, a scientific psychopathology *must engage with practices, not patients.* It would then require modern researchers to recognize as their "therapist colleagues" those who have mastered those practices.

However, the path followed does not stop at the question of a psychopathology that is finally scientific. Rather, this question leads to others, which are far more speculative. If the Yoruba "father of the secret" must be recognized as a colleague, such recognition cannot be conceived unilaterally or we would be back at the situation we hoped to escape from, the one in which we always "know better." For, the most extreme asymmetry would arise wherever there was a possibility of symmetry. However, this "colleague" can't be separated from the group to which he belongs. The question then becomes one of learning how "Yoruba colleagues" can be addressed in a way that confirms their belonging to the group that authorizes them to speak, by making those groups "present." A scientific psychopathology, therefore, raises the question of a "group-to-group" relationship, and the threat of asymmetry now arises from the radically different makeup of the groups in question. In other words, the problem now becomes "how to live with the Yoruba." Once again the question of "modern practices," including ethnopsychiatry, cannot be limited to its own concerns. We are not yet done with obligations.

Until now, I have used the term "culture" without defining it, although the term is subject to multiple, antagonistic definitions. I want to address that definition because of the urgency of the question of "how to live with the Yoruba." I'd like to try out the definition proposed by Tobie Nathan, for what he defines is not an entity but a problem: "Culture tries to resolve two problems: closure and transmission. How can a group be enclosed so that it is impermeable to others, and how can this closure be transmitted to the next generation? To resolve this technical problem, every culture makes its own choices."[13] The closure Nathan speaks about is not physical but logical, in the sense that logic is, obviously, inseparable from practice and cannot be made explicit and formalized in logical terms.[14] Moreover, this closure, which characterizes what Nathan refers to as "ethnic groups," is explicitly associated with the question of exchange. "In Cairo there were thirty, maybe forty, ethnic groups who paid close attention to their reciprocal definitions: a Copt from one place was not the same as a Copt from another place. An Orthodox Greek was not a Catholic Greek, and so on. This resulted in clearly defined ethnic groups; but these groups were very familiar with one another and felt that it was a matter of their personal, individual survival to preserve those cultures because that was the condition of exchange."[15] Here, culture has nothing in common with a "closed" identity, in the static sense that is often associated with the notion of autopoiesis.[16] Culture is part of a technical problematic that is strictly ecological in nature: closure and the transmission of closure as "condition of exchange." What's more, it cannot, as such, be subject to a theory. Each mode of closure and transmission constitutes a unique solution that creates its own constraints, problems, requirements, and obligations in terms of exchange. In short, it creates the practical identity of what is called "exchange."

It is important to bear in mind that culture as it appears in the strictly ethnopsychiatric problematic, where the question of healing arises, and culture the way an ecology of practices presents the problem are distinct. The culture to which the therapeutic process actively refers is relative to this process and to the "disorder" whose meaning is to be constructed and situated. The process and those who speak on its behalf are not capable, as such, of representing what culture in general is and what it is capable of. Their evidence relates to the way in which it is tested and implemented through the practice of therapy, the way in which it participates, as a resource and ingredient, in that practice.[17] Consequently, it is in *technical* rather than scientific terms that the question of culture arises in ethnotherapeutic practice. The definition of culture is not, and should not be, the same whether it relates to the "exchanges" between cultures or to therapeutic practices.

But the perspective of the "finally scientific psychopathology" Nathan suggests not only presents the problem of obligations, to which the question of "exchange" responds—how to live with the Yoruba—it also provides a partial answer to the question put forth here: how can what I globally refer to as the social sciences be practiced without becoming subject to the curse of tolerance? Such a perspective provides the means to approach the practices of these sciences, like all the others, in terms not only of obligations but of *requirements*. We can now ask about the requirements associated with a scientific practice because those requirements will no longer be focused on the question of proof. Their construction follows from the obligation I proposed earlier: resisting the challenge of the "curse of tolerance."

Why would a science be possible "with the Yoruba colleagues" (and others) but not with patients? Because, unlike patients, who are there to explore, through a technique of influence, the obligations entailed by their illnesses, such "colleagues" will not be (overly) influenced but will be fully capable of putting at risk the way in which their practice is described. In what sense is a "social" science possible? Is it conceivable to create a psychology of the "housewife," the "adolescent," or the "TV viewer," for example? How, in such cases, can we claim that what psychology addresses is capable of putting at risk the arguments that refer to it? TV viewers, adolescents, and housewives are categories already defined as the targets of techniques of influence. Don't scientists take advantage of this social construction to secure a stable difference between those who ask questions and those who bear witness, through their beliefs, opinions, and presumptions, to the influence they are exposed to?

Just as laboratory practitioners require what they question to be capable of the purification needed to produce reliable evidence, practitioners of the social sciences, in order to construct reliable knowledge, need to encounter people able to resist their authority. This ability may be an individual trait, but requirements cannot depend on individual character. From now on, I will refer to those able to resist as "practitioners" in the broad sense, meaning people engaged in some activity or role that entails obligations and requirements. Didn't more reliable knowledge about what a housewife is find support for its conditions of possibility in the work and commitment of feminists, for whom such questions mattered? Likewise, a possible starting point for knowledge about those we call drug addicts can be found in the confrontation with associations of "unrepentant drug users" who challenge, practically, politically, and conceptually, the representations we have constructed about them.

At the beginning of this book, I stated that the question asked involved two entangled populations: "scientists" and "mediators," the latter referring to the authority of forms of knowledge that claim to be "scientific." If my hypothesis is correct, if scientific practices in the "social sciences" *require* that what they address has the practical ability to put the researcher at risk, its first consequence is the problematizing of this entanglement. We must conclude that when practitioners address what is commonly called a "socially disadvantaged" population, they can no longer refer to the authority of science because the category "socially disadvantaged" excludes the construction of knowledge that would be scientific. But when an intervention has something to do with science, when it is not to be cultivated as a technique of influence, it will always mediate between possibly conflicting practices, each of which is capable of presenting itself in a positive manner. Mediation will then imply, like the question "How can we live with the Yoruba?" the creation of the "practical identity of what will be called exchange." Which means that it is urgent that the question of the difference between scientific practices and technical practices be made to exist.

22

The Politics of
Technical Inventions

It may appear surprising that I feel the distinction between scientific and technical practices needs to be explicitly constructed. In fact, and even though the word *science* refers—as it does here—to the "modern" sciences alone, the distinction is so ambiguous that some feel it should be abandoned altogether for a more comprehensive concept: "technoscience." However, to treat technique as distinct is crucial if we are to resist the effects of fascination associated with discourse about the inexorable technoscientific redefinition of the world. To begin this construction, I'll begin with a minor asymmetry between science and technique. In *The Invention of Modern Science* and throughout *Cosmopolitics,* I took the risk of accepting the difference between the so-called modern sciences and other knowledge practices. But I did so on condition of presenting this difference, of making it present, in a way that differs significantly from the way it is ordinarily introduced. It is much more difficult, however, to talk about the "invention of modern techniques." Attempts have been made to define what a modern technique might be, that is, to politically define the "technician" in the community in the same way the "scientist" was defined at the moment the modern sciences were invented. However, no such attempt marked an "event," none was able to situate us as its heirs, whether we extend its consequences or resist it. In other words, the category of "technique" remains de facto available for mainstream use, regardless of the question of modern and nonmodern.

How can we approach the question of what distinguishes "science" from "technique"? To establish the scope of our discussion, we must first recognize the category, itself well defined, of "scientific technique." Perhaps the simplest example of scientific technique is the cycloidal pendulum invented by Huygens. From a scientific point of view, the law of isochrony associated with the pendulum when it is used as an instrument to measure time, the fact that the period of a pendulum (at uniform gravitational acceleration) depends only on the length of the cord, is valid only for oscillatory movements whose amplitude is small. So isochronism is an approximation. Huygens's technical invention is the invention of a device that transforms approximate isochrony into *exact* isochrony. This device consist of nothing more than two blades attached on either side of the pendulum's anchor point in such a way that, depending on the amplitude of the pendulum's motion, a portion of the cord will contact first one, then the other of the two blades. As a result, the motion of the pendulum is modified (without friction) and the apparatus behaves "as if" the length of the pendulum's cord varies as a function of its distance from the vertical. Of course, the shape of the blades (cycloidal) was calculated using the scientific law of oscillation in such a way that the modified motion is exactly isochronous, regardless of the amplitude of the pendulum's oscillation. This shape, therefore, embodies a mathematically well-defined transformation of motion. We have here a perfect example of technique as applied science. The general law of pendular motion is known and guides the required modification of pendular motion in such a way that the pendulum can become a reliable instrument for measuring time. Once this is done, the technically reinvented pendulum makes exact what the scientific pendulum expressed only approximately.

Note, however, that the above example is a bit too perfect. The ideal transparency of the invention of the cycloidal pendulum is matched by its limited scope. This pendulum, if placed aboard a ship, will be useless.[1] It requires a protected environment, as is true of most laboratory instruments whenever they continue to operate exclusively in their place of birth, in the hands of "competent" users, capable of using them and evaluating the conditions, limits, and meaning of their reliability. Strictly speaking, a technique is "scientific," that is, directly authorized by scientific practices, only within this framework. When the instrument migrates to "foreign" labs (for example, when "nuclear magnetic resonance" migrates to hospitals), the qualifier "scientific" incorporates a value judgment that hides the reinvention that gave it the power to leave its place of birth.

Although, strictly speaking, a scientific technique above all celebrates the satisfaction of a requirement—what is called a pendulum ideally satisfies (whenever friction is made to approach zero) a mathematical definition—the reinvention that gives a scientific technique the power to migrate must be expressed in terms of obligations. Understood in the practical sense of creating links, rather than the speculative, rhetorical, or political sense, the ecology of scientific practices is invented and negotiated primarily by means of such migrations. What links physics and biology is, first and foremost, technique: spectrography, radioactive marking, nuclear magnetic resonance, and so on. In this case, the obligations that technique must fulfill in order to migrate relate to the practitioners who accept it in their lab as a reliable ingredient for argumentation or proof. But whenever migration is toward places where answers are developed to questions that are not scientific but have been raised by other groups, or on their behalf, the difference is indicated by a radical reinvention. In this case, the obligations that preside over such reinvention do not only involve a change of users. Their primary characteristic is their heterogeneous and contingent nature. That is why the cycloidal pendulum is a superb example of applied science but a poor example of technical invention. It satisfies a single obligation associated with the practical values of its users; it makes isochrony "really" true. But it disregards all the other requirements a "timekeeper" worthy of the name should satisfy.

The "modern technical laboratory" (which includes all services, including market research and legal services, that are present during the design of a device) escapes the obligation that identifies the scientific laboratory, the distinction between fact and artifact. It produces something new rather than reliable witnesses. However, it must satisfy a very large number of disparate obligations expressing constraints that, in one way or another, will have to be integrated into the definition of what is produced. Under what conditions will the instrument function, what qualification will be required to use it, what sort of reliability will be necessary, what price can be charged, are there existing patents that might cause a problem, what safety requirements or standards will have to be respected? All these questions and many others share, along with the science that is supposed to "explain" it, in the process that brings a product or technical device into existence. The modern opposition between rationality and opinion has no authority here, at least not automatically. Everything depends on the evaluation of power relationships. A device can incorporate something judged to be a mere habit, or risk taking the opportunity to alter that habit.

To speak of obligations here is to underline that none of the constraints that share in the invention of a technical device has in itself the power to determine how it must be understood. For example, it is the "technical laboratory" that has to determine the meaning it will confer on the knowledge developed in "scientific laboratories." A scientifically established "property" can be exploited as well as avoided. The technical design then ensures that it has no consequence on how a device functions. In other words, the "technical laboratory" is obligated but not determined. It transforms obligations into determinations for each particular device, each determination then assuming a meaning and being articulated with others. The implications of the social, the legal, the economic, the scientific do not preexist the production of the technical device as such. If that device, once produced, can be described in terms of those different "points of view," it is because this production was also the invention of the meaning to be attributed to what matters from each of those points of view and the coherent articulation among those different meanings.[2]

Why, then, speak of a "modern technical laboratory" when the operations of disqualification, of opposition between the "rational" and the "illusory," the search for a fixed point from which the distinction between "objectivity" and "rhetoric" could be stabilized, have no authority there, when some of the obligations that participate in the production of a technical assemblage openly display that what they refer to has no fixed meaning, when the meaning of every constraint can have the instability of the pharmakon: poison or remedy, obstacle or opportunity? Whereas it is important to restrict the category of "scientific technique," which provides technique with a legitimacy that it would obtain from science, the category of "modern technique" must be maintained because the laboratory I have just described is emphatically not a general model. It is "modern" in two respects, one of which is constitutive, the other opportunistic.

The technical laboratory as we know it is modern primarily in that those who are active there must, ideally, satisfy an obligation that identifies them as modern. Each represents a dimension of reality, which is to say, the obligations that ensue, but *none can claim to transform an obligation into a condition.* They must all be ready for operations of reinvention, for the mutations of meaning by which the device will be invented.[3] In other words, the "representatives," or spokespersons, must strip what they represent of any claim to constitute, in itself, a "cause" or "condition" that must be respected as such. They have to "abstract" themselves and what they represent from the ensemble of relationships of reciprocal capture that matter for them as practitioners in order to make it available for entering into new, concrete arrangements. The modern technical laboratory, therefore, requires a "doubly modern" world: a world capable of providing it with "technicians" who actively and deliberately proceed with the reinvention of what they represent, and a world capable of accepting the reinvention of meaning and values that a new technical device or product entails and will try to instigate as soon as it leaves the laboratory.

Where the environment consists of other technicians, lawyers, standards auditors, even consumer associations with access to "counter-laboratories," the reinvention-betrayal is limited. Just as the properties that make gold gold have been multiplied and become more exacting throughout a history in which alchemists and *testatores* were at odds, the history of standards, legal requirements, and sales criteria constitutes the translation of multiple histories in which were negotiated the spectrum of differences between constraints, which technical invention can attempt to reinvent, and conditions, which it must apply as is. The difference between constraint and condition, therefore, expresses the power relationship between the laboratory and an environment that has the power to eventually impose a standard whose meaning cannot be reinvented, that is to say, the power to ensure that the laboratory satisfy clearly determined demands.

However, whenever the environment can be assimilated to the "public" of consumers and users, the technical laboratory, as we are familiar with it, can be said to be modern in quite a different way. It exploits the opportunity offered by the division between rationality and opinion. Even the demands of an industrial enterprise may be ignored as incompetent "dreams" when (as was the case in the early days of computer technology) it has no countertechnicians able to examine a technical proposal. The technical laboratory then makes active use of the rhetorical resources of modernity to claim that what it produces is accepted as such, to silence or disqualify those who do not have the power to force it to consider as obligations the conditions, requirements, and distinctions they would like to see recognized. It is then that the opaque argument of the "requirements of technical rationality" makes its appearance, or that a device will be presented based on the science that is supposed to explain it. At this point, social or economic "needs" with a given identity, which the device satisfies and which justify it, will be brandished. Stories flourish about rational, scientific knowledge becoming able, because of the discovery of the laws to which phenomena are subjected, to use those phenomena to satisfy those needs. Here, the "modern Constitution," in the sense Bruno Latour calls it into question, becomes functional, lending its syntax to the statements that shackle our thinking about hybrids by referring them to two agencies supposedly capable of being defined separately: nature and society.[4]

This second aspect of the modernity of technical laboratories obviously expresses a major political problem in which what we understand as "democracy" might be gambled or lost through derision. Unlike practices organized around the requirement of proof, technical practices do not, as such, contravene the obligations of democracy, and the possibility even exists of a convergence between these obligations and the requirements imposed on invention. The modern technical laboratory, if it is actively stripped of the rhetorical resources of modernity, communicates directly with a cultural-social-political question. How can those who are affected by what is being produced be "invited" to participate in its production? How can they become concerned parties, multiplying questions, objections, and requirements? Doing so separates what I have distinguished as "constitutively" and "opportunistically" modern. It complicates the technical invention, certainly, but in a way that does not violate its mode of functioning. To this extent, what I call the "modern technical laboratory" is, as such, an exemplary site of experimentation for what Bruno Latour has called, generically, the "Parliament of Things."[5]

Needless to say, the perspective of the Parliament of Things in this modern reading is utopian. It is nothing more than an empty dream if it does not function as a diagnostic vector for what makes it merely a utopia, and as a learning opportunity for resisting what today opportunistically frames our world. The multiplicity of practices that gather around issues and must invent their interrelations, implies an obligation to actively resist the hierarchy of knowledge, to make an active distinction between scientific and technical inventions, to engage in a deliberate process of clarifying the requirements and obligations of each practice. It especially implies the creation of new types of practices, of "interest" groups capable of promoting ignored dimensions of issues and situations. A single condition is placed upon participation in the Parliament of Things. Everyone may introduce themselves in terms of the requirements their practice brings into existence. But in a hypothetical trustworthy and reliably functioning Parliament, everyone must also agree that those requirements be expressed as obligations during the process of deliberation/invention. No one can introduce themselves by establishing conditions—take it or leave it— from which the possibility or impossibility of agreement would follow.

"Planck" would be welcome in the Parliament of Things, with the conservation of energy, which would apply "even to Martians." But the unified view of the world he defended would not resist the challenge of having to meet those it disqualifies. And the historical analyses of "Ernst Mach," who embedded the "energy" factish in the practices and questions of technicians, would also be welcome. But only to the extent that they celebrate the event in which those practices are gathered and do not disqualify factishes as simple logical generalizations. Galileo's billiard balls and "large Poincaré systems" would become legendary fixtures there, commemorating the risk of a definition that requires that description and reason coincide, and proposing to other protagonists, like the makers of "artificial life," a contrast that may help them present their own practices. But a scientific factish, regardless of the requirements and obligations whose satisfaction it celebrates, and its representatives, all those for whom this factish might make a difference, will not make law, for in this Parliament it is the meaning that can be given to such differences that counts, the way those differences can be represented, *become present for others.* That is why what is suitable for the Parliament of Things are not the requirements and obligations of the "scientific laboratory" but of the "modern technical laboratory," where tests lead to creation rather than proof, where proof is, depending on the circumstances, required or irrelevant, where what is constructed is not a consequence but a novelty.

However, one question must be asked that highlights the problem. The technical laboratory I'm referring to is modern before it is technical. And even within the framework of an established democracy, in which the difference between those who have the power to impose obligations and the "public"—consumers or users, who participate only in terms of their opinion—would not be ratified but actively opposed, the "Parliament of Things," as I have characterized it, would also be modern, would embrace only "modern" representatives of conflicting interests. Such representatives must be capable of "promoting" the constraints that, according to them, an innovative device, disposition, or product must satisfy based on commonly intelligible criteria alone. They must also be prepared to participate in the translation-betrayal-invention of what they represent.

In short, they must *cooperate* with the invention even when they complicate the process.

These limits are not obvious in our tradition for reasons that should not be confused with a blind belief in progress or a fatal submission to the argument of rationality. There is no need to define ourselves using terms that disqualify us—an apparent form of humility—but that also transform into destiny what we humbly recognize we are responsible for. If there is something in our tradition that positively singularizes us in the sense that it must be taken into account in an ecology of practices, it is the invention of "politics." The limits are so obvious because the condition that must be fulfilled by those who participate in the Parliament of Things is a political condition, one that is usually espoused when condemning "corporatism." Every "citizen," every concerned group or practice, in the very act of defending their interests, must be able to put the formulation of the interests in question to the test of general interest and acknowledge their participation in the polity and its collective project. The nature of the general interest, the collective project, the polity, the citizen, all can be questioned. The condition does not apply to the way in which the relationship between the group and the individual will be taken into account but to the obligation to share in constructing that relationship.

Everything we today judge to be normal, a synonym of progress, has been invented through struggle, resulting in the invention of new identities for the collective citizen, and transforming the modes of belonging, requirements, rights, and obligations that follow. But there is nothing neutral about this dynamic of invention. It defines *our* "ecology" in a way that is political, that requires that we accept *the test that distinguishes between condition and constraint.* No one can impose a condition because everyone is subject to the same condition—collectively inventing what the world we all have in common will be. And that is why this test cannot serve as a general obligation that defines the principle of an ecology of practices. Were that to happen, the conditions of the politics we accept and whose constraints we demand would change their meaning in the manner of the pharmakon. They would become conditions we would apply to others. Unless we are prepared to tolerate the fact that the categories of the "political" may not (yet) be obvious to them. And that would be our curse.

Therefore, we may have to accept, as a new starting point, that our political dynamic is in itself a "logical closure," as understood by Tobie Nathan, not an ideal that can be generalized by right. The ecology of practices, if it has to avoid being captured by the "great divide" between modern and nonmodern cultures, must certainly include the categories of the political, but it must not give them universal scope as part of "what is human in man."[6] Those categories must be *complicated,* considered along "with" other categories that "fold" the question of the decision differently. And among those other categories, we find those "supernatures," those multiple worlds that seem destined to create obligations and conditions that are foreign to the question of belonging—even when there is conflict within such belonging—to a polity.

23

The Cosmopolitical Question

If the question of politics were reduced to that of "nonmodern" practices, that is, if only those practices were to "complicate" the principles of an ecology of practices, the situation would be extremely dangerous. For it would expose me to a renewed temptation of "tolerance." This is what the multiplication of "musts" that populate the preceding paragraph of this text translates—we are the ones who "must." Consequently, I would have failed had those imperatives referred to an abstract norm our categories had to subject themselves to in the "name" of some duty toward universality. Here, as elsewhere, the question of tolerance challenges and puts at risk an overly seductive proposition.

If the question of "others" does not have the meaning of a normative imperative that must be obeyed, it is to the extent that it may become a vector of requirements and obligations for *practices* that can be represented by practitioners. And if practitioners are to complicate the model that relates the Parliament of Things to the model of the modern technical laboratory, the limits of that model must be made apparent from the viewpoint of modern practices. This means that it must be understood that the existence of "others" does not complicate our life but forces us to recognize that complication.

The figure of the modern technical laboratory had as its objective recognition of the difference between technical practices and scientific practices. But it leaves open the question of knowing whether all "technicians," even if we limit ourselves to those who claim to be modern, are equally at home in the modern laboratory.

Let's examine a body of techniques invented by a lineage of "constitutively modern" laboratories—metallurgy. This choice is partly based on the fact that few would question the high degree of success that has characterized this technical approach, one that has integrated a number of scientific ingredients but has also, on many occasions, paved the way for and guided scientific questions; and partly on the fact that metallurgy descends, quite recognizably, from techniques that incorporate a constitutive reference to supernature. Modern metallurgy is certainly modern, and not "modernist," in that the purification of supernature ingredients, although it has changed the nature of metallurgy, has not led it to define itself in contradistinction to its past, when the blacksmith was always something of a sorcerer. We could say that one technique has been substituted for another by reusing certain components, retranslated, of course, but recognizable in the new assemblage, and that the new technique does not exhibit in the form of symptoms the absence of components it has not reused. Modern metallurgy has no need to disqualify its ancestor, the blacksmith-sorcerer, and may even be passionately interested in a practice that is no longer its own. This is why it can retroactively be said that the art of metalworking has allowed itself to be "purified": it has not become an "impossible profession." The practitioner of metallurgy is a "natural" inhabitant of the modern technical laboratory, for the translation-betrayal of the conditions of his practice constitutes the very history of the invention of that practice, which is robust with respect to the redefinitions affecting it. The same description could also apply to chemical technicians, who have learned to incorporate in their creations questions as disparate as the accessibility of resources, patents, workplace safety, and, lately, the problem of pollution and the need to recycle.

My intention is not to review techniques but to create an appetite for the distinctions among them that need to be constructed. Philippe Pignarre has proposed a suggestive cartography of what we call a drug, that is, what binds together those who refer to it, from chemists and biologists in pharmaceutical laboratories to the doctor who prescribes it and the obedient patient who purchases and consumes it.[1] The typical case is already more complicated. Modern drugs can be presented as being "purified" of the irrationality that condemned the pseudo-drugs of the past, but the great singularity of present-day connections, their codified nature, the sophistication of the procedures of "socializing" what is born as a "chemical molecule" and becomes, rarely, "a drug," clearly illustrates that purification in this case no longer has the same meaning as it does in metallurgy or chemistry. It is not only a case of production, but one of precarious production, for it constantly alludes to what it must defend itself from and is forced to incorporate unsettling tests (clinical trials using a placebo) that it would willingly forgo (one day, rational pharmacology will be able to . . .). The technicians of "modern" drugs today act on behalf of rationality by disqualifying others types of drugs, those that cannot be separated from a practice of influence. In the Parliament of Things, these technicians will be asked to negotiate with the representatives of patients' associations, who will no longer be impressed by this mode of presentation. The issue here is not to oppose modern drugs, presented as the "only rational" drugs, to some form of "good medicine" to which we must return. What is in question is the future of such modern drugs, which face the challenge of political questions from which they are currently protected by their qualification as "rational." The modern technicians of drugs, including prescribing physicians, also have their place in the Parliament of

Things, but the utopia of this Parliament brings about the challenge of a modification of their "psychosocial" type.

But the question of drugs leads us indirectly to the question of Freud's so-called impossible "profession": psychoanalysis. As a technique of influence, psychoanalysis raises the question of its modernism. Psychoanalysts singularly lack a sense of humor about nonmodern "psychotherapies," and it appears that their training does not prevent them (which is the least that can be said) from presenting themselves in ways that equate with the condemnation of every other therapeutic practice. The master words that cannot be disentangled from a dream of eradication, the somewhat cynical tolerance toward predictable alloys, between "suggestion" and the gold of analysis, the analytic diagnosis of "resistance to analysis" all create a political problem. Whereas modern metallurgy can accept that the "purification" of its profession has in fact been a reinvention, whereas pharmaceutical technicians can claim that their definition of a drug, as partial as it may be, obligates them to reject the enormous majority of candidate molecules, psychoanalysts seem to require that we acknowledge their technique as "scientific," in the sense that it is said to satisfy what the human psyche requires and would therefore have a relation to truth as opposed to fiction.

Let us assume that the analytic scene, like the divinatory scene, be considered as an "influence laboratory" in the technical sense, that is, as foreign to the obligation that defines scientific laboratories as having to distinguish between the two meanings of an artifact. Now, the problem is to determine if this type of laboratory shouldn't be recognized as *positively nonmodern*. And in that case we could claim that, contrary to metallurgy and the art of manufacturing modern drugs, psychotherapy cannot be subject to the "political" requirement that defines a constitutively modern technique. Like ancestors or djinns, the unconscious would not be compatible with the processes of reinvention because reinvention does not only imply active translation but translation that betrays. All of them would refuse to let themselves be represented by "citizens." Psychoanalysts' lack of humor and the daunting nature of their claims would then imply *by default* that, in their case as well, the "conditions" of exchange have to be invented. How will the technicians of psychotherapy be represented in the Parliament of Things?

Once a possibility exists, aside from the "we musts," to treat "modern" and "nonmodern" symmetrically, the problem of the ecology of practices can finally become worthy of the awe-inspiring word that gives its name to this series: *cosmopolitics.* For the word signals the path along which the question is to be constructed, that of the (re)invention of politics, and the unknown that causes this path to bifurcate. Preserving the "political" character of the construction means that I do not have a choice, that we do not have a choice. Every proposition, no matter how utopian, if it is part of our tradition, draws from the inventive resources associated with that tradition. But those resources do not allow us to become "angels," authors of a utopia that is valid for every inhabitant on Earth. They do not give us the ability to meet and recognize those who should be the coauthors of such a utopia. The prefix "cosmo-" indicates the impossibility of appropriating or representing "what is human in man" and should not be confused with what we call the universal. The universal is a question within the tradition that has invented it as a requirement and also as a way of disqualifying those who do not refer to it. The cosmos has nothing to do with this universal or with the universe as an object of science. But neither should the "cosmo" of cosmopolitical be confused with a speculative definition of the cosmos, capable of establishing a "cosmopolitics." The prefix makes present, helps resonate, the unknown affecting *our* questions that our political tradition is at significant risk of disqualifying. I would say, then, that as an ingredient of the term "cosmopolitics," the cosmos corresponds to no condition, establishes no requirement. It creates the question of possible nonhierarchical modes of coexistence among the ensemble of inventions of nonequivalence, among the diverging values and obligations through which the entangled existences that compose it are affirmed.[2] Thus, it integrates, problematically, the question of an ecology of practices that would bring together our cities, where politics was invented, and those other places where the question of closure and transmission has invented other solutions for itself. Cosmopolitics is emphatically not "beyond politics," it designates our access to a question that politics cannot appropriate.

Cosmopolitics is, of course, a speculative concept, and its effects will first of all affect the way in which we understand ourselves and understand others in contrast to ourselves. To avoid identification with the uninteresting notion of a "return of the religious," in the sense criticized by Marcel Gauchet, it assumes that the dramatic opposition between the choice of the religious and the choice of the political, as presented by Gauchet, will be consigned to the order of justifications that are, as such, part of our history.[3] Returning to history choices that are presented as transcendent and mutually exclusive ("either submission to an order received in toto, determined before and outside our will; or responsibility for an order accepted as originating in the will of individuals who themselves are supposed to exist prior to the bond holding them together")[4] is obviously something I have been doing throughout this book. But this return to history now imposes a number of new ingredients. The case of psychoanalysis shows, from my point of view, that if we are to understand ourselves, we need "others." More specifically, we need to construct ways of escaping the power of slogans and master words that lead us to create dramatic, disjunctive disparities between "ourselves" and "others." For psychoanalysis, like those other impossible professions of governing, teaching, and transmitting, is not only a vector of the opposition between the so-called enchanted character of the "nonmodern" world and the equally so-called disenchantment of the modern world. It is also its victim, condemned to demand a status, that of a "modern" technique, that is unsuitable to it and that reduces it to a caricature of itself.[5]

It is to Tobie Nathan and Bruno Latour that I want to turn in seeking the instruments to complicate this opposition. With regard to the "nonmodern," both the distinction between logical closure and self-contained identity and the very practice of "ethnopsychiatric" consultation can be used to introduce a bit of humor into the grand theme of submission to an integrally accepted order, a humor that those affected apparently do not lack, for they mock the way we judge them.[6] Tobie Nathan's initial experience of what he refers to as logical closure did not occur with some "archaic" tribe but in 1950s Cairo. To take this experience into account, it is not enough simply to abandon, as Marcel Gauchet has, the representation of the poor, powerless savage who mistakenly believes he can subject natural forces to his magic. We must also accept the full weight of the *technical* terms used by Gauchet regarding what he calls religious logic, especially when he speaks of "the supposed conformity of collective experience to its ancestral law" (32) and the understanding of the actual in terms of "an established state of affairs it is supposed to legitimate" (14). "Supposed" is a profoundly ambiguous term, which certainly indicates a refusal to "psychologize," in the form of belief, what Gauchet attributes to logic, both the others' and our own, since we, as individuals, are "supposed to exist prior to" the bonds that hold us together (11). But this refusal is still a half measure. If we follow Nathan's proposition that this logic is itself the operant, both the position and the solution of a technical problem, the opposition between "religious" and "modern" loses much of its dramatic character. The adjectives "ancestral" and "established" no longer designate references that establish identity, but characterize logical operators that are integral to a practice, one through which closure or territory is defined and transmitted. What's more, these operators are fully capable of operating according to different modalities depending on the different practices associated within a culture. They bring about a logical and syntactic constraint on thought without expressing any sense of group adhesion. They have the same robust character as culture itself. In other words, they do not justify the fatal opposition between a religious coherence based on an intangible past and a political coherence that is entirely tied to the future, but problematize our disqualification of robust nonmodern techniques. To refer, as Gauchet does, to the grand theme of disenchantment in order to describe the "supposed" dramatic choice between two mutually exclusive logics is a way of "sugarcoating" a rather different kind of historical process of exclusion, one that includes prohibitions, persecutions, rules, and professional monopolies. In short, violent destruction rather than choice.

With regard to the "modern," Latour's Parliament of Things opposes what he calls the "modern Constitution," the very expression of the disenchantment of the world. In a way that is typically antifetishistic, a nonintentional nature is "constitutionally" opposed to a society that must "decide" without a transcendent model or norm. This Constitution, Latour shows, dooms us to let proliferate, without further consideration, the hybrids or factishes we produce and which produce us. It enabled David Cohen, for example, the French promoter of the need to map the human genome, to defer to "politics" decisions that would have to be made about the use of the data and techniques produced in the new labs: that use will be whatever "we" decide it should be. But this "we," purely human and apparently decisional, will intervene in a situation that will already be saturated with decisions made in the name of technique, science, and rationality. And this "we" will consist of new, or supposedly new, hybrids, because politicians will demand that experts tell them who "we" are from the scientific point of view, that is, what the genome is, what a genetic disease is, what a group at risk is, what they have to fear, and what they are duty bound to hope for.[7]

As for the Parliament of Things, it corresponds to a delocalization of politics. It can be characterized in terms that derive from Leibniz: "Not everything is political, but politics is everywhere." And wherever there is politics the challenge is to explicitly abandon the founding reference of our politics, which ties it to the "will of humans supposed to exist prior to the bond holding them together," and to loudly proclaim that what holds us together are bonds and hybrids that produce us as we produce them.

The cosmopolitical question is not about the "reenchantment of the world" but the coexistence of disparate technical practices corresponding to distinct forms of reciprocal capture, characterized by different logical constraints and different syntaxes. Is it enough, for the resolution of this question, to reduce disparate, distinct, and different to a minimum? This appears to be what Bruno Latour had in mind when he presented what at first seems to be a merely quantitative contrast between "short," nonmodern networks (in which hybrids belong to a group and do not try to recruit the maximum number of humans or nonhumans) and "long," modern networks (which can range from an experimental protocol resulting in a genetic map to the redefinition of what it means to be human, depending on whether or not a person belongs to a given risk group).[8]

Long, rhizomal networks are, according to Latour, what characterize our modernity.[9] And they impose upon those they connect the condition that characterizes both the modern technical laboratory and the Parliament of Things. Under this condition, whatever appears to be a "condition" must accept being a constraint whose meaning may well have been stabilized but can always be transformed. The continuous change of meaning is mandatory whenever the network, as it grows, connects, fabricates, and reinvents increasingly disparate elements of what we call nature and what we call society. The difference between long and short would then not be merely quantitative. It would be an active difference, the "rejection of the modern condition" then expressing the short network's active opposition to its extension. And this difference would then become expressive of the contrast not only between "us moderns" and nonmodern "others" but also between our different technical practices, for example, between the long networks in which "metallurgy" participates and those networks that need to remain short, which would include "psychoanalysis."[10] The distinction between long and short networks would then present a cosmopolitical problem of coexistence among practices that not only are different but very differently define their "conditions of exchange" with other practices.

Nonetheless, in *We Have Never Been Modern*, Bruno Latour argued for the symmetrical treatment of the long networks constructed by modern humans and the short networks of nonmoderns. And he was right to do so as symmetry implies the absence of "long-range causalities" such as modernity, objectivity, rationality, and so on, which are designed to identify what short networks "lack." "We should refer to them the way de Gaulle referred to stewardship: 'objectivity will follow.' Give me long networks, and I'll also give you all the objectivity you'll need. Conversely, objectivity is not some cold monster that would endanger all traditional cultures. Its presence alone cannot break the similarities among communities."[11] Latour's symmetry argument is made against the radical asymmetry of the great divide and its denial of any similarity. As such, it makes it possible to address the problem of representing "nonmodern" practices in the Latourian Parliament. But symmetry does not resolve this problem.

More specifically, we can state that the symmetrical treatment of networks is perfectly legitimate as a requirement of a scientific practice, that of the "symmetrical anthropology" that

Latour defends in *We Have Never Been Modern*. Like every other science, symmetrical anthropology states its requirements and brings its obligations into existence; and here symmetry holds the status of a requirement: "I require the ability to describe and compare without having to recognize any intrinsic obstacle to this comparison." But a requirement creates a problem and does not guarantee a response that will satisfy it. It constitutes the very risk of "symmetric anthropologies" and obligates them to closely evaluate what others present as an "intrinsic obstacle." Symmetric anthropology can be said to be a highly productive approach whenever it puts our own practices to the test of symmetry, because in doing so it openly clashes with our claims to be different, rational, objective. In doing so, symmetric anthropology even puts us on the path of a new type of protagonist that might be produced by a social science that would not be modernist, that is, that would only address those who are fully capable of putting at risk anyone who attempts to represent them. Symmetric anthropology tries, at its own risk, to represent "us," to represent our so-called modern practices, in a way that is distinct from the way that the practitioners it represents present themselves. Modern practitioners, represented in this way, retain their singularity but lose the power to disqualify, to present themselves in a way that is fatal to "others."

It is only normal that the concept of a Parliament of Things resulted from symmetric anthropology. Every new undertaking seeks and invents the place where it will be able to make a difference, a place that should take into account the requirements and obligations it brings into existence. But the cosmopolitical question insists upon the construction of words that slow the problem down, that prevent it from rushing toward a solution. What must be resisted is the perspective of "sorting" among practices that would have become comparable, a perspective that can be associated with the dangerous figure of a "gardener" carrying out comparisons and establishing sorting criteria.[12]

To the requirement of anthropological symmetry there corresponds the obligation to recognize that this requirement situates the anthropologist. The "we" who symmetrizes doesn't speak "for everyone," using words that also apply to us and to those we will encounter. They may *allude* to the necessity of "sorting," as an interesting variant of the concept of testing, if we refer to chemistry rather than gardening. For in chemistry, sorting is never a judgment. It is always relative to differential properties that make a difference in a particular process (heating, reacting, dissolving). But the allusion doesn't resolve the problem since it is silent about the eventual mode of presence in the Parliament of different types of protagonists, which conditions an effective exchange. Here, we return to the question "How do we live with the Yoruba?" not in the sense that the problem would be resolved but in the sense that it can begin to be explored. If sorting is not the business of science, but again relates to a technical problem that can only be introduced "with" the Yoruba, how can we construct this problem in such a way that it does not assume we are "angels," capable of understanding everything, sharing everything, and, therefore, of sorting everything?

We, who are not angels but think in political terms, must therefore create obstacles that prevent us from rushing toward others while requiring that they resemble what we might become, obstacles that prepare us to wonder about their conditions, the conditions they might establish for eventual exchange. This is *our* problem. Its construction in no way ensures that the "Yoruba" will meet with us (any more than the construction of an experimental device ensures that the being we wish to mobilize will agree to show up). Our words are relative to our practices and we now ask that they tell us which obligations will guide us where angels fear to tread.

24

Nomadic and Sedentary

To construct the words that bring into existence the obligations of what I refer to as "cosmopolitics," I will make use of a dangerous distinction; dangerous because it is loaded with existing meanings and susceptible to misunderstandings: the distinction between nomadic and sedentary. What is interesting about these misunderstandings is that they operate in both directions. We know that sedentary populations have turned their back on nomadic peoples. But, conversely, when the distinction becomes metaphoric, it is the sedentary populations that become the object of scorn, peoples who cling to existential, professional, or cultural territories and reject the challenges of modernity. The sedentary belongs to popular opinion, while the nomad is willing, in Popperian terms, to detach herself from her beliefs in order to challenge them and to follow a problem wherever it may lead.

I have made this choice because the ecological problematic in which I situate myself provides the example of an interesting use of this type of distinction-opposition. Models of population growth in a milieu with limited resources introduce two parameters on which the "success" of a population depends. The "r" parameter reflects the rate of reproduction of the population.

The "K" parameter expresses the "carrying capacity" of the milieu, that is, its ability to satisfy the needs of the population, which depends on the population's ability to exploit the resources in its milieu. We speak of an "r" strategy whenever a population reproduces itself rapidly; we speak of a "K" strategy whenever it consists of individuals capable of inventing or profiting from the opportunities offered by the milieu. Two things are interesting about this method of ecological description. First, the fact that the strategies are in tension with one another. For example, we cannot increase both the rate of reproduction and the extent of parental care that is required for children capable of learning, that is, of conferring a meaning upon the concept of opportunity.[1] Second, the fact that the difference between strategies has only comparative meaning, typically when two species interact. For instance, it appears that, in general, interaction among predators and prey, when it achieves a certain level of stability, implies that predators are "more K" and prey are "more r."

Using ecology to implement the distinction between nomadic and sedentary implies two constraints: first, avoid negative definitions for categories, the way opinion is defined negatively when compared to reason, and create two positive categories in tension with one another; second, define both categories relative to a well-defined interaction. In other words, it is not a question of identifying "nomadic" and "sedentary" individuals but of identifying them only in relation to a given interaction, of creating a contrast whose scope does not exceed that interaction. Those who appear as "sedentary" on the basis of a given interaction may well take risks in an elsewhere foreign to the "nomads" who judge them. Within such spaces, these nomadic judges may be seen as hopelessly sedentary, desperately tied to a territory that assigns limits and conditions to the risks they boast of.

Naturally, experimenters will claim to be nomadic. Are there any risks they aren't willing to take? For they are prepared to "delegate" invisible electrons, "activate" impalpable fields, and have adopted the oldest terms in our tradition—particle, element, atom—and mutated them. And what of theoreticians, who have denied the obvious fact of simultaneity at a distance and now speculate about multiverses connected by wormholes? Their theories are proof that they will not recoil before any risk, any obstacle associated with common sense. But if you were to suggest that they take seriously the question of irreversibility, they would be scandalized, or disdainful, as if you had committed a breach of etiquette. When relativist sociologists investigated the risks taken in laboratories and the offices of theoretical physicists, they produced an image of a sedentary population, apparently defined by beliefs and confidences that perhaps should be respected but certainly not taken seriously. As for those relativist sociologists, who define themselves as nomadic compared to the sedentary scientists they study but whose "beliefs" they do not share, they can be identified by their confidence in the determining character of social causality, by the categories of the "sociological" territory by which they judge the territorial and conditional nature of other scientific risks. I could continue with the series of disqualifications. The Popperian scientist laughs at the psychoanalyst who is not subject to the risk of refutation, while the psychoanalyst snickers at the limited nature of the risks taken by the physicist, who brackets her subjectivity, but eventually lands on the analyst's couch to acknowledge that, just like everyone else, she is subject to the most archaic fantasies.

The advantage of the contrast between nomadic and sedentary is that it reminds us that the opposition between modern and nonmodern practices is paralleled by a relationship of generalized conflict among so-called modern practices. And the only kind of community formed among modern practitioners when confronting nonmodern practitioners is based not on the risk of nomadization they would share but on a judgment they all accept. According to that judgment, during an encounter, the most modern and rational practitioner is identified as the one who can designate the other as being more sedentary than herself. To the extent that modern practitioners accept as sites of encounter only those that, in principle, eliminate any reference to invisible, nonhuman intentionality, this principle ensures the only unanimous judgment they are susceptible to, namely, that they agree to define nonmoderns as "territorialized by the invisible."

Correlatively, the contrast I have established allows us to avoid the opposition between modern and nonmodern. The cosmopolitical question does not begin at the borders of modernity but within the definition of politics we have invented, with the "nomadism" that is required by a political process that demands abstracting the territorial categories of participants. From the cosmopolitical viewpoint, such a situation presents a serious weakness. For it fails to take into account the danger that nomadism might become a norm. Practitioners who pride themselves on being nomads will feel perfectly at home in the Parliament of Things. Nothing in the Parliament of Things obligates them to pay attention to anything other than conflicts, rivalries, and negotiations through which others create new risks for them and constrain them to new formulations. This means that, although lucid in the sense in which politics or the modern technical laboratory requires, although capable of reinventing the meaning of the territory of their practice, the researchers who approach their practice in terms of the values of nomadism are in no way prepared to meet someone who refuses to cooperate, to play the game, to take an interest in the challenges associated with those values. Except by "tolerating" the delay that such a refusal sometimes imposes—curse!

"Your challenges don't concern me, nor do they interest me." In making such a statement, through its mocking indifference or its silence, practitioners who present themselves as sedentary are not present by default, in the sense that they would *not yet* be active, political representatives of their practice. They are present by way of affirmation and present the nomadic with a dilemma: either continue along their path, take advantage of the power relationship and lack of articulated, reasoned resistance; or stop and wonder what they are dealing with. This latter alternative illustrates the split between politics and cosmopolitics, between the affirmation of a universal principle of political capture and the equally universal question of the values of an ecology of practices as such.

When, in "The Science Wars," I introduced the challenge met by what I refer to as a speculative ecology of practices, I excluded the possibility that it might become a source of values everyone would be required to submit to, and in whose name everyone's place and relationships could be determined. I am not betraying that constraint here, for the question is not one of submission to "cosmopolitical values" but of temporary cessation, a recoiling before the temptation to shift from "he refuses to cooperate" to "he has nothing to show, nothing to say, no obligation to offer, no requirement to promote."

One well-known example is the mapping of the human genome. I use this example because of its obviously political dimension. The most glaring scandal is the void that appears where we should see a crowd of protagonists, representing the interests affected by ongoing technical innovation. What effects will the genetic tests being prepared have for employment and insurance? What will happen to those who belong to a "risk group"? What about parents who have been told that their child, once it matures to adulthood, will have a given probability of a given disease? How will the definition of diseases for which a test will become available be negotiated, what kinds of passions will be aroused, what apprehensive demands, and with what sort of economic or political pressure? All the while, the number of possible tests will calmly continue to grow, together with the number of statistical studies that establish new correlations that can be used to assign meaning to new tests. Today, silence and arrogant propaganda prevail, and along with them a reference to a "politics that will decide," although this would imply that propaganda has given way to collective thought in the Parliament of Things. The first consequence of this would be the identification of the "void" as the fundamental question. How can protagonists capable of complicating the problem be empowered? How can the presence of those who might share in the associated risks, choices, and decisions be ensured?

Yet, even if the political question was, utopically, resolved, if the Parliament of Things, delocalized across the community, transformed into a collective adventure what today is the material for professional and industrial strategies, there would still be silence to be heard, a different kind of silence. Who are we to evaluate what we are in the process of deciding among ourselves? Are we fully aware that we are about to connect, and thus transform through multiple and partly unpredictable acts of reciprocal capture, histories that, on Earth and until now, were distinct? I associate this silence with a response to what philosopher Alfred North Whitehead called Oliver Cromwell's "cry," which "echoes down the ages: 'My brethren, by the bowels of Christ I beseech you, bethink you that you may be mistaken.'"[2]

This silence does not imply a prohibition, a limit before which everyone must bow. In itself it expresses no power, provides no response. It only has meaning in giving the question its cosmopolitical dimension, creating the space in which the voice of those who are silent becomes present. And this silence might make other voices audible, for example, the voice of an African mother afflicted with AIDS, who refuses the therapeutically recommended abortion: "I have AIDS and I'm not dead. Therefore, I am protected and my child will be as well."[3] What we call "probability," or "risk group," has no meaning for her, and not because of ignorance. The risks this mother assumes are not defined by probabilities but by the nomadism characteristic of her culture, which implies other spaces in which our "scientific" definition of disease has no meaning or, more specifically, no genuinely interesting meaning.

The immediate consequence of the political, rather than the cosmopolitical, definition of a question such as the technical, political, industrial, or legal historicization of the human genome is that everyone is forced to argue on so-called common ground, which means that all the arguments employed will have to assume the mode of existence of negotiable constraints, while respecting political rules. And those who wish to express their conviction that the question is not "only political" will be tempted to employ all-purpose slogans, to have recourse to prohibitions they claim to be universal, such as the sanctity of life or the intangible character of the separations instituted by nature. It is at this point that the opposition I wish to avoid will arise, against which I introduced the distinction between nomadic and sedentary as relative to an issue. This opposition, one that is catastrophic, would be between the nomadic values of modernity and the sedentary values of tradition. For tradition, presented as purely sedentary, associated with nonnegotiable prohibitions, cannot be a source of problems, distinctions, intelligence. Where the hybrids in relation to which we assume the risk of transforming ourselves are negotiated, tradition will appear as an obstacle, one that has value only if a power relationship requires that "interlopers" must be tolerated, must even be listened to courteously.[4]

The cosmopolitical touchstone, therefore, is the possibility of a deterritorialization of the Parliament of Things, without which it risks becoming the territory of those who think of themselves as "nomads." It represents the possibility that those who present themselves, for a given problem, as sedentary and who refuse to play the political game, will be respected in the sense I have associated with William Blake. They are cursed who demand of the other that he "express himself like everyone else" and promote the constraints he wishes to see recognized, and who subsequently listen patiently or openly ridicule the arguments they have extorted from him. And the curse holds even if the power relationship requires that certain consequences of the argument be reinvented and included in the proposed settlement.

As I have indicated, the cosmopolitical question does not begin at the edges of modernity. It is everywhere, although not everything is cosmopolitical.[5] And, in particular, it is present at the very heart of modern nomadic practices. If this weren't the case, the distinction between nomadic and sedentary wouldn't belong to the ecology of practices. It could never be anything more than a restraint that moderns imposed upon themselves, a form of tact that would have to prevail in their dealings with nonmoderns. The cosmopolitical question and the distinction between the nomadic and sedentary dimensions of a practice, which allows me to put it to work, are present, but they are generally presented in the form of hostility and discomfort.

When sociologists of science decipher "beliefs" that are presumed and sustained by scientific practices, why should scientists feel they are being attacked? More specifically, why do they feel that they are not so much being attacked as misunderstood? It is because they know that the term "belief," when spoken by those who describe them, has an ironic ring to it, whereas for them, what is called belief is the actualized possible, or the possible in the process of actualization, that the existence of their laboratories celebrates and depends upon. Laboratory practice can be fully expressed only in the duality of a nomadic requirement and sedentary affirmation. But this affirmation has no public expression and appears only as the pretension to truth and objectivity. It is then reduced to the arrogant vocation of discovering, beyond illusion, a truth not susceptible to exchange, which everyone would have to celebrate in the same terms.[6] Kuhn's paradigm accurately expresses how the sedentary component of these practices becomes mute and dogmatic when they are experienced as cut off from the outside, without being associated with what Tobie Nathan would call practices of exchange, the capacity to encounter those who come from elsewhere in a civilized manner.

It is in order to allow for a more "civilized" encounter that Bruno Latour introduced the term "factish," intended to prevent the critical visitor from characterizing what laboratories bring into existence in terms of "belief." Reference to the factish should counteract polemical intent, which no academic courtesy in the world can dissimulate, and which leads ironic sociologists to visit laboratories. Of course, they travel there as nomads— constrained by other risks, animated by other requirements, they are only passing through. But why, as civilized visitors or as good (symmetric) anthropologists, don't they linger for a while around their hosts' fire, or coffeepot? Why don't they listen to the hopes and doubts, the dreams and fears, that are expressed in strange idioms that narrate both the fabrication and the vocational autonomy of factishes?[7] Why don't they also relate, in the strange idiom they speak among themselves for their own celebrations, the doubts and hopes, the dreams and fears they experienced, here or elsewhere, during their peregrinations? For such things must be expressed in the language of the practitioner who experiences them, whose obligations force her to experience them. The idiom and the factish affirm the territory. We can never fully understand another's dreams, hopes, doubts, and fears, in the sense that an exact translation could be provided, but we are still transformed as they pass into our experience. The experience is one of a deterritorialization that is ignored by the byways of criticism, a "transductive" experience without which all criticism is a judgment and a disqualification.[8]

The term "cosmopolitics" introduces what is neither an activity, nor a negotiation, nor a practice, but the mode in which the problematic copresence of practices may be actualized: the experience, always in the present, of the one into whom the other's dreams, doubts, hopes, and fears pass. It is a form of asymmetrical reciprocal capture that guarantees nothing, authorizes nothing, and cannot be stabilized by any constraint, but through which the two poles of the exchange undergo a transformation that cannot be appropriated by any objective definition.

◦As I noted in Book I, requirements and obligations belong to the regime of justification. It is on their terms that a practice can express its own singularity or that a constructivist approach can explore the singularity of individual practices, the way in which they are justified, the values they bring into existence, without using the customary words that assert rights, claims, hierarchical presentations and disqualifications. In this sense, requirements and obligations have been, for me, markers during an exploration in which the possibility arises of an ecology of practices that has not been reduced to a state of generalized conflict, to a reductionist pyramidal order, or instrumental apportionment. However, with the introduction of the cosmopolitical idea, the "justification" of practices must be transformed and may indeed be transformed when the sedentary components of another practice pass into the experience of a practitioner and resonate with her own sedentary components. This is an experience of deterritorialization, but it is also and inseparably an experience of "feeling" one's own territory, and as a result a practitioner, *modern or not,* may realize that she cannot present herself as a nomad, may discover how her practice situates her, resulting in a nonequivalence that she is meant to actualize, that makes her hope, doubt, dream, and fear.[9] This is not a matter of admitting that "she too" is attached, but rather a matter for celebration. She must then also recognize that, for her as well, there are "political limitations," issues that should not be challenged through the game of politics. She needn't apologize for this, but her first obligation is to recognize it.

According to the ecological terms I am proposing, the only

one who is dangerous, irremediably destructive or tolerant, is someone who believes himself to be "purely nomadic," because he can only define his practice in contrast to all the others and, regardless of his good intentions, can only define others in terms of tolerance. From the point of view of someone who feels he is purely nomadic, only tolerance can protect the sedentary from conquest, destruction, or the slavery to which confrontation with the nomad condemns them. This is why nomads' experience of the fact that they too have a territory is the very condition for an answer to the cosmopolitical question. The obligations to which the nomad subscribes are then no longer relative to the explicit characteristics of practice alone, but communicate hopes and doubts, dreams and fears, with the event that brings about the coming into existence of the territory in question, with a creation that affirms, through the obligations it entails, its autonomy from its creators. Together with that event there resonates an unknown for which every practice brings into existence a new dimension.

25

The Betrayal of the Diplomats

I want to return now to the question of closure as a *condition of exchange.* In what sense does the distinction between nomadic and sedentary components transform the problem of exchange? More specifically, does this transformation cause us to abandon the field I defined at the beginning of this series of essays, that of an ecology of practices, and the corresponding problem of "psychosocial" types likely to assert, rather than endure, that ecology? After all, hope and doubt, dream and fear, seem to refer to a "purely subjective," individual dimension of life, a dimension that the concept of a "type" should eliminate. When we speak of the "physicist" or "field biologist," haven't we decided to overlook what affects Max Planck, Albert Einstein, and Stephen Jay Gould as individuals? But, conversely, in treating dreams or fears, for example, as ingredients of a strictly "private" or "psychological" life, aren't we overlooking the fact that in other cultures dreams are messages that involve more than one individual and that fright is a crucial ingredient in relationships among distinct universes, the sources of constraints that confer upon doubts and hopes the matrices of their meanings? A division between affective "private" life and the "public" type is one of the dimensions of the "great divide" that the cosmopolitical question needs to problematize.

The trial that Tobie Nathan's co-therapists must undergo, the divination, prescription, and fabrication of active objects, the "authoritarian" reading of the other's dreams (not the "proposition" of a possible interpretation), is here exemplary. This trial does not express the necessity of identifying with the culture in which such gestures, such readings, are practiced. It is not an initiation. What is at stake is the experience of "passing through." What for the other is a condition must be experienced by the co-therapist in a way that does not transform it into a constraint whose meaning and scope he would be free to reinvent. This does not mean empathy, the sharing of the "same" experience. This does not refer to an experience of vertigo, or radical doubt, or the *mise en abyme* of territoriality. I am not claiming to deny that such experiences can occur, but I do wish to deny that they serve as a reference, title, privilege, or vocation. The vertigo of the blank page and the proximity of philosophy and madness are thematically somewhat too flattering, too devoid of humor not to make us smile, and the plunge into chaos always occurs with qualifications, without which neither the artist, nor the philosopher, nor the scientist will "bring back" anything.[1] What "passes through" during the trial is first and foremost the impossibility of reducing to belief what is a condition for the other, and the impossibility of adopting a position of tolerance toward the other's inability, in this case, to make a distinction between condition and constraint.

For Nathan, as we have seen, this trial must be required of every practitioner who is likely to participate in the creation of a "scientific psychopathology." Therefore, it belongs to the "psychosocial type" of researcher capable of escaping the modern passion for disqualifying any practice that fails to accept the existence of a unique world. But from the cosmopolitical perspective, it is reasonable to ask the extent to which the question of "passage" does not arise everywhere, the extent to which the very condition of exchange, including exchange among "political representatives," does not have as its condition some form of "passage."

In fact, we are already familiar with a "psychosocial type" whose practice involves exchange and who incorporates the tension between territoriality and deterritorialization I associated with the experience of "passage." This is the "diplomat." What is difficult and interesting about the practice of diplomats is that it frequently exposes them to the accusation of betrayal.[2] The suspicion of those whom the diplomat represents is one of the risks and constraints of the profession and constitutes its true grandeur. For what is demanded of the diplomat is characterized by an irreducible tension. On the one hand, diplomats are supposed to belong to the people, to the group, to the country they represent; they are supposed to share their hopes and doubts, their fears and dreams. But a diplomat also interacts with other diplomats and must be a reliable partner for them, accepting as they do the rules of the diplomatic game. Therefore, the diplomat cannot be one with those she represents. If she were, diplomatic activity would be devoid of meaning or reduced to that other strategy invented by humans to avoid the risk of all-out warfare, hand-to-hand combat between two "heroes" whose destiny everyone agrees to abide by. But cynicism, indifference, or treason, even if such accusations are frequent, do not constitute the truth of the diplomat's practice but its negation. They point to the risks inherent in any practice, the risk of negating the obligations that bring it into existence. For the experimental researcher, this would be an indifference to the distinction between a reliable witness and an artifact, for the field researcher this would be the search for causes having the power to cause, thus turning her field into a scene of proof that would be valid for other fields. As for diplomats. they must translate, and thus betray, what for them are also and at the same time "conditions," which marks them as sedentary. Afraid they will betray but translating nonetheless.

Perhaps the diplomat is also someone who can help us formulate what the question of science entails whenever it is a question of the "social sciences," fields populated by techniques that belong to Freud's three impossible professions, techniques for which the political requirements defining a "modern" technique are inappropriate. In *The Invention of Modern Science*, I had suggested that we consider the practitioner of theoretical-experimental science as a unique hybrid of judge and poet, whereas the art, and the risks, associated with the field scientist lead her to resemble a detective or a "sleuth" hot on the trail. But at the time, I had found no analogue for what I had referred to as sciences "of a third kind," because they are addressed to beings who, by definition, are interested or could be interested, or are capable of being interested, in what is required of them, in the way they are addressed. It now seems to me that the diplomat is such an analogue. Like the diplomat, the practitioner of a science for which "the conditions of the production of knowledge for one are, inevitably, also the conditions of the production of existence for the other" should situate herself at the intersection of two regimes of obligation: the obligation to acknowledge that the dreams of those she studies, their fears, their doubts, and their hopes, pass through her, and the obligation to "report" what she has learned from them to others, to transform it into an ingredient in the construction of knowledge.[3] Thus, as a therapist, that is, *a technician of influence,* Tobie Nathan must connect with the conditions of his clients' culture, learn from them how to help them. But as a "scientist," he must "bring into existence" what he has learned, construct the means for resisting universalist fables and for interesting people like his colleagues, or myself. He must learn how to transmit and, in so doing, risks betraying. Like the diplomat afraid of betraying, he must continue to translate.

If the analogy holds and if this dual dimension—passive when it is a question of letting themselves be marked by this "passage" and active when it is a question of "reporting" and, therefore, possibly betraying—characterizes the psychosocial type of practitioner whose first obligation is not to rely on the indifference of those she describes, certain consequences might follow. Those consequences specify the scope, the meaning, and the risks of those scientific practices, and may help explain why the "social" sciences have so many difficulties in inventing practices that correspond to what I will from now on refer to as the *sciences of contemporaneity*, characterized by the fact that those who produce knowledge and those about whom knowledge is produced share the same temporality.

The first consequence of my analogy is that for such sciences the problem is not individuals but practices in the broad sense. I had already reached this conclusion when I suggested that we consider that, to the extent that a scientific statement has to be able to be put at risk by what it addresses, the so-called social sciences cannot hope to produce reliable knowledge unless they stop addressing "humans" and begin to address beings they know are capable of having a position about the relevance of the questions addressed to them. The figure of the diplomat also includes this constraint. Diplomacy exists only among "powers." The diplomat knows that she is obligated to report to those who have commissioned her, and that what she has to report, a proposed agreement, may be associated with a form of betrayal. But, except in the case of unconditional surrender (for which she has no use), she also knows that those she reports to have the power to reject her proposals. The inherent grandeur of her practice is tied to this risk. And it is this risk that the anthropologist of the sciences would avoid in visiting a "mad scientist," or one of those unfortunate and illuminated autodidacts who construct for themselves a new theory of the unification of forces, or cook up a splendid quantum explanation for telepathy. To the extent that she works with impunity, because what she has chosen as an object is socially defined by ridicule or inadequacy, what the anthropologist "reports" will be an insult, the pure and simple exploitation of a position of weakness. Similarly, when compared to the grandeur of one who accepts the risk of betrayal, the social psychologists who screen the opinions of individuals about issues that are practically foreign to them are "petty." They exploit those individuals' goodwill or their willingness not to disappoint a "scientist." The verification of the pettiness of this exploitation is the fact that the categorization process loses its relevance as soon as an individual is questioned whose practice in fact incorporates the issue for which the process gathers "opinions." The categorization process expresses a power relationship, the presumed difference between the person who developed it and those who are subjected to it.[4]

But there is a second, less obvious, consequence. The diplomat returns to those who sent her with a proposal that, for them, entails risk, but she herself cannot participate in the evaluation of that risk even though it may result in the conclusion that she has betrayed her mandate. Similarly, the production of knowledge that takes a practice and its practitioners as its object relates to their sedentary component but cannot authorize the least conclusion about the nomadic risks these practitioners are liable to accept.

To make this clearer, let's return to the case of ethnopsychiatry. What ethnopsychiatrists, as therapists, experience is indeed the sedentary component of the other's life, the experienced significance of active objects, and the multiplicity of universes, that is, the effect of logical constraints whose ingredients are active objects and multiple universes. But they experience this in their own way, not by sharing in the other's fears and obligations. That is why, unless an ethnopsychiatrist has been initiated and recognized by other initiated therapists, who would then be his "peers," this experience in no way authorizes him to participate in the nomadic risks of these therapists when they test the ingredients of their practice. Similarly, the anthropologist visiting a laboratory will experience the sedentary conditions of its inhabitants, the way they manipulate their devices or construct their interpretations—this way and not otherwise—the way they hesitate, grow excited, or indignant in the face of a proposal, shrug their shoulders at a suggestion, a question, or an objection that expresses the incompetence of their interlocutor, or suddenly change their style and mumble, in some impoverished epistemological jargon, some all-purpose rationalization. But he knows, or should know, that unless he is recognized as a colleague, his visit does not allow him to formulate a judgment about the controversies, innovations, or negotiations the laboratory participates in, it does not allow him to share in the risks taken there even if those risks follow from conditions he has experienced.[5]

In place of the obsession of the power relationship established between interrogator and interrogated, there would be an asymmetry. What interests the anthropologist of science is not what interests the scientists she visits. What interests her is what for scientists often goes without saying, what is transmitted implicitly and is, nonetheless, an essential element of these scientists' practices. For it results in the difference between someone who is self-taught and scientists capable of recognizing the risks they can take and what would disqualify them were they to propose it. Which brings us back to the requirement I formulated earlier. The researcher who extends his practice of translation/representation to whomever he chooses would commit extortion, an abuse of power in the name of science. Only practitioners, never individuals as such, can authorize such a practice because only in their case is the sedentary dimension of experience already the object of a practice of transmission. Only they can forcefully object to the way they are represented, not feel humiliated, fascinated, dispossessed; only they can avoid identifying with what is reported about them, avoid attributing to the one who interrogates them the position of "subject supposed to know." They may accept it with a smile or take an interest in the interest of which they are the object. They may even learn from their guest something of the uniqueness of their own practice, just how different it is from the general models of rationality or justification they thought they had to use. But the eventual consequences of this are not the business of the guest.

There is, therefore, nothing neutral about the analogy between the sciences of contemporaneity and diplomatic practices. To take but one example, it allows us to conclude that there will be no "pedagogy" in any scientific sense as long as teachers do not have the means to define themselves as collectives of practitioners, and as long as the educational specialist feels free to tell individuals "who happen to teach" how to go about their business, confirming thereby that those he addresses are "only individuals," have not produced, together with others and within actual groups, the means to evaluate the relevance of what is put before them. More generally, the temptation and the possibility of having someone submit to the requirements of the laboratory can now be understood as a trap par excellence. For, in this case, the power of the laboratory defines the fact that the interrogated being can be a practitioner in the sense of an obstacle to be eliminated or avoided. To subject Albert Einstein, Bruno Latour, Tobie Nathan, Jacques Lacan, or Martin Heidegger—assuming they agree—to a reliable laboratory protocol means you will be engaging only with purified human artifacts, characterized by mere quantitative differences.

In contrast, the operation of representation the researcher-diplomat must risk should never authorize the researcher to recognize what can be generalized beyond any differences. That is why certain basic concepts of our "social sciences," which are supposed to ensure their autonomy, such as "motivation," "interest," "suggestion," "symbol," "cognitive processing," and "unconscious," change their meaning. They do not lose their value, far from it, but their value has nothing to do with the risks of producing scientific knowledge. For the researcher who honors them is thereby made to require of those he encounters that they manifest their submission to science, and that they confirm the autonomy of those categories. Motivation, suggestion, value, or the unconscious are ingredients of a *practice of influence* (marketing, teaching, human resource management, psychoanalysis, publicity, etc.).[6] As categories that have scientific allure, such concepts address the "making" of the modern individual, one who, for example, would be prepared to assume that her behavior is subject to what others may identify in terms of scientific categories. Addressing someone in terms of their "motivation" or "lack of motivation" could then be compared to activating a logical constraint of which motivation is an ingredient.

What is at stake in the practice of the diplomat are the effective conditions of an "encounter," not the recognition of submission. That practice is always situated "between" two groups. Following this analogy would extend to all the sciences of contemporaneity Tobie Nathan's argument concerning what a scientific psychopathology should be, a science that focuses on psychotherapists rather than patients. As we have seen, such a science raises the question of a "group-to-group" relationship; it implies collaboration with "nonmodern" practitioners whom the practice of collaboration must actively acknowledge as belonging to an actual group. Therefore, it is not an individual that must be addressed but a practitioner qualified to require that the conditions of her practice are recognized. However, the proposed extension creates the possibility of hesitation. Is it still a question of "science" or a question of "politics" in the broad sense, that is, the invention of means for "living together" within the same community?

Are the sciences of contemporaneity, therefore, quite

simply "political"? If so, my attempt at characterization will have failed. The fact that this characterization has direct political consequences—the connection between the production of reliable knowledge and the existence of actual groups having the means to construct and claim a practice that cannot be reduced to opinion—would merely be another way of saying that there is no proper "place" for these sciences, that they have been entirely dissolved in the general problem of negotiation among practices. I will have managed to turn my diplomat into the translation of my own interests, a practitioner of the ecology of practices I have been trying to construct.

The term "represent" always has a dual meaning, related either to the scientific or the political register. The laboratory scientist must be able to construct the representation of what her factish "says" in order to be able to represent the constraints and possibles resulting from that saying elsewhere, before others. Leaving the laboratory, more specifically, the network of laboratories where "competent colleagues" have tested and acknowledged this representation, does not put into question this representation; the only question is the meaning that will be assigned to it. The scientist who has left the laboratory is "autonomous" in the sense that, wherever she goes, her factish has made her autonomous, it has satisfied the requirements that allow her to represent it before others. In contrast, we must recognize that the translation produced by the diplomat cannot allow a meaning to be assigned to the term "representation" independently of the problem presented or the vocation assigned to the "group-to-group" exchange that is risked.

To resolve this problem it is important to distinguish two "types" of practice that follow from the diplomatic analogy. I refer to these as the "diplomat-researcher" and the "diplomat-technician," both of which are distinct from the expert (for example, a scientist who leaves the laboratory) or the "nonmodern practitioner" (who would agree, *as a practitioner authorized by a group,* to collaborate in the construction of a "finally scientific psychopathology"). The expert is defined as representative of what allows her group to exist and what her group brings into existence, but she also defines that group as actively interested in the construction of an issue, a device, an innovation that would require its representation. The expert is, in any event, considered "autonomous," because the vocation assigned to the exchange in which she participates implies that the "sedentary" conditions of her group are not questioned. Rather, they are recognized not as conditions of exchange but as conditions for the contribution of the expert to this exchange. But we can also say, whenever a diplomat-researcher reports to her colleagues what she has learned from a visit, that she too is considered "autonomous," although for very different reasons. The relevance of her translation, the fact that it has been accepted as neither ridiculous nor insulting for those who have made it possible by their hospitality, is crucial. But its aim, a contribution to collective scientific knowledge, is not the business of her hosts, just as their own risks are not the business of the diplomat-researcher. Consequently, the group-to-group relationship is "located," restricted to a successful exchange. The problem of the "diplomat-technician" is quite different, however, for, like any technician, she must produce something "new"—in her case, the possibility of "group-to-group agreement." She is "delegated" by those she represents to promote their interests and their conditions in negotiations in which those interests and conditions *can be challenged.* In this case, the touchstone of the enterprise is no longer the construction of knowledge that might incorporate what she brings back. Here, the problem is indeed (cosmo)political and the touchstone of the diplomat-technician's enterprise is situated, at the time of her return, with those she represents.

The important point here is, as usual, the absence of confusion. When he is "scientific," Tobie Nathan, speaking of culture, speaks about it from the point of view of therapeutic techniques he has experienced and which he represents with the autonomy that his experience as a therapist has given him. And he can then say that those techniques, and the thought they constrain and generate, presuppose a cultural identity that presents itself as closed and stable, from generation to generation. But whenever he "acts as a mediator," which is to say addresses those whose task, in one way or another, for better or worse, is the definition and implementation of rights, duties, requirements, and obligations that will affect the modalities of life for populations that originate outside France, he becomes a technician in the sense that the common measure of technique is a creation that nothing authorizes. And in this case, cultural closure should certainly not be denied, but it becomes an ingredient of a very different problem related to the very existence of the group. The problem associated with closure then becomes one of the risks and becomings such closure will make possible for those who confront the question of what is proposed to them.[7] As I have pointed out, those risks do not directly concern diplomat-researchers, in the sense that they do not address the nomadic component of a practice. But for the practice of the diplomat-technician they are a *touchstone:* the touchstone that makes the difference between successful translation, which invents a modality of peace, and the proposal that will be rejected and will turn the diplomat into a traitor.

26

The Diplomats' Peace

The question associated with the diplomat's return—What is a culture capable of? What does its closure imply?—has neither a representative nor a theory. It finds an answer, on a case-by-case basis, only in the event of this return: hopes and doubts, dreams and fears. We might well ask, then, What commitment is needed to make the risks taken by the diplomat-technician, or the true diplomat, conceivable? And what autonomy does the diplomat have when she must report to those who delegated her and who await her return? The problem is important because it involves the construction of a "psychosocial type" and not the heroic or ascetic image of someone who might be engaged in critical lucidity, forced to renounce, unlike all the others, "being situated," "having their own site," bringing their own requirements into existence.

The commitment of the diplomat-technician associates her with what I called the cosmopolitical question. Her autonomy binds her to the unknown of this question. Not to the unknown in the sense that it would transcend all situations, but in the sense that it can find a response only in terms of a given type of situation, according to the terms of a given encounter. It is an answer to the question of knowing whether a form of commerce can be invented in this case and on these terms.

That is why it is important to emphasize that the invention of diplomacy precedes that of politics in the sense we have given to this term since the ancient Greeks. The diplomat was invented so that peoples, nations, groups could, if necessary, success- fully coexist without the destruction or the enslavement of one by the other being the sole outcome of their relationships. This implies that the peoples, nations, or groups in question accept not the necessity of peace but at least *its possibility, its eventuality.* This corresponds to the constraint I associated with the ecol- ogy of practices. The diplomats' peace is not a norm with which everyone must comply, which everyone must recognize as tran- scending their own interests and values. It is not the negation of war but a constraint, an ingredient of a belligerent regime, which distinguishes it from other belligerent regimes. After all, while peace constitutes a relatively hollow goal when it is associated with slogans of goodwill or tolerance, the multiplic- ity of belligerent regimes is filled with interest. From total war to experimental controversy, "war" can refer to a multitude of polemical situations, a multitude that is, in itself, material for invention and experimentation. The "diplomats' peace," there- fore, is another name for a belligerent regime that is singular- ized by peace as a possibility. The diplomat's commitment, the requirements her practice assumes, the obligations that put her at risk, make her the representative not of a general and hollow ideal of universal peace, but of possible peace, always local, pre- carious, and matter for invention.

If the diplomat-researcher and the diplomat-technician must be distinguished in terms of the risks that commit them, one characteristic reunites them because it conditions their activity: they both depend on the way in which those they rep- resent define their environment. For in both cases, representa- tion means, first and foremost, that those one represents agree to be represented, that is to say, agree to assume the risk of translation-betrayal.

Fear is one of the components of what "passes" during an

exchange, when the nomad experiences what sedentary experience entails, what it brings into existence, the unknown it
causes to resonate. But this operation of *passage* assumes hospitality. Hospitality is always more or less conditional. Seen as
a condition of possibility for the construction of knowledge, it
implies that this practice of construction is possible only if the
"diplomat-researcher" is welcomed as such, only if she can
be accepted without being treated as a hostage, recruited for
some bellicose purpose, required to identify with the cause that
mobilizes a group, or to join in the vindication of an identity.
The failure here may express the incompetence of the visitor,
her anxiety, her intolerance of fear, the nonactivation of her own
sedentary component, her own practical commitment. It may
also express the fact that the moment is not opportune, which
should not be surprising and is part of the landscape of risk associated with any science. The possibility of a scientific practice is
never a right. It always depends on requirements that must be
satisfied but whose satisfaction the scientist cannot impose. For
sedentary experience to be "knowable," the exchange must be
accepted. The hosts may instill "fear" in their visitors but they
must not "want to frighten" them, they must not present themselves as "frightening."

Similarly, if the diplomats' peace is to be possible, the diplomat must accept the fear associated with the risk of betrayal
implied by her mandate, but she must not be frightened by those
who have commissioned her.[1] The "diplomat's return" must be
a challenge, but this challenge must be shared, although asymmetrically, by those who await this return. The unknown who
engages the diplomat, the question concerning possible commerce, must be, in another form, accepted by those who have
sent her. She must understand that the proposals she brings
back will not be evaluated in terms of their static fidelity to the
explicit definition the group assigns to its requirements.

Technician or researcher, the diplomat-practitioner
depends upon the conditions of exchange with those she intends

to represent, either diplomatically or scientifically. But in this capacity she has her own hopes and doubts, dreams and fears. And her existence and the question that existence embodies— "What does she want of us?"—bring about, for those she meets, the risk of new conditions of exchange, or, more specifically, new modes of actualizing their own conditions. While the risks cannot be shared, it is crucial that the knowledge of the risk both must experience, each in their own way but through the *presence* of the other, is recognized and accepted. We could say that this is the principle of an asymmetric becoming, a becoming that could be said to be transductive, and it is this principle that can expose the belligerent ecology of practices to a regime that makes peace possible.

Here, it may be appropriate to introduce the question of "sorting" that Bruno Latour once associated with symmetric anthropology. The unanswered question involved determining how the sorting was done, according to what criteria and what methods of encounter. How can we "sort" without reference to a transcendent agent of arbitration before which all must bow?

In this case, the sorting might affect the way a practice itself, *faced with a particular problem,* defines its relationship to its environment. Is it a question of delegating experts? Do we need diplomats? Is it war? Sorting doesn't affect the solution to the problem but its construction, for the very fact of accepting the need to answer such questions brings the cosmopolitical question into existence. It implies that the group is capable of recognizing its sedentary dimension, that is, capable of recognizing that "not everything is of equal value" for it and experiencing this difference not as a right whose legitimacy must be recognized by everyone but as a creation whose possible destruction can only be expressed in terms of distress or fright.[2] And it also implies that a distinction between the expression, nonnegotiable as such, of this difference and the static identification with a particular formulation of this difference be conceived as possible.

As I noted previously, only those are incontrovertibly destructive or tolerant who believe they are "purely nomadic," who are not subject to distress or fright, and the group that is so identified can only delegate experts. It is the quintessential characteristic of practices I have characterized as "modernist" to be key suppliers of such experts. Again, with respect to psychoanalysis, it is in these terms that I framed the problem. The psychoanalyst, when she lives as a "modern practitioner," also lives as a "nomad," detached from the illusory connections that attach others. The analyst can be considered to be "at home" anywhere, for her practice defines all forms of "territoriality" or sedentariness as susceptible to "analysis."[3] The unconscious brought into existence by the analytic scene may authorize certain practitioners to play the role of experts in the community but it will never let them hesitate about the need for diplomacy. The same is true for the human genome, when it gives Daniel Cohen the power of a nomad who knows, wherever he may be, he has the right to promote the relevance of his practice and delegate to "political" decisions the responsibility to limit this universality on behalf of "human values that must be respected."

Sorting is an operation of immanence. "Do we need diplomats?" is a question that can be asked only by practices and groups able to delegate experts who are not defined as purely nomadic, who know that their risks are dependent on particular values and commitments and are not guaranteed by a right that only circumstantial obstacles might limit. "Can we risk the possibility of peace that the use of diplomats assumes?" is an entirely different question. And this question, always relative to a specific problem, brings about the cosmopolitical unknown expressed by the eventuality and risk of a possible commerce.

Sorting is not a judgment, then, but a test whose meaning refers to the *speculative* question of the ecology of practices. From the matter-of-fact point of view of a "neutral" ecological description, "modernist" practices are an enormous success because they are creative of all-purpose slogans, authorizing a crowd of stakeholders guaranteed to possess the ability to distinguish, without fear or doubt, without contact, what justifies such sorting and what serves as an obstacle, a resistance to be eliminated or an illusion that must be tolerated. Similarly, the choice of war, in the sense that it excludes the possibility of peace, cannot be denounced on moral grounds, and certainly not by those who intend to be "modern" and never send diplomats but experts, agents of modernity. Sorting is a test that does not have the power to judge but on which the possibility of ecological creation is dependent.

My field, says the anthropologist at the moment she presents it in terms of the knowledge she has gained. My country, says the diplomat even when she has been disavowed by it. The possessive indicates something quite different than a right of ownership or knowledge that would have been confirmed. It expresses the creation of a relationship of belonging that neither the experimenter nor the field scientist nor the modeler have to know. A Galileo who referred to "my inclined plane" would be ridiculous, for the very ambition that brings its factish into existence is valid for any situation where a falling object is involved. An entomologist who said "my ants" would be indulging in a somewhat indefensible sentimentality, for the ants are observed as bearing witness for their species; whether they are here or someplace else is indifferent. On the other hand, there is nothing ridiculous about Shirley Strum referring to "her" baboons, for she has made their study a science of contemporaneity, and even played the role of a diplomat when it became a question of saving "her" group from destruction.[4] But perhaps it was Leibniz, the philosopher diplomat so often reviled by those he represented, who accepted the risks imposed by the possibility of saying "my body" even though he denied the concept of a body causally subject to the soul that "possesses" it. This paradoxical sense of belonging expresses the highly speculative *vinculum* between the body as an interacting crowd and the soul that can be said to be "dominant" only because there passes into it something like a collective echo of the crowd that transforms it into an affirmation—"I have a body."

Two floors are always needed, two floors that are inseparable but actually distinct because of the asymmetry of their relations. The Leibnizian *vinculum*, Deleuze writes, connects two expressions of the world, the soul that actualizes this world and the body that realizes it. "The upper floor is folded over the lower floor. One is not acting upon the other, but one belongs to the other, in a sense of a double belonging. The soul is the principle of life through its presence and not through its action. Force is presence and not action. . . . But the belonging makes us enter into a strangely intermediate, or rather, original, zone, in which every body acquires individuality of a possessive insofar as it belongs to a private soul, and souls accede to a public status; that is, they are taken in a crowd or in a heap, inasmuch as they belong to a collective body. Is it not in this zone, in this depth or this material fabric between the two levels, that the upper is folded over the lower, such that we can no longer tell where one ends and the other begins, or where the sensible ends and the intelligible begins? Many different answers can be made to the question, *Where is the fold folding?*"[5]

The very principle of the cosmopolitical Parliament is to determine, for every question, at each conjuncture, where the fold is folding. How does it, on the one hand, distribute the crowd of experts—who accept the risks by which their practice will find new fields of realization—and, on the other, the force of the cosmopolitical question as actualized by the presence of diplomats?[6]

27

Calculemus

"What would a man be without an elephant, or a plant, or a lion, without grain, without the ocean, without the ozone or plankton, a man alone, much more alone than Crusoe on his island? Less than a man. Certainly not a man. The ecological regime nowhere states that it is necessary to go past humans to nature. . . . The ecological regime simply says that we do not know what man's common humanity consists of and that possibly, yes, without Amboseli's elephants, without the meandering waters of the Drôme, the bears of the Pyrénées, the wood pigeons of the Lot, or the groundwater in the Beauce, he would not be human."[1] "We don't know." The expression marks the grandeur of the seventh regime, that of ecology, which Latour suggests adding to the other six regimes (inspiration, domestic, opinion, civic, commercial, industrial) previously characterized by Luc Boltanski and Laurent Thévenot. Each regime has its own definition of grandeur and pettiness, its own justification. Ecological grandeur is associated with the suspension of certainty in which Cromwell's cry can again be heard: "bethink you that you may be mistaken."[2] His cry does not demand proof that we are not mistaken. It begs for hesitation, for fear to enter into the logical enclosures in terms of which all practices define their ends and means, and what remains outside the scope of their concerns. It does not object but seeks to induce in the territory associated with such definition the experience of a deterritorialization that makes the hopes and doubts, dreams and fears, of others present.

In Latour's ecological regime, the scientist is "petty" who leaves her laboratory without fear to assume the role of an expert who does not care about what the purification required by her practice has led her to ignore. "Petty" as well is someone who demands that those who are an obstacle transform themselves into "experts," accept the political obligation of participating in the negotiation. Latour's ecological regime, therefore, sorts not so much practices as the ways in which those practices are liable to present themselves, to be present for others.

What becomes of the Parliament of Things if the grandeur it celebrates and which justifies it is not the political grandeur of constructing ever longer networks, but the ecological and cosmopolitical grandeur that subjects the ever-renewed relationships between ends and means those networks invent to the test of the "dreams, fears, doubts, and hopes" of others—those whom the network being constructed does not interest, but who know that, if the Drôme stops meandering, if we banish the bears, if we ban the veil from our schools, or even if we expect that the desire to conform will cause the veil to disappear "spontaneously," something of the "common humanity of man" risks being destroyed?

Here, it is a question of cosmo*politics,* not the art of healing, of which dreams, fright, and transgression are an integral part, or the art of transmitting and teaching, where it is a question of creating risks, requirements, and the obligations of a practice as such. In one way or another, the question of representation is central and can only be asked if those who come together accept the risks and challenges to which that coming together obligates them. In the political version of the Parliament of Things, it is through what I have called the nomadic component of their practices that practitioners were required to participate, for the political challenge is addressed to this component. But in its cosmopolitical version, the need to suspend the rush to reduce an issue to the political register alone creates the necessity for the presence of the sedentary as such. The question, therefore, is that of the presence of what is, by nature, absent from political representation, of the mode of presence in the Parliament of those whose "sedentary" conditions might be endangered by the imperative principle of the common formulation that is to be invented.[3]

In Book I, I accepted the constraint of sticking to the question of "psychosocial types." That is, I speculated on what practitioners situated by their territory might be capable of. When it is a question of "politics," even cosmopolitics, this constraint is crucial if we are to avoid the trivial dream of an angelic future: souls, now without bodies, would assume a relationship of perpetual peace. In adopting this constraint, I have held to a determinate aspect of what we call thinking, thinking as both situated and empowered by stabilized, practical, collective modes of existence.[4] And, as is always the case when politics is involved, I arrived at the question of the separation of power and the nonconfusion of roles. The cosmopolitical Parliament must respect, and even actively bring about, *relative to each issue it addresses,* the distinction between experts and diplomats.

Just as the distinction between nomadic components and sedentary components only has meaning relative to an encounter, the distinction between experts and diplomats is always relative to an issue. It does not refer to a stable difference defining groups as such, and it establishes no procedure for decision making. It has to do with the position of a problem, not its solution. Does this issue actively interest a group, in the sense that it will contribute constraints that should be taken into account (yes, but . . .) while accepting as legitimate the common account that is to be invented? In this case, experts would be delegated, authorized to take risks for the sake of the eventual invention. Does the solution to a problem, or even the fact that it is raised, risk affecting a group with respect to what conditions the values and logics of its mode of existence? In this case, its diplomatic representation is necessary, and diplomats, far from being authorized to contribute constraints, will have to introduce conditions and requisites, and will have to inform those they represent about an eventual "compromise" they may or may not accept.

Not only is the distinction between experts and diplomats relative to an issue, it is relative to the way in which that issue is formulated, the way in which the problem is defined, that is, above all, to how its different components are evaluated. The cosmopolitical definition of a problem, therefore, is always situated in time, relative to a present that new diplomats or new experts will bring to pass.

For instance, formerly, when the question of illegal drug use was discussed, experts felt free to represent "victims," whether of drugs, the society that leads to drug use, or the suffering of the subject for whom drug use was reputed to be merely a symptom. Today, those who have asserted their political existence as "unrepentant drug users" have forced experts to redefine the scope of their expertise. Political parliaments remain ignorant of this but the cosmopolitical Parliament already exists where the problem of knowing how to live with drugs invents new questions brought about by the presence of both drug-using experts and drug-using diplomats. It also exists because the necessity for still more diplomats is being revealed, diplomats who would bring into existence the mute voice of people for whom the drug problem was never primarily political, or subjective, or medical, or scientific. And with them the dreams and fears, doubts and hopes, pass, to create the experience of a deterritorialization of our own categories, in terms of which the private right to do-whatever-doesn't-harm-anyone-else struggles with the state's right to enjoin citizens from any conduct identified with the destruction of the social bond citizenship assumes.

During the last few years, a number of pseudophilosophical sophistries have shown, to their authors' great satisfaction, that an animal cannot be a "legal subject" because it is not capable of the reciprocity that one should be able to expect from a legal alter ego. During that same period, a number of elements were gradually put in place that could be used to begin to present the cosmopolitical problem of what we inflict on animals. The industrial use of animals in the cosmetics industries, as well as industrial livestock production, are scandalous, and the experts in this case have to communicate to those who delegate them, to those whom the experience of this scandal places on a war footing, the announcement of partial victories and information about future strategies. But when animal experimentation introduces the problem of measuring the conflicting legitimacies of sacrificing mice, rats, or even chimpanzees, and of alleviating the suffering of human beings, we move from a scandal to an undecidable question. The cosmopolitical pathway then becomes one of inventing ways of actively, deliberately bringing this undecidability into existence for all the protagonists.

The requirement imposed in England for all those who claim that they cannot do without animal experimentation then assumes its fullest meaning. Researchers must incorporate into their publications a complete description of the living conditions of their laboratory animals and their possible consequences for the experiment, an evaluation of the suffering inflicted on the animals by the experiment, and proof that such suffering was strictly necessary. And if the article is to be accepted, all of these elements must be presented in accordance with guidelines that force scientists not to confuse what is convenient and what they can risk asserting. It's a double-acting trigger because one can legitimately claim that this requirement is scientifically as relevant as the technical characterization of a measuring instrument used for an experimental protocol, but that it also requires the experimenters to accept the challenge of taking an interest in what they would define without it as "means." The experimenters find themselves forced to experience the discomfort or anxiety to which their practice exposes them. This is in no way a punishment or compensation for what is inflicted on the animal. Rather, it is meant to bring about a deterritorialization within experimental territory and the correlative possibility that experimenters might understand, not simply tolerate, the questions their practices raise, might meet the diplomats of the opposing party and invent propositions that make a precarious peace possible.

It is not only the conditions of those who do not define themselves in political terms that pass into the cosmopolitical Parliament made present by their diplomats. Even within our modern tradition, where we dare to state that "the dead must bury the dead," we can also say that "the dead call out to the living so they might wake the dead."[5] If other peoples know how to keep watch over their ancestors and restore their voices through the words they create, the history we have invented for ourselves is haunted by the ghosts of those it has crushed, vanquished, or bowed, and by the shadow of everything our reasons, our criteria, have destroyed, or reduced to silence or ridicule. The past cannot be measured in terms of remorse or loyalty. And, in our tradition, it does not correspond to any requirement that we might satisfy. That tradition, precisely because it has given time the power to bury the past, because it does not give us the words, gestures, and actions by which the living might think, or create, their obligations to the dead, needs this unknown. The Parliament, because it is *our* invention, because it is cosmo*political,* must acknowledge that none of its negotiations will ever produce the "right" account or write off the past.

If our past doesn't have the words to express what it wants from us, it is nonetheless capable of forcing those who wish to be vectors of possibles designating the future to think "for, or rather, in the presence of" the past.[6] Some will object that we are not guilty for everything—the disappearance of the dinosaurs, for example. But we are responsible even "for" them, for the way we relate that disappearance, and especially the way in which we have, for a long time, made it the normal and quasi-moral consequence of the progress that leads directly to ourselves. Even the pedagogue is not guilty of the crimes committed in the name of pedagogy, but she is their heir, and must think and speak "in the presence of" the multitude of those who have never understood, or understood only too well, what education is about. Those who have to answer to diplomatic representatives must agree to think and speak "in the presence of" the past, "in the presence of" the judgments and disqualifications that were defined as so many victories by the history they inherit.

The cosmopolitical Parliament is not primarily a place where instantaneous decisions are made, but a delocalized place. It exists every time a "we" is constructed that does not identify with the identity of a solution but hesitates before a problem. I associate this "we" with the only slogan Leibniz ever proposed: *Calculemus.* Let us calculate. It's an odd expression, constructed to conceptualize the possibility of peace during a time of war. But Leibniz was a mathematician, not an accountant or statistician. For him, calculation was not a mere balance sheet contrasting homogeneous quantities, calculations of interest or benefits that were presented as being commensurable. For a mathematician, the accuracy of a calculation and the validity of its result are relatively simple questions, "trivial" in the language of mathematics. What is important, and which is not in the least trivial, is the position of the problem that will, possibly, allow it to be calculated, the precise creation of relationships and constraints, the distinction between the various ingredients, the exploration of the roles they are liable to play, the determinations or indeterminations they engender or bring about. There is no commensurability without the invention of a measurement, and the challenge of Leibniz's *calculemus* is, precisely, the creation of a "we" that excludes all external measures, all prior agreements separating those who are entitled to "enter" into the calculation and those subject to its result.

There is nothing contingent in the fact that the cosmopolitical Parliament accepts a "mathematical slogan." Mathematicians' practice commits them to transforming the conditions of a calculation, once these have been identified, into material for new risks, to transforming the unfolding of any distinction as an occasion for a nomadism through which new mathematical beings will come into existence. In Book I of *Cosmopolitics,* I wrote that, for me, mathematics is not a "modern practice" because requirement and obligation are literally and inseparably constitutive of the definition of mathematical beings. I would now add that, in their case, "sedentary" experience and "nomadic" risk cannot be separated, as expressed by the mode of existence associated with the mathematical beings they bring into existence. Mathematicians confer upon them, almost irrepressibly, a type of preexistence some might compare to Platonic ideas, but they also proudly emphasize the boldness required by their construction.[7] In other words, mathematicians can give their *calculemus* to the cosmopolitical Parliament for, among all practices, mathematics is unique in that, regardless of the question, it has no need to be represented by diplomats. For mathematicians, no issue can endanger mathematics; rather, issues provide them with a challenge, an opportunity for creation.

Calculemus, therefore, does not mean "let us measure," "let us add," "let us compare," but, first and foremost, let us create the "we" associated with the nature and terms of the operation to be risked. It is not a question of acting in the name of truth and justice, but of creating commensurability. It is a question of knowing that the "truth" of the created common measure will always be relative to what such creation will have been capable of, knowing also that a radical heterogeneity preexists such creation, the absence of any preexisting shared measure among the ingredients to be articulated. For Leibniz, reference to the best of worlds created by God in no way ensures the justice of human calculations, for we know neither the terms of the problem nor the quantities assigned by God to each of the terms. As Leibniz stated in the *Theodicy* (section 118): "It is certain that God sets greater store by a man than a lion; nevertheless it can hardly be said with certainty that God prefers a single man in all respects to the whole of lion-kind. Even should that be so, it would by no means follow that the interest of a certain number of men would prevail over the consideration of a general disorder diffused through an infinite number of creatures."

Of course, the Leibnizian *calculemus* does not derive solely from mathematicians. Through Leibniz it reveals the affinity between mathematics and philosophy as practiced by Leibniz, a speculative philosophy. What does the insistent presence of speculative philosophy throughout the pages of *Cosmopolitics* imply? Why, for example, have I employed, implicitly or explicitly, a speculative concept like "transduction" without fear that it would lead me to judge or to disqualify? What singular relation have I constructed between speculative philosophy and the "cosmopolitical Parliament"?

"Every science," writes Whitehead, "must devise its own instruments. The tool required for philosophy is language. Thus philosophy redesigns language in the same way that, in a physical science, pre-existing appliances are redesigned."[8] "In the same way," that is, according to the respective requirements and obligations of each. But philosophy in Whitehead's sense, and Leibniz's as well, can require nothing of the world in the sense that this requirement would create a hierarchy between the essential and the illusory. According to Whitehead, philosophy must be obligated by everything that "communicates with immediate facts," knowing that these have nothing in common with facts that have been purified or produced to serve an argument but refer to anything we can claim to be a matter of experience (which is why "what does not so communicate is unknowable, and the unknowable is unknown").[9] To be a matter of experience does not belong to the register of proof (I have proof that animals suffer) but to the affirmation that demands to be heard (the suffering of animals inhabits my experience).

Mathematics is unique in not taking into account what it does not bring into existence. Since the ancient Greeks, the "material" triangle, materialized by an actual drawing, has been a sign, signifying the requirements the ideal triangle satisfies and the obligations it imposes. Not taking something into account is not the same as denying or disqualifying. Mathematical practice does not need to deny. The material triangle is not the victim of the ideal triangle. The mathematician, because she will go as far as the power of words to judge can take her, because she pushes this power of judgment to its limit, where it becomes the power to create, knows the difference between ordinary judgment and mathematical creation. That is why she can, with less fear than others, more specifically, *with a fear that is constitutive of her practice,* create problems where the ordinary use of words provides answers. But speculative creations have to answer a rather particular obligation, the obligation to "save" from habits that judge and hierarchize all that we experience. For instance, it must specify both the material and the ideal triangles, each according to a mode of existence, but without hierarchy or dependence. It does not do so in the name of "truth" or to save the "victims" of our judgments. Or else, this truth and those victims are related to the unknown cosmopolitics alludes to. It is rather "in the presence of" the victims of all our inconsistencies, the radically unknown victims produced at every moment by the power to judge that arms the words we use, that speculative thought is risked. It proceeds in the immanence of the obligation it brings into existence, accepting that the tools it designs and redesigns will be recognized as defective if they can be used to establish a position of power, to hierarchize or disqualify.

Like mathematical practice, the practice of speculative philosophy, as I have interpreted it, has no need of diplomacy, nor does it need to impose any conditions whatsoever on others.[10] Rather, it must accept and understand on its own terms, that is, translate-betray every condition that has been expressed. But the operation of translation-betrayal does not make "speculative philosophers" experts. We can understand the temptation of the philosopher to imagine that the world would be better if she were a "counselor of the Prince," but any relationship with power and its responsibilities is poisonous for philosophy. No more than anyone else do philosophers have access to a definition of the "common interest." Nor are the words and uses of words the speculative philosopher creates those of diplomats. For, their own challenge—to minimize requirements and maximize obligations—results in a delocalization of the issues, detachment (not indifference) from the particular "calculations" by which history is carried out. Speculative philosophers are obligated by everything to which they may be sensitive, but they are never obligated by the claims individual languages make. Just as mathematicians "take no account" of the material triangle, speculative philosophers take no account of specific languages, or rather of the specific privileges those languages give to their reasons and requirements. That is why speculative thought is present in the Parliament of Things, but present without interaction. It can be compared to the "soul" united with the cosmopolitical body by means of the Leibnizian *vinculum*, for it must fabricate concepts that actualize what a given epoch realizes, the disparate ensemble of the *calculemuses* that may be realized during that epoch. It alone can say "my epoch."

To highlight speculative philosophy in this way compared to all other philosophical "genres" (moral philosophy, philosophy of art, law, history, science) is to recall that those genres derive their apparent necessity primarily from the conviction that what they designate (the question of morality, art, history, etc.) "needs" to be represented by other means than those of their corresponding practices. I maintain that this conviction, however legitimate, primarily expresses the problem presented by the practices in question. In other words, whoever comes to philosophy to study history, consciousness, the work of art, madness, or physics should be welcomed, but as exiles, with the words that express a hoped-for future in which they themselves or their descendants will be able to return to the regions they should never have been forced to leave in the first place.[11]

The cosmopolitical Parliament creates the words to express this future. The experts and diplomats who populate it are not philosophers, but they cannot limit themselves to representing the point of view resulting from their practice. They must make present both what it means to belong to such a practice and the singularity of the point of view it produces. It is the possibility of presenting oneself in this way that I have explored throughout *Cosmopolitics*. If I have been led to "disqualify" as "modernist" certain practices pertaining to what are referred to as the "social sciences," it is because, in their case, "belonging" cannot be acknowledged. But the speculation I have offered also implies that such practices, if they escaped the curse of tolerance, would crucially matter for the very possibility of the *calculemus*. For, *calculemus* depends on the question of the possibility of peace, as accepted by those groups it may endanger, while the practical requirements I have associated with the social sciences also depend on peace. As knowledge-producing practices, they depend on the fact that other practices are "contemporary," knowable because they have accepted the kind of peace that is the condition of their knowability. Correlatively, the knowledge they produce, the narratives they construct, are needed for encounters that escape the curse of tolerance. When a scientist leaves the laboratory because she wants to promote a possible that has been born there and must encounter those for whom her proposal matters, she must turn to experts or diplomats who have learned and been empowered, through the translations-betrayals they authorize, to make her feel that what they represent must be taken into account.

The cosmopolitical Parliament "exists" today occasionally, but precariously and improvisationally, without memory and without any long-term consequences, the way a microbubble forms in a liquid just below the boiling point. It cannot be stabilized, cannot exist, in the ecological sense, without the active, engaged, risk-filled production of practices that create knowledge about practices, capable of narrating the way they differ and matter, thereby enabling histories to be made with them the way we learned to make histories with plants, chemical reagents, and atoms.

However, there remains an insistent unknown, namely, what we think we are familiar with. What will become of politicians as such who, today, are the only officially recognized mediators, officially allowed to participate in decision making, or, more specifically, who are allowed to participate in what experts, and the powers those experts represent, allow them to decide? We should not be surprised by this unknown, this too well known that has become enigmatic, for the unknown here expresses the difference between the *image* I have been trying to bring into existence and a *program*. The image is a challenge as it takes literally the claims to universality by which we have invented ourselves, which must be combined with their corresponding obligations if there is ever to be a civilized future for those claims. Consequently, the image is silent about politics as it is currently defined, rather bringing about politics where it is currently clandestine, where political problems find "solutions" without their ever having been presented as problems in the first place. But telling about these problems as "political" does not answer the question of politics as a practice.

That politics and its specific obligations are today an unknown, that the nomadic visitor looks in vain for the fire around which she might listen, where the hopes and doubts, dreams and fears, of those who welcome her might pass into her experience, is not, however, a positive achievement of my analysis. It is a limit and an appeal. For that reason, it is important to emphasize that the challenge of the cosmopolitical Parliament, like Latour's Parliament of Things, is not a new beginning, purged of old political conflicts, which would be relegated to the warehouse of useless gadgets. But it implies that a practical contrast has been substituted for programmatic oppositions, such as the opposition that is supposed to differentiate the Left from the Right. It is such a contrast—which the "Left" could bring into existence and which could bring it into existence independently of programs, promises, and ideals, independently of a future in which it "would be in power"—that Gilles Deleuze proposed in two pages that deserve far greater attention than they have been given.[12] If "the Left really needs people to think," if its job, "whether in or out of power, is to uncover the sort of problem that the Right wants at all costs to hide," it is because the distinction between Right and Left satisfies a key contrast: "Embracing movement or blocking it: politically, two completely different methods of negotiation." That is why the Right has available to it "direct mediators, already in place, working directly for them. But the Left needs indirect or free mediators, a different style, if only the Left makes it possible."

28

The Final Challenge

At the start of *Cosmopolitics,* I said that my project was to bring into existence the question of an ecology of practices, not as a solution but as a learning process, the creation of new ways of resisting, in the present, a future that derives its plausibility from our powerlessness as well as from effective power relationships through which that future is established.

The widespread parasitism I associated with the capitalist redefinition of practices that must, first and foremost, be resisted may coincide with the capitalist axiomatic Deleuze and Guattari describe in *Anti-Oedipus* as intrinsically different from the old codes and the territories associated with them. The capitalist axiomatic in itself cannot be identified with an ensemble of axioms, it exists only in the invention of axioms, their mutations, their rearrangements, and in the invention of factitious reterritorializations that will be made, unmade, and remade depending on circumstances. "It is with the thing, capitalism, that the unavowable begins; there is not a single economic or financial operation that, assuming it is translated in terms of a code, would not lay bare its own unavowable nature, that is, its intrinsic perversion or essential cynicism. . . . But in point of fact it is impossible to code such operations."[1]

It goes without saying that the invention of axioms is the antithesis of any "parliamentary" process because it is precisely the function of axioms to substitute nonnegotiable imperatives for the negotiation of a problem. There is no "we" in the capitalist axiomatic, certainly not the "we" of "capitalist individuals" who would be opposed to the "we" of those they exploit. In this sense, capitalism is indeed "the limit of all societies," and Deleuze and Guattari seem to encourage us to move toward that limit.[2] "But which is the revolutionary path? Is there one? . . . To go still further, that is, in the movement of the market, of decoding and deterritorialization? For perhaps the flows are not yet deterritorialized enough, not decoded enough, from the viewpoint of a theory and a practice of a highly schizophrenic character. Not to withdraw from the process, but to go further, to 'accelerate the process,' as Nietzsche put it: in this matter, the truth is that we haven't seen anything yet."[3]

I have introduced this terrible statement to conclude this essay because it apparently dissolves any possibility of the Leibnizian *calculemus,* because it serves as a test, challenging the very idea of an ecology of practices. No one really knows what "to go further" really means, and the authors of this statement do not know themselves. What they do know and claim is that we cannot hope to resist capitalism by insisting on maintaining the old territories, or by attempting to bring them back as they once were. To do so is to risk releasing the nightmare of monstrous reterritorializations, the return of an archaic that has never existed. We can also state that "to go further," "to accelerate the process," does not mean proceeding in the same way as capitalist axiomatization, accelerating it using the same coordinates. And the "terrible curettage" and "malevolent activity"[4] that respond to the "we haven't seen anything yet" are entirely directed toward the imaginary reterritorializations capitalist axioms continuously generate, and have nothing to do with a "modernizing" crusade. But what is of the utmost importance is to avoid the trap of using Deleuze and Guattari's terrible statement as an instrument of judgment about who is entitled to belong to the "revolutionary path."

This may be the trap that the majority of Marx's heirs fell into when they relied upon this aspect of his analysis, the only one that provided any reassurance, namely, that the class of wage earners was securely privileged in that it was the only class whose objective interest was to abolish the wage relationship, and that its point of view may serve to construct the only genuine revolutionary theory. For, the stability claimed by this theory on behalf of the privilege of the wage-earning class makes it indifferent to any "ecological" concern. All calculations must inevitably amount to the same thing, they must confirm, in the final instance, the legitimacy of class analysis, which is alone capable of reliably, realistically leading and organizing minority struggles (feminist, ecological, gay, drug users, etc.). If the capitalist axiomatic is indeed the antithesis of the Leibnizian "calculemus," then accepting as the privileged "we" what would first be defined by its position in this axiomatic may be the trap held out to those who attempt to fight against it. It is a trap because any such position is relative to the ongoing process of axiomatic invention. That the practitioners of class struggle wish to promote a mode of analysis capable of sparing others any number of inanities, any amount of moralizing or naïveté; that they are seasoned fighters, obligated by their practice to identify and analyze the various forms of pseudo-consensus, arguments of misguided common sense, or the obscene platitudes that anesthetize us, is one thing. But that this practice gives them the power or the right to judge, to hierarchize, to organize all minority struggles, is something quite different. To quote Deleuze and Guattari's analysis, "the flows are not yet deterritorialized enough."

If we cannot "wait" for the depredations of capitalism to demonstrate the truth of the class struggle and produce class awareness, we must also actively, diplomatically avoid constructing a common measure by which the appropriate methods of accelerating the process, of pushing further, can be unilaterally evaluated. And this includes the measure of the "schizophrenic character" of production, which we might be tempted to evaluate based on the arguments of Deleuze and Guattari. For, such a character, should it claim to transcend any process of calculation/negotiation, is liable to become a new law against which each of us would be inadequate, necessarily inadequate.

"And then . . . and then . . ." "Either . . . or . . . or." "And therefore . . . !" For Deleuze and Guattari, the crux of the matter is to prevent these operations from being subjected to and organized by some "either . . . or" that would claim to mark decisive choices among immutable terms.[5] But to escape the "either . . . or," more accurately, to dissolve it, does not necessarily require the too fascinating risks associated with the figure of "the" schizophrenic. We are surrounded by "either . . . or's," and primarily the one that enables us to condemn fetishes. We have already encountered a number of them—"knowledge production" versus "simple opinion," fabricated artifact versus autonomous being, "modern" versus "nonmodern," and so on. And it may well be that the figure of the diplomat, always suspect, accused of evading "the" issue ("it depends . . .") or betrayal, always trying to escape static disjunction ("you're either with us or against us"), is interesting precisely because it avoids any romantic attraction for an "elsewhere" that might inspire the hope of escaping from "here" once and for all.

Chemists wanted to know what characterizes gold. They learned, step by step, how to prepare the reagents that made their question answerable by dissolving everything but gold. I've tried to operate as a chemist, to select a reagent that doesn't dissolve what interests me, one that respects practices but attacks the way practices use the disjunction of "either . . . or" to reterritorialize themselves in terms of ancient codes derived from (political) philosophy. A critical operation such as this does not correspond to judgment in the ordinary critical sense. As with chemical tests, crisis recovers its etymological meaning—separation, or discrimination. It is relative to an issue and does not refer to some general form of legitimacy. If all that glitters is not gold, one may nevertheless have good reasons to appreciate what glitters for its glittering.

I have, in fact, addressed questions that seemed to lead inevitably to the alternative disjunction of "either ... or." Did the neutrino exist at the origin of the universe or is it purely a human fabrication? Is "reality" subject to laws or is it veiled from us? Does matter consist of atoms or are these simple, conventional references? Does life "emerge" from physical-chemical processes in the sense that it would find its explanation in them or, rather, in the sense that it could be considered an irreducible novelty? Are djinns relative to a therapeutic process or does the supernature they are part of have, in itself, the power to impose its existence upon us? If my efforts have not met with failure, their effect will be not to resolve but to dissolve such disjunctions, which only appear fundamental because they serve as an ever-renewed appeal to judgment, the settling of accounts in the name of some predetermined calculus. If my effort has not met with failure, its effect will be to generate interest in other questions and allow them to proliferate along with the practices they are associated with—relaying the risks of those practices rather than objectivizing them. As for the dramatic alternatives we began with, they will have adopted the regime of "and then ... and then," "either ... or ... or," and "therefore ... !" This "therefore ... !" can be associated with the characters I have presented—Ilya Prigogine designing and testing his various functions, Tobie Nathan inventing therapeutic strategies in which "either ... or" has no place, Stuart Kauffman at the edge of chaos, or even Chris Langton hunched over his keyboard. It is not an insult but a form of homage to claim that their shared grandeur lies in the fact that they resemble far less the "objective scientist" than Joey, Bruno Bettelheim's child-machine, which *Anti-Oedipus* immortalized.[6]

In other words, in asking questions about the "psychosocial type" to which a practice corresponds, I have gambled that
here too it is possible that "we haven't seen anything yet." Possible, but not probable. Speculation about a possible has nothing to do with the calculation of probabilities. It is possible that
the notion of "type," which until now seemed to justify a static
judgment (Max Planck authoritatively defining the "true physicist's" passion), might become an endogenous process of self-
affirmation, one that is relative to the commitment of a practice
but deciphers its obligations under the ever-renewed tension
between the necessity of translation and the risks of betrayal.
Planck would then no longer refer to "we physicists," as if the
"we" preexisted the conclusions he will draw on their behalf.
Rather, he would say "This is what it takes to be a physicist and
these are our factishes," thereby discovering what his refusal to
abandon realist enjoyment, his refusal to accept Mach's critical
lucidity, commits him to. That is the first meaning of *calculemus*:
the "us" does not preexist the invention of the calculation, the
negotiation of the "and . . . and," of the ingredients to be taken
into account, and the "either . . . or . . . or" of their articulations.
It is this very invention that produces this "us."

The cosmopolitical Parliament is a speculative idea, a consequence of which is the crucial character of the "schizophrenic
character" of practices,[7] the way in which they escape the "either
. . . or." It is what enables them to participate with other "us's" in
the calculation, and satisfy what Tobie Nathan has referred to as
the "conditions" of exchange, an exchange of a "highly schizophrenic character." For the general obligation that governs this
disparate ensemble, the principle of immanent sorting to which
it corresponds, is precisely—but in different ways depending on whether one comes as an expert or as a diplomat—the
active, inventive exclusion of "either . . . or's." It should not be
claimed that capitalism has been vanquished by some miracle
wherever the cosmopolitical Parliament exists, no matter how

precariously or evanescently. But we can say that it then encounters an inventive power that is capable of resisting its axiomatic power. We can also say that the question of the difference between a political parliament, even a Parliament of Things, and a cosmopolitical Parliament is specifically the result of this inventive exclusion of "either . . . or." The least "either . . . or" respected or ratified produces a fault line through which everything starts all over again.

"Either" you negotiate with us "or" you have nothing to say. How can we fail to recall those "upright men," more Roman than Greek, who, in France, are always referred to when speaking of the "Republican ideal" (so oddly distinct from democratic practices). To which the "others" can retort, once the cosmopolitical Parliament exists: Please, sit down and behave yourselves.

However, we can't just stop there. For the cosmopolitical Parliament itself refers to something that cannot be included in any calculation, something that causes distinct calculations to resonate together, as it resonates in each of them.

What Makes Nature Tick? is the title of yet another book on the popularization of contemporary physics, with its neutrinos and quarks, its big bang and its interactions, unified or not. How can we tell the author that this title, which he hopes is appealing, is genuinely obscene? It was more than a century ago, in 1872, that Emil Du Bois-Reymond uttered his resounding *Ignorabimus*: we do not know, we will never know, even if we attain the knowledge of Laplace's demon, "that which is," meaning both what it is that we know (that which "inhabits space" wherever there is matter) and what it is that we are, we the knowers (the essence and origin of consciousness). From a cosmopolitical point of view, these eternal "limits" of knowledge have no more interest than the image of knowledge by which they are defined as limits. Nor is it in terms of limits that we can determine the obscenity of a "theoretical" response to the question of what makes nature "tick." Perhaps the genius of the English language with which this title was invented and its appeal might provide a way out. For the title implies, although implicitly, "What is it that 'drives' nature?" "What makes nature 'desire'?" And the question proliferates and diversifies wherever someone succeeds in constructing an answer, in producing a new factish. *What makes you tick?* If the question cannot be a matter of calculation, it is not because it would address a unique transcendental for all calculations but in the sense that it designates what every calculation affirms and requires in the very act in which it is engaged.

"Now, it's up to the two of us!" says the mathematician faced with a complicated problem, or the experimenter faced with a promising but somewhat unreliable detection device. And the "two" thereby designates a doubled disparate multiplicity. As for the "us," it does not preexist the eventual coming into existence of "the two of us," the mathematician and her solution, the experimenter and her finally reliable device. What exists are the dreams and nightmares in which are mixed, permuted, and combined bits of machinery, processes, or equations, and bits of humanity, gestures, or reasoning, in which operations that alter signs, bodies, meanings, assemblages are experienced. Neither the mathematician nor the experimenter relies on anyone to construct the terms of the solution, and the "two of us" to come is nowhere inscribed or prescribed. And yet, reference to the possible "two of us" exists as a requisite, without which neither the mathematician nor the experimenter would risk the adventure. *It's what makes them tick.*

What "drives" diplomats and experts toward the cosmopolitical Parliament? As noted, this Parliament has one, and only one, requisite: the same one that makes the ecology of practices a form of speculative thought rather than a matter-of-fact situation. What it requires and affirms is peace as possible. Not the convergence of all calculations as realized or even realizable, but the production of convergence as a possibility. This is the only reference diplomats share, but it is a reference without which the adventure would not be undertaken. This requisite does not belong to cosmopolitics as such, for the question of peace does not have politics as its privileged site. It does, however, correspond to the transition from politics to cosmopolitics, for it expresses the insistence of the "cosmos" within politics. However, the fact that it can be expressed in terms of "peace"—in the sense that peace is not the goal, but that the reference to peace, as possible, is required—indicates the danger that is exposed when we overlook the pharmacological character of an appealing solution.

When Leibniz attempted to consider the question of damnation and introduced Beelzebub's refusal to "ask forgiveness," which God had decreed to be the condition for his salvation, he also showed, perhaps in spite of himself, the extent to which apparent generosity, whenever it is one-sided, can exacerbate the rage of the one it was supposed to benefit. The cosmopolitical calculus will always remain exposed to rage, to Beelzebub's despair: "Poison penetrates our limbs, and already rage rushes furiously / Through all our joints; wicked deeds must be piled upon wicked deeds. / Thus we are purified. The only offering for the enraged / Is the affliction of the enemy. It pleases me to scatter him to the winds, / Mangled alive, torn into a thousand pieces, / Made into so many examples of my own sadness, / So that, when the trumpet calls for the resurrection, / He is deprived of flesh."[8]

A particular syntax may be needed to stabilize the *pharmakon* of this peace. Beelzebub's venomous rage is the rage liable to invade the victims of crimes to whom the risky commitment of the *calculemus* is proffered. It may also be necessary for the physicist to free herself, without feeling she is sacrificing herself for the sake of peace, from the obscenity found in the very idea that her practice might appropriate the question *What Makes Nature Tick?*. And it is probably crucial for stabilizing that other *pharmakon* of our tradition, the one that condemns Beelzebub to

damnation. For it is the essential goodness of the God of mono-
theism that makes him unable to be anything other than the
one who "forgives." We need such syntax in order to "dream the
dark" and celebrate each event where heterogenesis is risked
for itself and not as the result of some essential goodwill or the
desire for peace.

That syntax *is not, and cannot be,* a religious syntax in the
sense that religion requires, in the sense that it creates a reference to a unity recognized to have the power to unbind, to
undo the affiliations it transcends. This syntax can, however,
be defined by the vocation of having to civilize that unity, having
to make it present not through the simultaneous collapse of all
the calculations into the incalculable, but as the "unknown" that
every calculation requires. Because it is a question of cosmo*politics,* we should not be surprised if it is within the tradition that
invented politics that I have risked identifying the lineage that
might serve as the practical vector of that unknown—a constitutively political vector like the lineages of science and philosophy,
and involved in a thousand and one crimes, just like science and
philosophy. But the fact alone of referring to it may serve as a
thought hammer for the heirs of the other two approaches, proud
of having eliminated the third. For, if the elimination of theology is for many a "triumph" of reason, doesn't that elimination
stabilize the definition of a belligerent reason, one that identifies with the triumph found in the destruction of the other?[9]

But what if it was neither religion nor belief that we lacked
but theologians capable of expressing in terms of the *logos* rather
than conviction that which brands all our calculations and judgments with the seal of uncertainty and risk? This is the final
unknown of the cosmopolitical question, a *practical* unknown
whose primary interest is its ability to compromise those who
allude to it, the way "magical" gestures compromise ethnothera-
pists. It is the fear that one day we may have to acknowledge that
we are not yet finished with a past we were so proud to have put
behind us.

NOTES

PREFACE

1. [The present translation is based on the updated two-volume French edition of *Cosmopolitics* published in 2003. The contents of that two-volume edition were compiled from an earlier edition of *Cosmopolitics* published in 1997 in seven volumes. *Cosmopolitics I* includes volumes I, II, and III of the original edition; *Cosmopolitics II* includes volumes IV, V, VI, and VII of the original edition. Volume references throughout the text follow this numbering scheme. For example, Book IV refers to volume 4 of the original edition, which is included in volume 2 of the present edition. Please note that the current English-language edition of *Cosmopolitics* has been revised by the author and varies slightly from the 2003 French edition. All translations of quoted French works are my own, except where otherwise noted. —*Trans.*]

1. ATOMS EXIST!

1. Bruno Latour, *Petite Réflexion sur le culte moderne des dieux faitiches* (Le Plessis-Robinson: Synthélabo, "Les Empêcheurs de penser en rond," 1996).

2. Jean Perrin, *Atoms,* trans. D. L. Hammick (New York: D. Van Nostrand Co., 1916), 105–6.

3. See Mary Jo Nye, *Molecular Reality: A Perspective on the Scientific Work of Jean Perrin* (London: Macdonald, 1972).

4. For this and many other aspects of Duhem's work, including Duhem's relationship with his colleagues, see Paul Brouzeng, *Duhem, Science et providence* (Paris: Belin, 1987).

5. This renewal, which I will return to in Book V, connects the, now

positive, question of pathology to that of nonintegrability in Henri Poincaré's sense (discussed in Book II).

6. Pierre Duhem, *The Aim and Structure of Physical Theory* (Princeton, N.J.: Princeton University Press, 1991), 143.

7. Ibid., 140–41.

8. Ibid., 43.

9. Ibid., 330.

10. Ibid., 335.

11. Ibid.

12. Ibid.

13. Perrin, *Atoms*, 82.

14. Ibid., xiv.

15. See Isabelle Stengers, "Le médecin et le charlatan," in Tobie Nathan and Isabelle Stengers, *Médecins et sorciers* (Paris: Éditions Synthélabo, 1995).

16. See Bruno Latour, *Pasteur: Une science, un style, un siècle* (Paris: Institut Pasteur et Librairie Académique Perrin, 1994). I can't recommend this book too highly. Intended for the "mass market," it presents an approach to scientific practice that we would today have to be blind to confuse with any form of relativism. See also Bruno Latour, "Les objets ont-ils une histoire? Rencontre entre Pasteur et Whitehead dans un bain d'acide lactique," in Isabelle Stengers, *L'Effet Whitehead* (Paris: Vrin, 1994).

17. See any of the books published by Bernard d'Espagnat since 1981.

2. ABANDON THE DREAM?

1. Roger Penrose, *The Emperor's New Mind* (Oxford: Oxford University Press, 1989), 480.

2. The connection between two isolates yields an isolate. The fact that the question of the unification of interactions, the Holy Grail of high-energy physics, is also considered crucial by theoretical cosmology with respect to the universe on the so-called Planckian scale (very, very, very young, hot, dense, etc.) does not break the isolation but merely provides the issues with the grandiosity of a *nec plus ultra*.

3. Stephen Brush, "The Chimerical Cat: Philosophy of Quantum Mechanics in Historical Perspective," *Social Studies of Science* 10 (1980): 432.

4. Max Jammer's *The Philosophy of Quantum Mechanics: The Interpretation of Quantum Mechanics in Historical Perspective* (New York: Wiley Interscience, 1974), already assembled a variety of more or less disparate interpretations. These have continued to proliferate with the appearance of Bell's theorem and experiments on hidden local variables.

5. Léon Rosenfeld, "Misunderstandings about the Foundations of Quantum Mechanics," in *Selected Papers of Léon Rosenfeld*, ed. Robert S.

Cohen and John J. Stachel, Boston Studies in the Philosophy of Science, vol. 21 (Dordrecht: D. Reidel Publishing Company, 1979), 495–502.

6. See Gerald Holton, "Mach, Einstein, and the Search for Reality," in *Thematic Origins of Scientific Thought* (Cambridge: Harvard University Press, 1973), 219–59.

7. Causal measurement, a principle of dynamics, will be integrated into the space-time curve that general relativity employs: the space-time trajectory is then replaced by a four-dimensional geodesic that the body, formerly subjected to gravitational force, now follows as if it were animated by a purely inertial movement. Special relativity does not address the problem of the forces and accelerations of rational mechanics. That is why, although general relativity introduces "free-falling elevators" populated with observers, the thought experiments employed by special relativity make use of "trains," that is, mobile bodies inhabited by observers and traveling at a speed whose justification is not dependent on physics (but, in this case, on their motors).

8. The mathematical regime of probability amplitudes is different than that of probabilities. Probabilities are mutually exclusive (either this, *or* this, *or* this). Probability amplitudes are additive and interfere with one another (this *and* this *and* this).

9. Jammer (*The Philosophy of Quantum Mechanics,* 120–21) relates that the evening before his death, November 18, 1962, Bohr had once again drawn on his blackboard the diagram for the "photon box" that Einstein had challenged him with in 1930, during the sixth Solvay conference, and that Bohr had succeeded in using as an argument in favor of quantum mechanics. Bohr's success in challenging Einstein on the very field he had invented had a significant effect in the adoption of the Copenhagen interpretation, but a vanquished Einstein continued to inhabit his conqueror's thought.

3. NIELS BOHR'S LESSON

1. This point is made with rare lucidity by Aage Peterson in *Quantum Physics and the Philosophical Tradition* (Cambridge: MIT Press, 1968).

2. Gilles Deleuze, *Difference and Repetition,* trans. Paul Patton (New York: Columbia University Press, 1995), 212.

3. Here, I am using the term "excited atom," although the intuitive representation that follows from Bohr's model leads us to speak of an excited electron. As we will see, quantum formalism connects Bohr's atom with the cyclical variables of Hamiltonian physics, which implies that the electron has no meaning independent of the atom, which is the only "subject" in the strict sense of the formalization.

4. Léon Rosenfeld, "Niels Bohr in the Thirties," in Stefan Rozental, ed., *Niels Bohr: His Life and Work as Seen by His Friends and Colleagues*

(Amsterdam: North Holland Publishing Co., 1967), 114–36.

5. The question of (local) hidden variables has been considered a closed topic ever since the experiments of Alain Aspect (1962). The possibility of trying to resolve the question by experimentation is directly inspired by the EPR article, but satisfies Bohr's requirements. Its starting point was a thought experiment attributed to John S. Bell. Bell invented a device and a measurable property (Bell's inequality) whose value would differ depending on whether the quantum reality is determined in terms of local hidden variables or is not, as assumed by quantum mechanics. The fact that quantum mechanics allows us to predict an inequality value corresponding to the nonexistence of local hidden variables shows that this nonexistence is not part of the interpretation of formalism alone. Denying the existence of these variables is not based on some positivist prohibition that has been added on, but is clearly stated by the formalism itself, something Bohr had always maintained. On the other hand, the possibility of presenting quantum mechanics with a *new* experimental challenge on this front refutes von Neumann's demonstration of the impossibility of conceiving another theory compatible with the experimental results, one that makes use of the quantum formalism and itself includes a reference to hidden variables. See Trevor Pinch, "What Does a Proof Do if It Does Not Prove?" in Everett Mendelsohn, Peter Weingart, and Richard Whitley, eds., *The Social Production of Knowledge, Sociology of the Sciences* 1 (Dordrecht: Reidel, 1977), 171–215. Aspect's experiments confirmed the quantum prediction. If this had not been the case, it is quantum mechanics itself rather than its interpretation that would have been called into question. Typically, those who speak of this experiment are silent about the fact that Niels Bohr's position is consistently the prevailing one (including his rejection of the "question of interpretation" as separable from that of the formalism). Some of them are now turning to the question of "nonlocal" hidden variables, about which the formalism, along with all of physics, has remained silent.

6. See David Bohm, *Wholeness and the Implicate Order* (London: Ark, 1980).

7. Bohm's thought is one of great, and seductive, poetry, and his interest in Oriental spirituality is certainly genuine. Nonetheless, the poetry in question is that of a physicist. The implicate order is directly inspired by the wave function that is to be transcended, and the unfolding explication enables us, as in general relativity, to make the practice dependent upon and subordinate to some transcendent truth, which it translates but also degrades. The connotation of degradation is further reinforced by the opposition between our mechanistic Occident, attached to its explications, and the mystical Orient, which "knows" that things are inseparable. In other words, the hierarchy of our practices, which makes physics the spokesperson for reality, is confirmed and celebrated by the fact that the

question of the interpretation of quantum formalism culminates in the "salvation" of Western thought, which is finally able to escape its limitations and join Oriental wisdom.

8. *Cosmopolitics, Book VIII,* "The Curse of Tolerance," discusses this generalization of complementarity as it appears in the remarkable work of Georges Devereux.

4. QUANTUM IRONY

1. Stephan Körner, ed., *Observation and Interpretation in the Philosophy of Physics, with Special Reference to Quantum Mechanics,* Proceedings of the ninth symposium of the Colston Research Society held at the University of Bristol April 1–April 4, 1957 (London: Butterworth Scientific Publishers, 1957), 183–86.

2. See Bruno Latour, "Les objets ont-ils une histoire? Rencontre entre Pasteur et Whitehead dans un bain d'acide lactique," in Isabelle Stengers, *L'Effet Whitehead* (Paris: Vrin, 1994).

3. See especially Bruno Latour, *La clef de Berlin et autres leçons d'un amateur de sciences* (Paris: La Découverte, 1993).

4. Shirley Strum and Bruno Latour, "Redefining the Social Link: From Baboons to Humans," *Social Sciences Information* 26 (1987): 783–802.

5. Of course, today, it's the computer and not an erector set that "represents" the model. And it does so in a manner that is much subtler, integrating data other than the "angles" and "lengths" of chemical bonds.

6. See it on a screen, naturally. Although any teaching laboratory is equipped to "duplicate" emission and absorption tests, which are now incorporated in a large number of standardized instruments, the encounter with "an" atom requires an extremely sophisticated laboratory. When an experimenter talks of "seeing" with emphasis, it is always with respect to a situation in which nearly all the technical and experimental resources of physics have been mobilized to create this "seeing."

7. Hans Christian von Baeyer, *Taming the Atom: The Emergence of the Visible Microworld* (London: Viking, 1993), xix.

8. See B. Harvey, "The Effects of Social Context on the Process of Scientific Investigation: Experimental Tests of Quantum Mechanics," in *Sociology of the Sciences, Yearbook: The Social Process of Scientific Investigation,* ed. Karin D. Knorr, Roger G. Krohn, and Richard Whitley (Dordrecht: Reidel, 1981).

9. Von Baeyer, *Taming the Atom,* 178.

10. This return cannot be "determined" but it can be inhibited. In the "new laboratories" where the habitat of the atom is to be found, spontaneous transitions are defined as due to the resonance between the atom and the electromagnetic field the atom induces. The atom is then placed in a cavity that absorbs those components of the electromagnetic field that are

likely to resonate with the excited atom. The result is precarious, for the slightest imperfection in the performance of the absorbing cavity restores the atom to its "natural behavior," but "excited" atoms are able to remain in such a state for a long time. In its "habitat," the atom "interacting with its field" suddenly seems to exist in a very "real" sense.

11. I'll return to this below, but I want to point out that a quantum analog of what we measure as lifetime can be defined but has no precise physical meaning in quantum mechanics. In effect, the definition of what we measure corresponds only to the first term of a calculation that proceeds by successive approximations (a description of the atom "perturbed" by the field it induces) and becomes muddled if we are misguided enough to make further approximations. In such a case we find deviations from the law of exponential decay, which gives lifetime its *kinetic* meaning.

5. THE PHYSICISTS' DOUBLE STANDARD

1. Nancy Cartwright, *How the Laws of Physics Lie* (Oxford: Clarendon Press, 1983).

2. For the relationship between von Neumann's axiomatic interpretation and that of the Copenhagen physicists, see Léon Rosenfeld, "The Measuring Process in Quantum Mechanics," in *Selected Papers of Léon Rosenfeld*, Boston Studies in the Philosophy of Science, vol. 21 (Dordrecht: D. Reidel Publishing Company, 1979), 536–46. "Intelligent beings on other planets would have . . . recognized the complementarity of position and momentum. They might have been spared axiomatics, however" (539).

3. This illustration was taken up by Georges Devereux in *De l'angoisse à la méthode* (Paris: Flammarion, 1980) and provides an eloquent transposition to the behavioral sciences. The tightly held stick corresponds to the situation that requires of the subject of experience an unequivocal response to the question asked. The loosely held stick implies that the way in which the response obtained affects the questioner is part of the meaning of that response. This is what occurs whenever the questioner accepts the fact that the situation she has created so the question can be asked does not have meaning for her alone but also, although quite differently, for the being that is questioned.

4. Cartwright, *How the Laws of Physics Lie*, 165.

5. Rosenfeld, "The Measuring Process in Quantum Mechanics," 539.

6. Cartwright, *How the Laws of Physics Lie*, 174. Much ink has been spilled about the question of the relationship between measurement and preparation. In strict terms, as Cartwright reminds us, "preparation" (such as the "sorting" carried out through the use of a field) has no effective physical meaning, if some interaction similar to a measurement does not make it irreversible. Although only the final measurement is supposed to be

capable of actualizing the separation "prepared" by the sorting operation, in practice the physicists reason as if the sorting had effectuated the
separation. They are forced to do that when they deal with devices where
the measurement, strictly speaking, is the final act of a "history" staged
by those devices. In such cases, as Cartwright points out, if the physicist
doesn't postulate that "someplace in that history" the reduction has actually taken place long before the final measurement, the ability to indicate
the meaning of the device, what it does, and why it is capable of doing so
would vanish.

7. Here, Cartwright picks up the realist argument used by Ian Hacking in *Representing and Intervening: Introductory Topics in the Philosophy
of Natural Science* (Cambridge: Cambridge University Press, 1983). In
"The Self-Vindication of the Laboratory Sciences" (in Andrew Pickering, ed., *Science as Practice and Culture* [Chicago: University of Chicago
Press, 1992], 29–64), Hacking returns to the same argument, and uses
the term "symbiosis" to describe the mutual stabilization between laboratory equipment and the properties attributed to theoretical entities that
the equipment brings into play. As Hacking points out, this stabilization
serves as a limit to the scope of Kuhnian scientific revolutions and the
theories of incommensurability often associated with them. In order for
it to be accepted, a new type of theoretical staging must preserve previously stabilized symbiotic productions. The "helium nucleus" will be able
to participate in problems that were inconceivable for Rutherford when
he used them as projectiles. But our knowledge concerning the composition of the helium nucleus must include the fact that it can behave as a
projectile "bombarding" a target.

8. Cartwright, *How the Laws of Physics Lie,* 194.

9. Ibid., 181–82 (my emphasis).

10. In Book II, I describe the technique of perturbation, which can be
used to construct a problem in dynamics by starting from an "integrable"
case corresponding to the Hamiltonian H_o, and by redefining the case
actually being studied as this "integrable case" perturbed by an interaction. From the classical perspective, the ideal is to end up by "incorporating" the perturbation in a new Hamiltonian, which will make the case
under study a new "integrable case." This is, generally, impossible. In this
sense, quantum mechanics is doubly unique. On the one hand, the postulate embodied in the Schrödinger equation amounts to claiming that any
quantum system should be able to be represented as a superposition of
stationary states (the quantum analog of representing a system as integrable). On the other hand, the need for a plausible connection with experimentation strongly suggests to physicists not to actualize this possibility.
In fact, with respect to "de-excitation," this possibility has indeed been

actualized in the case of a hypersimplified model of the quantum atom, known as the "Friedrichs model." Here, it has been possible to move from the representation of an atom (with two energy levels, fundamental and excited) perturbed by a field (represented by a continuous spectrum of frequencies) to the representation of a single undisturbed system. But the result is quite astonishing, for any possibility of speaking of the excited state and its lifetime has disappeared. It is as if, from the initial moment, the energy of the transition was "already" transferred to the field. The system is, in effect, represented by a new continuous spectrum of frequencies. The "atom," that is, the discrete energy levels, has disappeared, now "incorporated" in the new definition of the field. Temporal evolution is limited to describing the "gradual dissolution" of a "wave packet" initially concentrated in space.

6. THE SILENT DESCENDANT OF THE QUEEN OF HEAVEN

1. It was only in the 1950s, following Pascual Jordan, that Bohr came to realize that the "choice" of measurement is translated, in physical terms, as an "interaction" that has no quantum meaning because it *irreversibly* creates a permanent mark. To the extent that he attributed no physical meaning to the wave function independently of its reduction, this came to mean that irreversibility was a fundamental feature of physics. But for those who wanted the wave function to supply the secret of its reduction, the problem was formulated in the following manner: "In what situations could Schrödinger's equation justify the approximation that enables us, in classical mechanics, to speak of an irreversible change?" It is this statement of the problem that was taken up by the Italians Adriana Daneri, Angelo Loinger, and Giovanni Maria Prosperi (see Book V, chapter 2, "Boltzmann's Successor").

2. This axiom allows us to claim an equivalence between the two ad hoc languages constructed independently of one another around 1927, that of Schrödinger, which I use here, and the language of Heisenberg's matrices.

3. Usually, the elements of a space are vectors defined by their scalar products and their "length" (the product of a vector by itself). Functions in Hilbert space are defined so they have analogous properties. The scalar product of two functions, however, is not commutative: the scalar product of a function and another function is not equal to the product of the second function and the first. The "length" or norm of a function (the square root of the scalar product of that function with itself) must be finite.

4. The complementary nature of the measurements is reflected in the fact that their corresponding operators do not have the same eigenfunctions. This can be immediately derived from the noncommutative nature of the scalar product of functions in Hilbert space. We can just as easily say

that the operators correspond to those measurements that "do not commute" with one another.

5. Here, I am following the customary approach, which focuses on the question of eigenstates of the Hamiltonian. Schrödinger's equation leads directly to this question, but it is important to note that the approach proposed by Heisenberg, although it is indeed equivalent to Schrödinger's in those cases where the latter is operational, allows for a more "pragmatic" treatment of certain problems. In that case, we "avoid" the Hamiltonian question and, in so doing, can no longer assign an unambiguous physical meaning to energy, position, or momentum. Today, this comment is purely technical because the so-called fundamental equations of quantum mechanics make use of the Schrödinger equation. But it reflects the possibility of another pathway, one that could have been taken, and which, centered on matrix representation, would have emphasized transitions rather than states (in the case where the approaches are equivalent, transitions can be used to identify states, and states can be used to calculate transitions). As Cartwright notes (*How the Laws of Physics Lie*, 127–28), it is the transitions that play a genuine "causal role" in quantum mechanics.

6. Except, and this restriction will assume its fullest meaning in Book V, at the "continuous limit." In this case, the Hilbertian axiom I presented loses its power of self-consistent definition.

7. No doubt this is also one of the reasons why the problem of Poincaré nonintegrability had so few repercussions until recent decades. Beings belonging to Hilbert space assert, through their mathematical identity, that the problem doesn't exist for them, that no qualitative difference can characterize them, that they are all entitled to claim the same type of representation.

7. THE ARROW OF TIME

1. I have been paraphrasing a statement from *The Critique of Pure Reason* in order to point out that Kant extended (rightly?) the singularity of the dynamic object to all phenomena.

2. It is the movement of the hands that "tells" the time, and every history of clock making, caught between the ideal of a perfect, and purely dynamic, pendulum and the sophisticated development of methods of connection (nondynamic and involving shocks and drive mechanisms) without which there would be no clock, reveals the tension between the two senses of time-as-measurement. See Isabelle Stengers and Didier Gille, "Temps et représentation," *Culture technique* 9 (1983): 21–41, translated as "Time and Representation" in Isabelle Stengers, *Power and Invention: Situating Science* (Minneapolis: University of Minnesota Press, 1997), 176–211.

3. My narration takes "Prigogine" as its subject, but the name

represents a collective and, what's more, a collective that has continued to mutate. The different sections of this story express the different physical-mathematical approaches to the object as well as a new group of researchers familiar with the corresponding approach. "Prigogine" refers to the characteristic of continuity, one that requires and obligates, that is, pursues and reinvents the same problem in an ever-changing technical and human landscape.

8. BOLTZMANN'S SUCCESSOR

1. Ilya Prigogine, *Non-Equilibrium Statistical Mechanics* (New York: John Wiley & Sons, Interscience Publishers, 1962), 2.

2. Ibid.

3. Ibid.

4. Ibid.

5. Ilya Prigogine, *The End of Certainty: Time, Chaos, and the New Laws of Nature* (New York: Free Press, 1997).

6. The "abnormal" collision produced by the reversal of velocities in the case of a dilute gas can be defined in terms of the destruction of correlations alone. For, if we bring about a new velocity reversal after a certain period of "abnormal" evolution, bringing the gas further from equilibrium, the gas will start evolving again toward equilibrium. Therefore, its evolution is once again compatible with the hypothesis of molecular chaos, which means that the "reverse" collisions have not created (post-collision) correlations that the second reversal of velocities would make determinant.

7. The volume of the ensemble corresponds, in somewhat more technical terms, to its "measure."

8. Gibbs also defined a "canonical" ensemble that corresponds to a system interacting with a heat reservoir (at a given temperature). In this case, the energy is not invariant, but all the systems can be characterized by the same temperature, that of the reservoir.

9. The first response to this problem is found in what is known as the "ergodic hypothesis." According to this hypothesis, during the process of dynamic evolution, the system will, sooner or later, pass through every point (or, according to the "quasi-ergodic hypothesis," in the neighborhood of each point) of the region of phase space occupied by the microcanonical ensemble. It can then be shown that the average value of the dynamic properties over long periods of time becomes identical to their average value for the microcanonical ensemble. The generality of this hypothesis was subsequently shown to be false: ergodicity is not a general property of Hamiltonian systems, but characterizes only some of them.

10. In the text that follows, except where necessary, I do not distin-

guish between the two fields. Recall, however, that the state of a quantum system can never be represented by a point, nor by its evolution along a trajectory.

9. BOLTZMANN'S HEIR

1. "Dynamical and Statistical Descriptions of N-Body Systems," the first article to discuss this new development, coauthored by Ilya Prigogine, C. George, and F. Henin, appeared in *Physica* 45 (1969): 418–34. "A Unified Formulation of Dynamics and Thermodynamics," which offers a general approach to their initial proposal, written by the above three authors together with Léon Rosenfeld, dates from 1973 (*Chemica Scripta* 4, 5–32).

2. In slightly more technical terms, this subdynamics corresponds to a kinetic evolution, and correlations are defined as linear functions of this "leading component" moving toward equilibrium. For other subdynamics it is the correlations that direct the evolution of the system.

3. Gilles Deleuze and Félix Guattari, *What Is Philosophy?*, trans. Hugh Tomlinson and Graham Burchell (New York: Columbia University Press, 1994), 79.

4. The effective construction of the transformation leading to the physical representation is quite laborious. It would be of no interest for those physicists who are "unsympathetic" to irreversibility and would therefore require of any innovation that imposes such irreversibility that it at least supply a new power of definition for previously inaccessible matters of fact.

5. We can see in this evolution a dynamic process wherein "bare" particles become "clothed" particles. The distinction between "bare" and "clothed" dates back to field theory, where the concept of interaction introduced the vacuum and virtual particles. We can also say that cyclical representation in Hamiltonian dynamics can be used to make the transition from interacting "bare" particles to "clothed" particles whose behavior is autonomous. The uniqueness of the 1970 physical representation is that the process of clothing occurs over time and is related to sorting correlations that dynamics has no reason, nor the means, to distinguish.

6. This is the same value judgment expressed by Bernard d'Espagnat's somewhat measured assessment of Prigogine's work in *Reality and the Physicist: Knowledge, Duration, and the Quantum World* (Cambridge: Cambridge University Press, 1989), 119–35. In discussing irreversibility, d'Espagnat maintains a stable opposition between, on the one hand, quantum and classical mechanics, whose laws are always valid, even though they are in practice inapplicable to macroscopic systems, and, on the other, "our" empirical reality of the "dissipative" properties of those systems, the object of an intersubjective understanding instrumental in

nature. According to d'Espagnat, this opposition allows us to position customary "coarse-grained" procedures as well as the more "sophisticated" procedures proposed by Prigogine. As a result, the idea that such sophisticated procedures would allow us to call into question both quantum and classical mechanics as idealizations would then seem untenable. For, doesn't their derivation from *classical and quantum languages* confirm, on the contrary, the legitimacy of their starting point? "If neither kind of mechanics describes for us the true laws which the constituent parts of macroscopic systems in some way obey, if both are *no more than* idealizations . . . it is hard to see how they could have formed a good starting point for the construction of any physics of these same macroscopic systems" (132). The value judgment inherent in this type of argument is always the same: whenever it's a question of dissipative properties, the fact that the starting point implies "saving" them by means of approximations does not challenge the value of that starting point. That such "merely empirical" properties might be saved more satisfactorily by Prigogine obviously does not give him the right to challenge what his derivation implies is legitimate. However, what this commonsense argument obscures is its assumption that macroscopic systems, as such, entail no obligations. For, what Prigogine is attempting to do is less a derivation that would save them than an exploration of the obligations those systems would entail if it were in fact not a question of merely saving them.

7. See Léon Rosenfeld, "The Measuring Problem in Quantum Mechanics," in *Selected Papers of Léon Rosenfeld,* ed. Robert S. Cohen and John J. Stachel, Boston Studies in the Philosophy of Science, vol. 21 (Dordrecht: D. Reidel Publishing Company, 1979), 536–46.

8. Léon Rosenfeld, "Statistical Causality in Atomic Physics," in Cohen and Stachel, *Selected Papers of Léon Rosenfeld,* 547–70. The quote appears on page 562.

10. THE OBLIGATIONS OF CHAOS

1. Pierre Duhem, *The Aim and Structure of Physical Theory* (Princeton, N.J.: Princeton University Press, 1991), 141.

2. The project for a qualitative mathematics of morphology is obviously associated with René Thom and catastrophe theory. For a generalization of this project and the introduction of a "morphological revolution," see Alain Boutot, *L'Invention des formes* (Paris: Odile Jacob, 1993).

3. Collisions and not "the collision." An isolated elastic collision does not create any particular problem for dynamics and there is no reason to describe it in terms of the correlations it produces or destroys.

4. For a more detailed explanation of binary representations, see Ilya Prigogine and Isabelle Stengers, *Entre le temps et l'éternité* (Paris: Flam-

marion, 1992), and Ilya Prigogine, *The End of Certainty: Time, Chaos, and the New Laws of Nature* (New York: Free Press, 1997). Here, I'll limit myself to pointing out that the binary numeric representation is the natural representation of the baker's transformation (or "Bernoulli shift"). Referring to this representation, we can immediately see why the expansion and contraction involve a coefficient of 2 and why the square has length 1 (the various coordinates, corresponding to the numbers comprised between 0 and 1, are then represented as a succession of *decimals,* 0 or 1).

5. Contrary to an initial state resulting from the "normal" reversal of velocities, this state does not result in an evolution that moves away from equilibrium over a finite period of time before returning to "normal," but one that does so over an infinite period of time.

6. Where, by reversing the velocities of a given state, the preparation would be characterized by an infinite "entropy leap."

11. THE LAWS OF CHAOS?

1. I. M. Gel'fand and N. Vilenkin, *Generalized Functions,* vol. 4 (New York: Academic Press, 1964).

2. Leslie E. Ballentine, *Quantum Mechanics: A Modern Development* (Englewood Cliffs, N.J.: Prentice Hall, 1990).

3. The kinetic systems relevant to the construction of the 1972 "physical representation" were "large systems" by definition; the laboriousness of the precautions imposed by this construction became a stumbling block.

4. I am speaking of the past for, today, a degree of skepticism makes this effort problematic.

5. For this and what follows, see Ilya Prigogine, *The End of Certainty: Time, Chaos, and the New Laws of Nature* (New York: Free Press, 1997).

6. Gilles Deleuze and Félix Guattari, *A Thousand Plateaus: Capitalism and Schizophrenia,* trans. Brian Massumi (Minneapolis: University of Minnesota Press, 1987), 412.

7. With regard to the baker's transformation, we again encounter the distinction between expanding and contracting fibers, and the principle of selection that excluded the application of the baker's transformation to a contracting fiber. Transformations that place equilibrium in the future can be applied only to continuous densities in x (expanding fibers), and those that situate equilibrium in the past apply to continuous densities in y (contracting fibers). To be complete, we should add that this condition implies a restriction on the definition of the system's observables. The definition of density, ρ, in phase space (like that of the wave function, ψ, in Hilbert space) is not an end in itself. It is used to calculate the observable values of physical properties associated with other operators. A consequence of the fact that the "new representation" of ρ implies

singular functions is that it can be used only in conjunction with the appropriate functions, often referred to as "test functions." In the case of the baker's transformation, the test functions for the expanding fiber must be continuous in y and, for the contracting fiber, continuous in x. Obviously, such a distinction has no meaning for the geometric model, but it is expressed, in classical and quantum dynamics, by the impossibility of requiring an answer to certain questions that are, in principle, admissible for integrable systems.

8. This change of representation makes use of the theory of Fourier transformations.

9. In this case, the resonances are found to be associated with contributions to the evolution over time of the density, ρ, which have no equivalent in the description in terms of trajectory, contributions that are responsible for the irreversible "diffusion" of this density. The density tends toward a uniform value in the entire space, which is to say that it adopts the behavior expected of the ensemble analog of evolution toward equilibrium.

10. Whenever I speak of an atom that is de-excited, I am speaking of a highly simplified model, the "Friedrichs model," to which I alluded in Book IV. This model corresponds to the special case where spontaneous de-excitation can have a spectral representation in *Hilbert space;* but the corresponding evolution is then exponential (characterized by a lifetime) only as a first approximation.

11. In particular, the "dissipative criterion" that makes physical representation possible satisfies the conditions that define the possibility of an irreducible probabilistic representation.

12. I am speaking in terms of possibilities. For mathematicians, large Poincaré systems continue to present the problem of the transition to the thermodynamic limit. The hope of mathematicians is that a definition of the properties of "large systems" as such (for example, an electromagnetic field not restricted to a container) could be constructed that would allow continuous spectra to claim their own habitat without having to resort to their finite homologues.

12. THE PASSION OF THE LAW

1. And they might even be happy to do so. I find it hard to believe that physicists (and there are a number of them who feel the need, when controversies arise or when meeting with specialists in the fields of social, cultural, or anthropological studies, to maintain that physics is not "relative to opinion") are satisfied with the position they feel required to adopt, given that, because of a dogmatism bordering on stupidity, it contrasts with the inventiveness of the arguments that they, as physicists, experience.

2. "Enfance," in *Abécédaire de Gilles Deleuze,* which was broadcast on the French television channel Arte on August 6, 1995.

3. Here, I am alluding to the ideas of Émile Meyerson. See Book I of *Cosmopolitics.*

4. I am using the term in the same sense used by Whitehead in *Process and Reality* (New York: Free Press, 1969). For Whitehead, there could be no absolute conformity, for this corresponds to a "vacuous actuality." The heritage of the past always implies a minimum of creativity regarding the way in which the past is inherited. Rejection of the dream of absolute conformity realized by the dynamic object is at the core of Whitehead's thought (29).

5. That is, in Whiteheadian terms, they would reduce all "social order" to an order integrally and necessarily corpuscular (ibid., 36), which is, in his schema, as aberrant as the concept of vacuous actuality.

13. THE QUESTION OF EMERGENCE

1. See Bernadette Bensaude-Vincent and Isabelle Stengers, *A History of Chemistry* (Cambridge: Harvard University Press, 1996).

2. In fact, it could not without the definition of the part, and the entire problem of emergence, disappearing. The reader may recall (see *Cosmopolitics,* Books II, III, IV, and V) that if the physics of laws has cultivated an art, it is the art of freely redefining what belongs to the whole and what belongs to the part. Moreover, spectral (or cyclical) representation eliminates any distinction between whole and part. Every independent mode, evolving autonomously, "represents the whole," just as the Leibnizian monad contained the world even in its smallest details. As for the challenge I introduced with respect to Prigogine's work, even though it assigns crucial importance to the argument that the "laws of chaos" are irreducible to individual representation, in terms of trajectory or wave function, it is in no way a question of emergence but of a correlative redefinition of the terms of the problem and its solution. Neither the laws of chaos nor reversible and determinist behavior "emerge." In both cases, the definition of behavior and the definition of the entities that behavior introduces are strictly inseparable. Correlatively, the "arrow of time" does not emerge from dynamics. It is incorporated into the problem of dynamics (posed by "large Poincaré systems") once this problem is defined by the requirement that it can be solved.

3. Richard Dawkins, *The Blind Watchmaker* (New York: W. W. Norton, 1987).

4. See Horace Freeland Judson, *The Eighth Day of Creation: Makers of the Revolution in Biology* (New York: Simon & Schuster, 1979), 470–80.

5. Let me point out what is self-evident. The event that celebrates the

creation or mutation of an experimental factish is not, in this context, instantaneous. It marks a recognized innovation and punctuates a history that can be as laborious as one wishes. The successful delegation is the result of what Andrew Pickering, in *The Mangle of Practice: Time, Agency, and Science* (Chicago: University of Chicago Press, 1995), 21–22, compares, not without reason, to a "dance" in which "active" moments on the part of the scientist, when he constructs, introduces, arranges, are followed by "passive" moments when what he has tried to introduce actively manifest themselves, sometimes satisfying, and often disappointing, the "desire" immanent to the arrangement, the desire to "capture" what is operative (*capture of agency*). I do not employ the terms used by Deleuze and Guattari arbitrarily here. Pickering is, to my knowledge, the only Anglo-American researcher in the field of the social study of science to explicitly make use of this reference.

6. See Isabelle Stengers, "Le médecin et le charlatan," in Tobie Nathan and Isabelle Stengers, *Médecins et sorciers* (Paris: Éditions Synthélabo, 1995).

7. In Book II I endeavored to show that no relationship exists between this definition of a state and what it so often relied on, namely, the convergence of the description of a state and the explanation of the behavior brought about by dynamics.

8. Karl R. Popper and John C. Eccles, *The Self and Its Brain: An Argument for Interaction* (Heidelberg: Springer-Verlag, 1977).

9. The mathematical physicist Roger Penrose seeks the answer in quantum mechanics. In *Shadows of the Mind: A Search for the Missing Science of Consciousness* (Oxford: Oxford University Press, 1994), he finds that consciousness alone requires that we think of time in terms of "flux." Time would be reducible to a four-dimensional static geometry, *except* in the case of the reduction of the quantum wave function. *Therefore,* the two problems are connected and the revolution in physics that will bring about resolution of the question of quantum reduction will *also* be the one that will allow phenomena of consciousness to be understood in physical terms. As a result, one can assume that purely quantum effects play an important role in the brain (not as an explanation but as a prerequisite). And Penrose identifies the possible site of such effects: the microtubules that are one of the ingredients of the cytoskeleton of every cell.

14. THE PRACTICES OF EMERGENCE

1. J. K. Feibleman, "Theory of Integrative Levels," *British Journal for the Philosophy of Science* 5 (1954): 61.

2. Here, I restrict myself to the types of situation that have inspired reductionist arguments concerning water. Today, the questions and practices relating to it, that is, to its identities, have obviously been multiplied.

3. See Bernadette Bensaude-Vincent and Isabelle Stengers, *A History of Chemistry* (Cambridge: Harvard University Press, 1996).

4. Controversial immunologist Jacques Benveniste unwisely went in this direction when he asked physical chemists to submit to the verdict of his cells. The biological argument I develop is operational only if the evidence attributed to cells is reliable. This is never realized without difficulty but, for one reason or another, Benveniste didn't make it a priority. That choice cost him a great deal.

5. In biochemical practice, the scientist can also "delegate" the identification of a molecule to a living organism.

6. The term "quasi" in Latour and Serres comes first in the genealogy of the object. But to the extent that the biological "quasi detector" is an experimental factish, we are able to speak of it, to assign it a "quasi end," only to the extent that the "cellular metabolic" object has also been stabilized by our research practices. The circle that allows us to speak of a quasi detector as being first, whereas only our experimental detectors allow us to speak of them, is not vicious; it indicates the specificity of experimental factishes whenever they are defined as being previously mobilized by the living organism's methods of self-production.

7. Jean Perrin, *Atoms*, trans. D. L. Hammick (New York: D. Van Nostrand Co., 1916), 208.

8. That is why molecular biologists have greeted with disgust the idea of the "self-organization of matter," to the astonishment of physicists who believed they would be welcomed with enthusiasm. If selection is not the only cause, molecular biology might not provide the privileged access to the living organism it was thought to provide. Thirty years earlier, a proposal like that of Prigogine would have found, especially among embryologists, a much more interested group of listeners.

9. In "The Sociology of a Genetic Engineering Technique: Ritual and Rationality in the Performance of the 'Plasmid Prep'" (in Adele E. Clarke and Joan H. Fujimura, eds., *The Right Tools for the Job: At Work in Twentieth-Century Life Science* [Princeton, N.J.: Princeton University Press, 1992]), Kathleen Jordan and Michael Lynch show that the preparation of plasmids, which involves multiple operations of delegation, is still "artisanal" in the sense that the difference between procedures that work and those that don't can't be reexpressed in terms of the properties of the corresponding actors. That this preparation is nonetheless presented as a simple routine may express the fact that contemporary molecular biology integrates what I distinguish: the technical performance of "delegation" satisfies such an intense aspiration that experimental obligations come to be, if needed, flexibly modulated.

10. I should point out here the remarkable exception known as the Internet. Its designers not only wanted to create a technical infrastruc-

ture, but also to institute constraints intended to obligate history, to counteract the possibility of centralized control. That is why the fact that the future of the Internet may appear to be uniquely indeterminate should be celebrated as something novel, expressing the appearance of an unprecedented psychosocial type of technical innovator.

11. See Isabelle Stengers, *The Invention of Modern Science*, trans. Daniel W. Smith (Minneapolis: University of Minnesota Press, 2000), 155–58.

12. See Vinciane Despret, *Naissance d'une théorie éthologique: La danse du cratérope écaillé* (Paris: Éditions Synthélabo, "Les Empêcheurs de penser en rond," 1996). Despret's book describes how the observer may discover that her interest in the power of fiction also subjects what she observes to the power of her own relativistic fiction. When irony is replaced by the humor of a shared problem, what is produced is no longer a confusion of roles but a singularization of interests. The dancing babblers of the Negev Desert, the conflicting strategies of ethologists addressing them, as well as the way in which the birds effectively create the possibility of such conflict, the way the observer can entertain different relations with each of the researchers as well as the way in which their respective strategies organize different styles of apprenticeship both with the bird and the observer—all of the intermingled elements in this situation overlap without exclusion, creating the possibility for judgment. And "not judging" does not mean blindly accepting everything but experiencing how the appetite for judgment destroys interest.

13. For fieldwork described and *implemented*, see Bruno Latour, "Le 'pédofil' de Boa Vista—montage photo-philosophique," in *La clef de Berlin et autres leçons d'un amateur de sciences* (Paris: La Découverte, 1993).

14. Stephen J. Gould, *Wonderful Life: The Burgess Shale and the Making of History* (New York: W. W. Norton, 1990).

15. DISSIPATIVE COHERENCE

1. Ilya Prigogine and Jean Wiame, *Experientia* 2 (1946): 451.

2. See the polemic between Thom and Prigogine in *La Querelle de déterminisme* (Paris: Gallimard, "Le Débat," 1990).

3. I use the term "realization" rather than "actualization" because, when it is a question of fluctuations, we are dealing with selection from among predetermined possibles. Fluctuation is merely a trigger. Moreover, it is remarkable that the topic of "order through fluctuation" should become of secondary importance in Prigogine's writings after the 1980s, when the topic of irreversibility as a source of order then predominates. The continuity of the requirement thus holds sway over the unexpected character of what helped satisfy that requirement.

4. Gilles Deleuze and Félix Guattari, *A Thousand Plateaus: Capitalism*

and Schizophrenia, trans. Brian Massumi (Minneapolis: University of Minnesota Press, 1987), 200.

5. We can see in this situation where the laboratory "denies itself," where the power of definition refers to the question of the change of meaning of the very concept of definition, a situation that is fundamentally dialectical. Providing we keep in mind that, in this case, as in that of emergence, dialectics does not refer to the situation as such but to the situation as qualified by the requirements that brought it into existence. It would be better, then, to speak of an "assemblage mutation" in the sense that the laboratory assemblage that allows experimental requirements and the question of their satisfaction to exist is replaced by a new assemblage that includes the question of what the concerned being may be capable of in what becomes *its* environment (and no longer the limit conditions created by the experimenter).

6. Ilya Prigogine and Isabelle Stengers, *Order out of Chaos: Man's New Dialogue with Nature* (New York: Random House, 1984), 280.

16. ARTIFICE AND LIFE

1. For the distinction between "states of things" (such as a dynamic system), "things" (a dissipative structure), and "bodies," see Gilles Deleuze and Félix Guattari, *What Is Philosophy?,* trans. Hugh Tomlinson and Graham Burchell (New York: Columbia University Press, 1994). I also want to point out that the attribution of a "body" to a living organism leaves unanswered the question of how the organism experiences "having a body."

2. Quoted in Steven Levy, *Artificial Life: The Quest for a New Creation* (New York: Penguin Books, 1993), 113–14.

3. In M. Mitchell Waldrop, *Complexity: The Emerging Science at the Edge of Order and Chaos* (New York: Simon & Schuster, 1992), 202–3, there is a report of that exciting night in the winter of 1971–72 when Langton, working on a simulation of the Game of Life, "sensed the presence of someone else in the room." There was something alive there on the screen.

4. Levy, *Artificial Life,* 346.

5. Stuart Kauffman, quoted in ibid., 128.

6. In *Cosmopolitics,* Book V, chaotic trajectories were seen not to be robust in this sense, but within the field of artificial life, it is Conway's Game of Life that serves as the typical example of a type of behavior "sensitive to initial conditions." The game is governed by a small number of simple rules that can be used to determine, from step to step, which squares on a checkerboard will be occupied and which will be empty. A rather extraordinary number of different evolutionary possibilities result from these few simple rules.

7. To the extent that I am unable to sympathize with the claims of

second-order cybernetics, I will limit myself to referring readers to Jean-Pierre Dupuy, *Aux origines des sciences cognitives* (Paris: La Découverte, 1994).

8. I have simplified the situation by contrasting Kauffman and Langton. The bottom-up approach was also followed by Langton, although in his case, inspired by the Game of Life, it emphasized diversity rather than robustness. Later on, I will introduce the notion of the "edge of chaos" that is based on Langton's work, an idea Kauffman felt he had "missed," that he "should have" formulated (see Waldrop, *Complexity*, 300–304). This edge, which is also generic, serves as a meeting place between the bottom-up situations involving Kauffman's stable attractors and those involving "chaotic" behavior, of which the Game of Life is an example. These two authors, both of whom worked at the Santa Fe Institute of "artificial life," make different use of the "edge of chaos." For Kaufmann's problem it was a crucial ingredient; for Langston's problem, a synonym for solution.

9. Judith Schlanger, *Penser la bouche pleine*, 2d rev. ed. (Paris: Fayard, 1983).

10. See Léon Chertok and Isabelle Stengers, *Le Cœur et la raison: L'Hypnose en question, de Lavoisier à Lacan* (Paris: Payot, 1989).

11. Andrew Pickering, *The Mangle of Practice: Time, Agency, and Science* (Chicago: University of Chicago Press, 1995).

12. Relativist sociologists of science often describe the development of detectors in this way: the world remains silent, it has told us nothing, but the machine confirms its builder's "way of seeing." The fact that novelty, here, confers a highly interesting and unique meaning to this would-be universal claim is not devoid of humor.

13. Waldrop, *Complexity*, 299.

14. Stuart Kauffman, *The Origins of Order: Self-Organization and Selection in Evolution* (Oxford: Oxford University Press, 1993).

15. Kauffman's network of randomly connected Boolean automata was, at the beginning of neoconnectionism, already a schematization of a "genome" exhibiting regulatory properties alone. Every automaton is a "gene" that, when activated, synthesizes a product that activates and/or inhibits other "genes." Activation and inhibition are generally associated with a selective logic and, therefore, have a utilitarian value. Here, they are what the problem requires. What Kauffman was interested in was the question of the stable cellular differentiation that occurs during embryogenesis, that is, the number of distinct "cellular operations" (the number of attractor basins) that can be obtained from a single genome.

16. In this way, the edge of chaos provides a model of behavior that calls to mind the way the living organism was situated by Henri Atlan "between crystal and smoke," and Karl Popper, "between the clock and the cloud."

17. Kauffman, *The Origins of Order*, 522.

18. Stuart Kauffman, *At Home in the Universe: The Search for the Laws of Self-Organization and Complexity* (New York: Viking Press, 1995), 243.

17. THE ART OF MODELS

1. François Julien, *La Propension des choses* (Paris: Seuil, 1992).

2. Stuart Kauffman, *The Origins of Order: Self-Organization and Selection in Evolution* (Oxford: Oxford University Press, 1993), 644.

3. The generic properties of interconnected networks are "universal" in the sense that they are "all-purpose." Similarly, once we are able to recognize, in any given milieu, a transformation operating far from equilibrium and subject to activation *and* inhibition by two of its products, those products being capable of diffusion throughout the milieu, the question of the appearance of spatial differentiation arises. If the inhibiting product diffuses more rapidly than the activating product, homogeneity will be broken (Turing instability) and give way to a succession of regularly distributed activity "peaks."

4. Stephen Jay Gould, "Judge Scalia's Misunderstanding," in *Bully for Brontosaurus: Reflections in Natural History* (New York: W. W. Norton, 1992), 460. Here, Gould is replying to Justice Anton Scalia, who had taken a minority position during a Supreme Court ruling in June 1987 against creationists' claim that creationism should be placed on the same footing as evolution in schools. Justice Scalia held that the reference to creation is distinct from religious belief, which claimed to explain creation by introducing a creator. The creation required by creationism can then be placed on the same footing as the origin of life from inanimate matter required by Darwinian evolution. But, Gould affirms, the theory of evolution does not need to make reference to the origin of life.

5. All discoveries are precarious, but irreversible. We can say that the former "vital force" has been irreversibly abandoned, along with its mission of attempting to "explain everything." And it is noteworthy that self-organization in Prigogine's or Kauffman's sense does not challenge "Darwinian discovery" but merely the neo-Darwinian narrative that postulates the omnipotence of selection. On the other hand, it is perfectly conceivable, although currently indeterminable, that *some* aspects of biological innovation might one day be recognized as irreducible to Darwinian scenarios (for example, see A. Portmann's work on ethomorphology) and that, correlatively, our concepts of what matter can do might be transformed. This may be an unlikely revolution, but it is not impossible. In any event, it would "follow" Darwin, for it is in contrast to "Darwinian discovery" that the problem will have been defined.

6. "Life is robbery," Whitehead wrote in *Process and Reality*, and this

lapidary characterization marks the disconnect (relative for Whitehead) made use of by Darwinian discovery. Life is a robber, and the adventures of living things are marked, before all other distinctions, by the risks encountered by anyone for whom robbery is their means of existence, who, in order to exist, must actively "exploit" a milieu. This definition can be used to identify the conceptual obstacle to the idea that Gaia is a living thing. The Earth bathes in the rays of the Sun, yes, but it is hard to say that it exploits them in the sense that a living thing exploits a milieu. The Sun is too reliable, it is closer to the "limit conditions" that are explored in a laboratory than to a being one might exploit at one's own risk.

7. Norbert Wiener, *Cybernetics: Or Control and Communication in the Animal and the Machine* (Cambridge: MIT Press, 1961), 24. We could see in Marx's *Capital* a "model" of history that satisfies Wiener's criticism. Marx did not intend to deduce history from a model. Marx's gamble, the prerequisite of his model, was that the history of capitalism could be understood on the basis of the question asked by a "universal," which was both necessary and relative to that history, namely, the tendency of the rate of profit to decline. It was the stability of the question—although how the problem was presented was constantly reinvented, that is, the way in which its consequences would be avoided—that allowed Marx to define a "historical logic" that united necessity and contingency. *Capital,* as the subject of that history, is both in a position of submission (to the question) and invention. See Daniel Bensaïd, *Marx l'intempestif: Grandeurs et misères d'une aventure critique* (Paris: Fayard, 1995).

8. T. W. Schoener, "The Controversy over Interspecific Competition," *American Scientist* 70:6 (1982): 586–95.

9. Isabelle Stengers, *The Invention of Modern Science,* trans. Daniel W. Smith (Minneapolis: University of Minnesota Press, 2000), 158. As will be seen in the following chapter, "tact" changes its nature whenever the power relationship that enables someone to identify the problem presented to the other becomes an abuse of power. That is why tact can characterize the teacher or the therapist, whose techniques have the ability to create a problem, but is in no sense an "all-purpose quality."

10. See Albert Goldbeter, *Biochemical Oscillations and Cellular Rhythms: The Molecular Bases of Periodic and Chaotic Behaviour* (Cambridge: Cambridge University Press, 1996).

11. And from time to time: what difference does it make if a theorem, made possible by such assumptions then serves as a reference for introducing the problem of unemployment? Of course, this doesn't concern the "real" economist, who knows the conditions to which the model was subject. See Jean-Paul Fitoussi, *Le Débat interdit* (Paris: Arléa, 1995).

12. J. L. Deneubourg and S. Goss, "Collective Patterns and Decision-

Making," *Ethology, Ecology, and Evolution* 1 (1989): 295–311.

13. The question raised here concerning ants is similar to the one asked by specialists of "distributed intelligence," who want to make intelligence "emerge" from simpler, modular behaviors. It should be pointed out, therefore, that the interest of the "ant model" is that it in no way entails the promise of a general theory of intelligence but, on the contrary, introduces the possible invention of a very specific mechanism, which only has meaning in terms of the "quasi-technical" definition of the problem: the collective-exploration-of-the-resources-of-a-milieu.

14. Gilles Deleuze and Félix Guattari, *What Is Philosophy?*, trans. Hugh Tomlinson and Graham Burchell (New York: Columbia University Press, 1994), 123.

15. That is why it is important to approach with caution Humberto Maturana and Francisco Varela's proposal to establish a general epistemology of the living based on the concept of autopoiesis. The concept is convincing when a "body" authorizes the modeling. The "informational closure" then expresses the fact that the "body" produces its own requirements. But autopoiesis does not explain the living thing. On the contrary, it is the living thing that, in some cases and with certain approximations, gives its relevance to autopoiesis and to the power relationship of the model it promises. In terms of this, consider the way Félix Guattari proposes complicating the concept by including the fact that the "model" is not only on the order of the practice of understanding (epistemology) but primarily refers to the effective fabrication (historical, political, machinic) of bodies: "Autopoiesis deserves to be rethought in terms of evolutionary, collective entities, which maintain diverse types of relations of alterity, rather than being implacably closed in on themselves. In such a case . . . when one considers them in the context of the machinic assemblages they constitute with human beings, they become ipso facto autopoietic" (Félix Guattari, *Chaosmosis: An Ethico-Aesthetic Paradigm*, trans. Paul Bains and Julian Pefanis [Bloomington: Indiana University Press, 1995], 39–40). But "modeling" then implies practical risks that have nothing to do with those of the field sciences, the risky becoming of the "emerging," which Guattari has explored through the question of schizanalysis.

18. TRANSITION TO THE LIMIT

1. This description of the transition to the limit is quite accurate, and is one of the points where physical-mathematical practice alludes to the problem most forcefully. In effect, it describes the gesture whereby the construction of an "object," the phase transition, is organized from some "critical point" whose very definition is the recognition that at this point we are required to abandon whatever it was that allowed us to differentiate

the phases. During a "second-order phase transition," that between a gas and a liquid, for example, we cannot determine what we are dealing with: droplets in a gaseous milieu or bubbles in a liquid milieu (cf. the "edge of chaos"). Correlatively, the definition of what is a gas, or a liquid, or a solid is transformed, for it implies making explicit whatever it is that makes the definition possible. In this sense, it is "critical." So, from the point of view of nonequilibrium phase transitions, we can say that the state of equilibrium is (ideally) characterized by correlations of zero scope and intensity.

2. Gilles Deleuze, *Difference and Repetition,* trans. Paul Patton (New York: Columbia University Press, 1995), 187.

3. Daniel N. Stern, *The Interpersonal World of the Infant: A View from Psychoanalysis and Developmental Psychology* (New York: Basic Books, 1985).

4. It is interesting to note how some have reduced Stern's argument to the reassuring status of a useful, but merely empirical, addition, while others have categorized him as a "cognitivist," thereby "normalizing" his opposition to conventional psychoanalytic interpretation.

5. Félix Guattari, *Chaosmosis: An Ethico-Aesthetic Paradigm,* trans. Paul Bains and Julian Pefanis (Bloomington: Indiana University Press, 1995), 94–98. See also Léon Chertok and Isabelle Stengers, *Le Cœur et la Raison: L'Hypnose en question de Lavoisier à Lacan* (Paris: Payot, 1989).

6. See Tobie Nathan, "Manifeste pour une psychopathologie scientifique," in Tobie Nathan and Isabelle Stengers, *Médecins et sorciers* (Paris: Éditions Synthélabo, 1995).

7. The contrast between before and after is also found among mathematicians themselves, in the difference between the laborious construction of a new mathematical being and the way in which that being, once constructed, imposes itself as a "discovery" in nearly Platonic fashion, self-engendered by the limpid coincidence between its definition, the requirements it translates, and the obligations it imposes.

8. Of course, this practice has nothing at all to do with the caricature that occurs when the teacher is convinced that mathematical definitions and rules *must* suffice. What is most astonishing is that, even in this case, transmission can sometimes occur, in spite of everything. Moreover, we can, I believe, parallel the selective character of knowledge, that is, the rarity of the success of the modeling operation, with mathematical transmission existence and virulence of this conviction, which can transform the misunderstanding into a violent confrontation. Whenever it is a question of "knowing how to walk" or "knowing how to speak," parents ignore the rules and proceed with rather remarkable tact, which may also be as "robust" as the knowledge whose actualization they induce and accompany. The multiple methods employed when learning to read also express a certain perplexed consideration for what the transformation of a series

of signs into a text assumes. The fact that mathematical transmission doesn't celebrate the assumption of meaning it depends on transforms the singularity its "categorical misunderstanding" constitutes into an instrument for issuing a verdict. Here, modeling includes a differentiation that associates this transmission with a nearly transcendent condition—one is "good in math" or one isn't.

9. Gilbert Simondon, *L'Individu et sa genèse physico-biologique: L'individuation à la lumière des notions de forme et d'information* (Paris: Presses universitaires de France, 1964).

10. See Gilles Deleuze, *Difference and Repetition*, trans. Paul Patton (New York: Columbia University Press, 1994), especially 129–68, for the natural illusion that consists in copying problems from propositions, and 222–62 for individuation.

11. Simondon, *L'Individu et sa genèse physico-biologique*, 18.

12. Ibid., 19.

13. In *Savoir-faire et compétences au travail: Une sociologie de la fabrication des aptitudes* (Brussels: Éditions de l'Université de Bruxelles, 1993), Marcelle Stroobants demonstrates that Simondon's thought is indeed a vector of resistance. She uses transduction to "denaturalize" the cognitive concepts of competence, know-how, or, worse, coping skills, which is to say, to reinvest them with their social and political dimensions. Stroobants thereby constructs the problem of "authorization," a concept inextricably cognitive and social. She shows how, in school and elsewhere, "competencies" are effectively transmitted, but in differential fashion. The process leads to modes of knowledge appropriation that may or may not "authorize," that is, which make a difference that is both subjective and social between knowledge that "qualifies" and knowledge that the specialist may identify in the other but which does not possess the resources to be valued and recognized outside the situation that gives it meaning.

14. See Gilles Deleuze and Félix Guattari, *A Thousand Plateaus: Capitalism and Schizophrenia*, trans. Brian Massumi (Minneapolis: University of Minnesota Press, 1987), 366–67. There, Deleuze and Guattari redefine the relationships between "innate" and "acquired." In the territorial assemblage they refer to as "natal," the innate is not "encoded" behavior autonomously unfolding, and the acquired cannot be explained on the basis of outside stimuli: "The natal, then, consists in a decoding of innateness and a territorialization of learning, one atop the other, one alongside the other."

15. For Simondon, individuation, regardless of its "field," is produced only once, and it is then a condition for a process of individualization that is itself permanent. This distinction is the axis of *L'Individuation psychique et collective* (Paris: Aubier, 1989), the second part of Simondon's doctoral

dissertation, which was published twenty-five years after *L'Individu et sa genèse physico-biologique*.

16. Thus, the processual and aesthetic paradigm proposed by Félix Guattari in *Chaosmosis* would imply putting any practice that banks on the Simondonian distinction between individuation and individualization at risk, while the therapeutic practice of Tobie Nathan approves this distinction. Speculative thought allows us to express this divergence, to make the terms resonate, to situate the respective risks (Guattari speaks of the therapy of the "hypermodern" individual and Nathan about the obligations associated with the therapy of migrant populations) but not to distribute its good and bad points.

17. In *Rewriting the Soul: Multiple Personality and the Sciences of Memory* (Princeton, N.J.: Princeton University Press, 1995), 224, Ian Hacking adopts a similar position. He quotes Whitehead, for whom the real problem is much less the possibility of "dissociated personalities," as shown by clinical treatment, than the unity of the "living person," whose precariousness is shown by the number of clinical cases. For Hacking, it is a question of a "desirable relationship" between philosophy and therapy: "Whitehead's philosophy has a ready-made slot for the multiple personality but can gain no support from it. His cosmology neither predicts nor explains any detail of the phenomena. Conversely, the clinical structure of multiple personality disorder is totally independent of Whitehead's cosmology."

18. Stephen Jay Gould, "Cardboard Darwinism," in *An Urchin in the Storm* (New York: Penguin Books, 1990), 50.

19. Stephen Jay Gould, "Triumph of a Naturalist," in ibid., 166.

20. Stephen Jay Gould, "Roots Writ Large," in Lewis Wolpert and Allison Richards, eds., *A Passion for Science: Renowned Scientists Offer Vivid Portraits of Their Life in Science* (Oxford: Oxford University Press, 1988), 143. Gould didn't realize just how accurate he was. Dorothy Sayers even had a theory of divine creation based on the experience of creative imagination. The Trinity is illustrated by the author's experience. There are always three books: the book you conceive of, the book you write, and the book that you and your readers read. See Barbara Reynolds, *Dorothy L Sayers: Her Life and Soul* (London: Sceptre, 1993).

21. Countereffectuation, as distinct from effectuation, refers to the practice of the event at the heart of Deleuze's thought. It runs through his *The Logic of Sense* (trans. Mark Lester with Charles Stivale [New York: Columbia University Press, 1990]), especially the twenty-first series, and appears in *What Is Philosophy?*, where the crucial distinction between functions and concepts is made. Functions, not only those that correspond to scientific propositions, but also those the organism actualizes

("the most elementary organism forms a proto-opinion on water, carbon, and salts on which its condition and power depend," 155) correspond to the effectuation of an event in a state of affairs, "but it is *counter-effectuated* whenever it is abstracted from states of affairs so as to isolate its concept. There is a dignity of the event that has always been inseparable from philosophy as *amor fati*" (159).

22. A slope that certain proponents of cognitivism descend with the monotonous candor of arrogance, but which also awaits those preoccupied with "minority" questions such as hypnosis. There is nothing like the experience of hypnosis to arouse the unwavering conviction that conscious experience is, in one way or another, "inauthentic," cut off from some more essential truth. At that point the field is ready for a revival, either informed or blindly repeated, of the romantic mise-en-scène that assigns truth to the depths of being and contrasts it with the repressive slogans of a soulless society.

23. The "dance" between the experimenter and a novel device is not really based on tact, for the finality of this dance, well described by Andrew Pickering in *The Mangle of Practice: Time, Agency, and Science* (Chicago: University of Chicago Press, 1995), is the stable distribution between end and means or question and answer. As for God, finding out if tact is all he requires is an interesting·theological question. I would be inclined to think that the God of Whitehead's cosmology, the God of speculative thought, does not possess, when interacting with the propositions of the world, the relationship of precedence that tact assumes. He may confirm them, but they are "new" even for him.

24. Tobie Nathan, *Fier de n'avoir ni pays ni amis, quelle sottise c'était* (Paris: La Pensée sauvage, 1993), 25–28.

19. THE CURSE OF TOLERANCE

1. Concerning the question of slogans, or master words, *(mots d'ordre)*, see Gilles Deleuze and Félix Guattari, *A Thousand Plateaus: Capitalism and Schizophrenia*, trans. Brian Massumi (Minneapolis: University of Minnesota Press, 1987), especially 75–80. [I have not used Massumi's translation, "order word," in this text. For an interesting discussion of the translation of Deleuze's work, see Jean-Jacques Lecercle, *Deleuze and Language* (New York: Palgrave Macmillan, 2002), 169.—*Trans.*]

2. Sigmund Freud, "Analysis Terminable and Interminable," in *The Standard Edition of the Complete Psychological Works of Sigmund Freud*, translated from the German under the general editorship of James Strachey in collaboration with Anna Freud, assisted by Alix Strachey and Alan Tyson (London: The Hogarth Press and The Institute of Psycho-Analysis, 1966), 23:209–53.

3. See Isabelle Stengers, *The Invention of Modern Science*, trans. Daniel W. Smith (Minneapolis: University of Minnesota Press, 2000), and Book I of *Cosmopolitics*, "The Science Wars."

4. Bruno Latour, *Petite Réflexion sur le culte moderne des dieux faitiches* (Paris: Synthélabo, "Les Empêcheurs de penser en rond," 1996).

5. This is not the place to discuss the different factishes I introduced during my exploration of the "laws of physics" (see *Cosmopolitics*, Books II, III, IV, V), for they are relative to the singular history of physics alone and the conflicts *among the physicists* who punctuated that history.

6. Recall that an electron whose motion "obeys" the laws that define it as "sensitive" to an electromagnetic field is a factish capable of autonomously confirming the law it obeys only because the physicist presumes its nonsensitivity to the power that law claims to have. The experimenter's joy when her factish, through its obedience, confirms its own autonomy, arises from her conviction that the being she investigates has not responded out of complacency, submissiveness, or abuse of power, and that it possessed all the means available to disappoint her.

7. William Blake, *Auguries of Innocence.*

8. Ibid.

9. See Bruno Latour, *We Have Never Been Modern*, trans. Catherine Porter (Cambridge: Harvard University Press, 1993), especially the chapter titled "Relativism."

20. THE CURSE AS TEST

1. Maurice Merleau-Ponty, *The Phenomenology of Perception*, trans. Colin Smith (New York: Humanities Press, 1962), xxii.

2. Ibid.

3. Ibid., xxiv.

4. Ibid., xxiii.

5. Georges Devereux, *From Anxiety to Method in the Behavioral Sciences*, preface by Weston la Barre (The Hague: Mouton, 1967).

6. Ibid., 23.

7. Ibid., 30.

8. Here, Devereux approaches what was probably a key source for Niels Bohr's thought. We know that Bohr had his colleagues read a novel by Poul Martin Møller about the anguish of a Danish student who was trying to think about his own thoughts, all the while failing to prevent the ensuing infinite regression. The story reveals the absence of a stable "demarcation." Every "ego" that observes the previous ego brings into being a new "ego" that observes the observer.

9. "A theory whose experimental strategy demands the *destruction* of that which it seeks to study . . . abrogates itself" (Devereux, *From Anxiety to*

Method in the Behavioral Sciences, 14).

10. Ibid., 304.

11. Ibid., 310.

12. Ibid., 216.

13. Ibid., 304.

14. Ethnopsychiatry or ethnopsychoanalysis? Devereux made use of both terms. The advantage of the first is that it reminds us that "-iatry," the practice of care, is not the monopoly of the physician. The advantage of the second expresses the fact that it is psychoanalysis that has risked making this claim. I have chosen the first because it is not at all certain that psychoanalysts are currently the privileged heirs of that risk. Maybe the question of what psychiatry can become is today more uncertain than the question of what psychoanalysis can become.

15. Devereux, *From Anxiety to Method in the Behavioral Sciences,* 312.

16. Ibid., 304.

17. George Devereux, *Mojave Ethnopsychiatry: The Psychic Disturbances of an Indian Tribe,* Smithsonian Institution Bureau of American Ethnology Bulletin 175 (Washington, D.C.: Smithsonian Institution Press, 1969), viii.

18. We could say that Devereux thus betrayed Freudian technique, whose intolerance required that the analyst retain no knowledge that would allow him to judge his patient. The patient was to be subjected to the test of the demarcations produced by the analyst in all their radicality.

21. ANXIETY AND FRIGHT

1. Tobie Nathan, *L'influence qui guérit* (Paris: Odile Jacob, 1994), 230.

2. Ibid., 237.

3. Ibid., 242.

4. Ibid., 18.

5. Edwin Zaccaï, "Entretien avec Tobie Nathan. Des ethnies à la psychiatrie. Allers et retours," *Cahiers de psychologie clinique* 4, *La Pensée magique* (Brussels: De Boeck Université, 1995): 163–75. See page 171.

6. Tobie Nathan and Lucien Hounkpatin, "Oro Lé, la puissance de la parole," in Isabelle Stengers, *L'Homme: La psychanalyse avait-elle raison? Rencontres de Châteauvallon* (Paris: La Pensée sauvage, 1994), 214.

7. See Tobie Nathan, *Fier de n'avoir ni pays, ni amis, quelle sottise c'était* (Paris: La Pensée sauvage, 1993), *L'influence qui guérit* (Paris: Odile Jacob, 1994), and "Manifeste pour une psychopathologie scientifique," in Tobie Nathan and Isabelle Stengers, *Médecins et sorciers* (Paris: Éditions Synthélabo, 1995).

8. Without considering that there is no reason to place all "nonmoderns" in the same boat. According to Nathan, the Yoruba of Benin, whose thought, he notes, managed to cross the ocean in the slave ships, would be

the real teachers of psychotherapy.

9. This is true for psychoanalysis as well—even if the psychoanalyst is happy and proud to follow Freud, who was able to "listen to" his hysterical patients, even if Lacanians are happy and proud to talk about analysands rather than the analyzed. An analyst always communicates what she has heard [from patients] to other analysts. It is always with competent colleagues that she tests the evidence her technique has been used to construct about the people who have come to her for treatment.

10. See Nathan, "Manifeste pour une psychopathologie scientifique," 21.

11. A less dramatic but equally significant example is given by the Rosenthal experiment, which gave two groups of rats to students (see Vinciane Despret, *Naissance d'une théorie éthologique* [Paris: Éditions Synthélabo, "Les Empêcheurs de penser en rond," 1996]). The rats in the first group, Rosenthal told the students, are from breeds selected for their intelligence; those in the second group, for their stupidity. Experimentation in a maze confirmed this. But Rosenthal had assigned his students rats with no particular features. What can we learn from this experiment, in which social psychology is more important than the psychology of rat mazes? Let us put aside trivial explanations. The students didn't cheat, and Rosenthal was able to show that the experimental control was inadequate, that is, the so-called objective and reproducible data resulting from the mazes were unreliable. But rather than asking where the "fault" lies, we may be better off describing how the three types of protagonist are "obligated" by proof, on behalf of science. The rats were subjected to the imperative of an experimental science that was being constructed, that much is obvious. But so was Rosenthal, who in the name of science allowed himself to *deceive* his students. And so were the students, who believed they were experimenters whereas they were the subjects of the experiment. Everyone, in their own way, obeyed the imperative that made their "behavior" intelligible, the imperative of a science under construction. Then what do observations and measurements in Rosenthal's experiment reveal? I would go so far as to say that they explicitly force upon us the fact that they reveal that *all* the behaviors found in behavior laboratories are *created on behalf of science*. Before all else, they reveal the price that must be paid to create a stable distinction between the one who asks the question and the one who answers it. All of them are obligated, each in their own way, by the reference to the scientific knowledge being constructed, the proof being established.

12. It is most likely the weakness of hypnotic techniques that reduces "magical gestures" to a relationship that lends itself to a psychological reading, and fright to the anxiety of having to assume a role liable to

inspire fantasies of omnipotence. See Léon Chertok, Isabelle Stengers, and Didier Gille, *Mémoires d'un hérétique* (Paris: Éditions La Découverte, 1990). No doubt this is why the history of those techniques leads us to witness the perpetual return to the evocation of a psychic agency (the Ericksonian unconscious, for example) sufficiently powerful to release the practitioner from her anxiety by allowing her to attribute to this "other" the responsibility of the effects produced.

13. Zaccaï, "Entretien avec Tobie Nathan" 171.

14. The paradigm, as understood by Kuhn, that effectively "closes" actual scientific groups, characterized by production procedures, in this sense corresponds to a culture. This confirms the fact that the "words" or theories it articulates do not have the power, in themselves, to make it explicit. It is only when practical "gestures" have been transmitted that the "theory" that appears to justify them assumes meaning, the whole corresponding to the specific mentality of the practitioner (a puzzle solver). See Isabelle Stengers, *The Invention of Modern Science*, trans. Daniel W. Smith (Minneapolis: University of Minnesota Press, 2000), 48–53. Note, however, that "paradigmatic culture" is relative to a form of participation that Nathan would consider secondary, in the sense that no scientist has paradigmatic language as her mother tongue. This distinction corresponds to the obvious distinction between the practice of scientific controversy as "role playing" and the practice of psychotherapy, which is a matter of life and death.

15. Zaccaï, "Entretien avec Tobie Nathan," 174. It is striking that Léon Chertok, who was able to practice hypnosis without protecting himself with a theory that justified its "influence," had experienced something similar to what Nathan went through in Cairo in his native "Litvakia" (currently Belarus). See Chertok, Stengers, and Gille, *Mémoires d'un hérétique*, especially 34–41, 147–48, and 166–68.

16. Autopoesis has two creators: Humberto Maturana and Francisco Varela. It is held in tension between a biological notion directly inspired by the study of vision in batrachians (Maturana), where informational closure enables the biologist to "judge" how the frog perceives the fly, and a speculative concept (Varela, but especially Félix Guattari in *Chaosmosis*), where this closure no longer authorizes the one who invokes it to make any judgment at all, no longer implies limitation, no longer communicates with the possibility of an ironical stance. "Viability" then becomes a purely descriptive concept a posteriori, and autonomy the simple reminder of the asymmetrical character of any reciprocal capture, a coupling that is always lateral and partial, and which does not establish any shared "logic." The ways in which one counts, because of this reciprocal capture, for the other and the ways the other counts for oneself are a priori

completely distinct, even if negotiation between those ways of "counting for the other" might be part of the outcome of that event (see *Cosmopolitics*, Book I, "The Science Wars").

17. According to Nathan, this same distinction is made by those directly involved. In the event of a disturbance, one does not at first turn to a mediator of the invisible. One first tries to make use of "Western" medicine or the available practical resources of the "lay" group (family counseling, parents, etc.). It is only when the disturbance is "serious," when it "resists," that the need to decode the "supernatural" message it necessarily expresses becomes imperative. The same is true for Nathan's own ethnopsychiatric work: those who are sent to him have already "benefited" from all the therapeutic and related resources our society makes available to those who suffer.

22. THE POLITICS OF TECHNICAL INVENTIONS

1. When Huygens was working, the question revolved around the ability to make use of reliable timepieces on board ships, a necessary condition for solving the problem of longitude.

2. See Bruno Latour's superb *Aramis: or the Love of Technology*, trans. Catherine Porter (Cambridge: Harvard University Press, 1996), where the consequences, in this case fatal, of the hierarchization of the obligations a technical device must simultaneously satisfy are explored.

3. See the description of the "alliances" that construct the lengthy networks of the modern world in Bruno Latour, *Science in Action: How to Follow Scientists and Engineers through Society* (Cambridge: Harvard University Press, 1988).

4. Bruno Latour, *We Have Never Been Modern*, trans. Catherine Porter (Cambridge: Harvard University Press, 1993).

5. Ibid.

6. Ibid.

23. THE COSMOPOLITICAL QUESTION

1. Philippe Pignarre, *Qu'est ce qu'un médicament?* (Paris: La Découverte, 1997).

2. Therefore, it corresponds to Félix Guattari's "axiological creationism." Perhaps I should have used the term "chaosmopolitical," with explicit reference to Guattari. But, independently of the problem presented by the additional wordplay involved, I prefer to follow Alfred North Whitehead in using the older term and, in so doing, highlighting the problematic, speculative dimension that remains associated with it.

3. Marcel Gauchet, *The Disenchantment of the World: A Political History of Religion*, trans. Oscar Burge (Princeton, N.J.: Princeton University Press, 1997).

4. Ibid., 11.

5. While I have emphasized the fact that our tradition is constitutively political, the reader will have noted that I have remained silent about the practitioners of the "impossible profession" of governing, known as "politics." That is because I know nothing of politicians' activity in terms of practice; I know nothing of their requirements, aside from the fact of their being elected, and I am incapable of identifying the obligations associated with that practice aside from the fact that they oppose legislation when they are in the opposition and bend to the necessity of the situation when they are in power. This empirical "I know nothing" expresses a situation referred to as the "crisis of politics." The question here is not so much to determine if political bodies are capable of resisting so-called economic imperatives. The crisis is associated with their inability to bring into existence the political problem constituted by the power of those imperatives, that is to say, to present themselves as practitioners confronting that problem.

6. Tobie Nathan, "Manifeste pour une psychopathologie scientifique," in Tobie Nathan and Isabelle Stengers, *Médecins et sorciers* (Paris: Éditions Synthélabo, 1995), 7 and 43.

7. The corresponding myth was invented by Stephen Hawking on the last page of his *A Brief History of Time*. When physicists have agreed on what the universe is, that is, *how* it is appropriate to describe it, everyone—philosophers, religious leaders, and so on—will be able to come together to debate the why of that *how*. This also brings to mind the way some people agree to treat human destiny from the temporal perspective of the death of the sun or the stars.

8. As I write these lines, Latour has decided to revisit the distinction, which his creation of the term "factish" clearly reflects. The critical discussion that follows is not so much addressed to Latour but to a position that is of interest because it appealed to him at a given moment.

9. The long network, always under construction, has the features of the rhizome (see Gilles Deleuze and Félix Guattari, *A Thousand Plateaus: Capitalism and Schizophrenia,* trans. Brian Massumi [Minneapolis: University of Minnesota Press, 1987]), except for the fact that—modernist rhetoric as an expression of a political problem—it is presented in the form of an ideal network, characterized by certain conditions and a finality (see the organism as claiming to constitute the aim of its organs' functioning in *A Thousand Plateaus*).

10. Whenever psychoanalysts enter a long network, as has been the case with drug policy, the result is politically disastrous. See Isabelle Stengers and Olivier Ralet, *Drogues: Le défi hollandais* (Le Plessis-Robinson: Éditions Delagrange/Synthélabo, "Les Empêcheurs de penser en rond," 1992). See also, for the role of psychoanalysis in managing the problems

introduced through technologies of "artificial" procreation, Michel Tort, *Le Désir froid* (Paris: La Découverte, 1992).

11. Bruno Latour, "Note sur certains objets chevelus," in *Pouvoir de sorcier, pouvoir de médecin, Nouvelle Revue d'ethnopsychiatrie* 27 (1995): 30.

12. In Bruno Latour, *We Have Never Been Modern,* trans. Catherine Porter (Cambridge: Harvard University Press, 1993), such a triage program appears: "Let us keep what is best about them, above all: the premoderns' inability to differentiate durably between the networks and the pure poles of Nature and Society, their obsessive interest in thinking about the production of hybrids of Nature and Society, of things and signs, their certainty that transcendences abound, their capacity for conceiving of past and future in many ways other than progress and decadence, the multiplication of types of nonhumans different from those of the moderns. On the other hand, we shall not retain the set of limits they impose on the scaling of collectives, localization by territory, the scapegoating process, ethnocentrism, and finally the lasting nondifferentiation of nature and societies" (133). Who is "we" here? Are we certain about the definition of what "we" are rejecting? What Latour does not wish to retain is assuredly a problem for "us." However, it could also designate conditions (stated in a language that disqualifies them) that nonmoderns might refuse to redefine as constraints, thus rejecting what I previously presented as a political test. In so doing, they would affirm the irreducibility of their short networks against Latour's generalized symmetry.

24. NOMADIC AND SEDENTARY

1. If we were tempted to transfer this type of analysis to human populations, we would need to use a Marxist approach and accept, for example, the definition of the salaried worker as a social relationship, for it is the only one in which the term "strategy" can be usefully transferred without blindly naturalizing adversarial relationships. The production and reproduction of the medical labor force are clearly more costly than they are for a nurse, who is trained more quickly, paid less, and has a much shorter working life (career). Correlatively, the physician is authorized to make decisions and fully exploit the resources of the medical environment, while the nurse is limited to knowing how to make use of those that have been assigned to her.

2. Alfred North Whitehead, *Science and the Modern World* (New York: Free Press, 1967), 16.

3. I want to thank Tobie Nathan for this example.

4. The situation is somewhat similar to the one in which I identified the "modern charlatan" in "Le médecin et le charlatan" (see Tobie Nathan and Isabelle Stengers, *Médecins et sorciers* [Paris: Éditions Synthélabo, 1995]).

The healer claims the ability to prove the validity of his technique by means of arguments resembling those used to demonstrate logical proof in the modern scientific sense. He then becomes a caricature that can be used to disqualify all "nonmodern" therapeutic knowledge.

5. Leibniz's distinction is important because, like everything we have inherited from him, it recalls the art of the problem, that is also the art of slowing down the rush of "obvious" solutions.. This type of distinction is practiced by nonmoderns. Not everything is treated as a message to be deciphered, expressing some invisible intentionality. It is only as a last recourse that a family will send one of its members to a master of the invisible. But the possibility of an invisible intentionality is nonetheless present in the "secular" management of the problem.

6. A somewhat recent example can be seen in the fallout from the Sokal affair and the resentment that was felt by physicists. That resentment, expressed in the form of disheartening prohibitions addressed to anyone who presumed to speak about physics without yielding to its objectivity, clearly indicates the disastrous effects of an ecology of practices dominated by nomadic values.

7. Fear and fright could operate as synonyms, but I have chosen, in the remainder of the book, to reserve for "fright" its privileged relationship with the intrusion of "supernatural others," where fear assumes a different technical meaning, one that is neutral with respect to any theraputic concern. This is primarily a question of breaking the "transference-countertransference" circle that, for Devereux, but also for Freud before him, enabled psychoanalysts to advance a bit too quickly from the question of "method" to that of "therapy." It seems to me that the words used to characterize controversies in the fields of knowledge should not communicate with words used by therapists, should not induce therapists to diagnose those controversies in terms of symptoms.

8. The concept of "transduction" was introduced in Book VI. It associates the question of creation (of a being or knowledge) with the problematic communication between two heterogeneous "realities."

9. For Tobie Nathan's Yoruba, the "modern physician" who lives in one world, which clings to the certitude that that one world contains every answer, is obviously sedentary.

25. THE BETRAYAL OF THE DIPLOMATS

1. See Gilles Deleuze and Félix Guattari, *What Is Philosophy?*, trans, Hugh Tomlinson and Graham Burchell (New York: Columbia University Press, 1994), 202.

2. Correlatively, the diplomat is also accused of lacking "principles." Doesn't she proceed on a case-by-case basis, carefully avoiding the

denunciation of the violation of general rights in order to focus on the construction of specific regulations? The diplomat is closely associated with the invention of jurisprudence (see Gilles Deleuze, "Gauche," in *Abécédaire* [Arte, 1995]. This broadcast was reissued as three videocassettes with the title *L'Abécédaire de Gilles Deleuze* [Paris: Éditions Montparnasse, 1997]).

3. Isabelle Stengers, *The Invention of Modern Science*, trans. Daniel W. Smith (Minneapolis: University of Minnesota Press, 2000), 166.

4. The situation is analogous to that of the modeler described in Book VI, who must determine how her field responds to the problem posed by the model and not verify the submission of that field to the problem posed by the model. Except for the fact, which here expresses the inadequacy of the "art of tact," that the problem presented by the categorization process does not express a knowledge that risks asserting that, one way or another, the problem also exists for the one who is interrogated. Here, the "problem" is expressed in terms that disqualify those who are interrogated because they express the difference that separates them from the one who interrogates (as a social psychologist, I know that opinion "must" be a function of variables I am going to identify).

5. Unless, of course, he makes the risky transition that ethnologists refer to as a "professional risk": *going native*. In the somewhat less risky context of sociological studies of parapsychological practices, this is what happened to Harry Collins and Trevor Pinch (see *Frames of Meaning: The Social Construction of Extraordinary Science* [London: Routledge and Kegan Paul, 1982]). Having identified a possible fraud, they cosigned an article informing the parapsychology research community how the fraud had been deterred.

6. The fact that I mention such techniques "wholesale" does not imply that, given the assumption that they cannot avail themselves of a science, they would be equivalent. My point is that in order to begin to create distinctions among them, it is crucial from the cosmopolitical point of view that those techniques are all recognized, and recognize one another, as "techniques of influence."

7. In Leibniz's *Confessio philosophi*, we find the model of a diplomatic enterprise whose failure is used to conceptualize the goodness of God and the damnation of Beelzebub. There is no need to point out that the terms of the peace treaty dictated by God to the hermit intercessor, the "simple" recognition by Beelzebub, the "pacified soul," of the fact that his evil alone was the cause of his damnation, assume the radical collapse of the "logical closure" that is Beelzebub's very damnation. No one "here below" can employ this terrible tautology for their own use, but it nonetheless expresses, hyperbolically, the test of choice between possible peace and certain war.

26. THE DIPLOMATS' PEACE

1. Where the distinction I have found it useful to make between "fright" and "fear" intersects the distinction I introduced between the definition of culture as associated with beings who refuse a political reinvention and its definition as associated with the ecological question of conditions of exchange. As such, fright belongs to a culture, fear expresses the *problem* of exchange.

2. This pair is intended to respect the difference, previously discussed with respect to "fear and fright." While the breaking of a "fetish" may be associated with fright, the questioning of a "factish" constructed by a practice whose vocation is the production of knowledge or the creation of new technical possibilities challenges the constructor's autonomy and is experienced as distress or disarray, rather than fright.

3. Lacan pushed this logic as far as it could go. See Léon Chertok and Isabelle Stengers, *Le Cœur et la Raison: L'Hypnose en question de Lavoisier à Lacan* (Paris: Payot, 1989), 186–92.

4. Shirley C. Strum, *Almost Human: A Journey into the World of Baboons* (Chicago: University of Chicago Press, 2001).

5. Gilles Deleuze, *The Fold: Leibniz and the Baroque,* foreword and trans. Tom Conley (Minneapolis: University of Minnesota Press, 1993), 136–37.

6. Leibniz, mathematician and diplomat, theologian and possible creator of the first "social science," the "first great phenomenology of motives" (ibid., 79), was able to work in the fold of two floors. He developed a conceptual system in which everything is a constraint and nothing a condition, where the evidence of common sense is reinvented with, as its sole imperative, the construction of coherence. But, a tireless letter writer, he continued to try to meet the varied conditions of his correspondents, to accept the variety of their requisites, which he would then betray, that is, actualize them in conceptual terms and, finally, because practice can never take precedence over the ideal, submit them to his correspondents. One floor has no meaning without the other. "His" correspondents were "present" to him, the way the body is present to the soul, in the operation in which they were translated-betrayed in the actualization of his concepts. And they were actually present, objecting, imposing their conditions during the test in which the possibility of the realization of those concepts within them was verified, the possibility that they might incorporate them without violence. Which is to say, to accept the becoming they proposed to them.

27. *CALCULEMUS*

1. Bruno Latour, "Moderniser ou écologiser? À la recherche de la 'septième cité,'" *Écologie Politique* 13 (1995): 5–27. The quotation appears on page 19.

2. Luc Boltanski and Laurent Thévenot, *On Justification: Economies of Worth*, trans. Catherine Porter (Princeton, N.J.: Princeton University Press, 2006).

3. This may relate to the question of Greek democracy, of which tragedy was a component. In *Le Théâtre des philosophes* (Grenoble: Millon, 1995), Jacques Taminiaux maintained that the Aristotelian reading of catharsis had to be understood not as a purification that eliminated the modes of *pathos* associated with fear and pity, but as an elucidation of the citizens' praxis through the experience of fear and pity inspired by the tragic fate of (noncitizen) "heroes." Taminiaux, coming to Aristotle via Arendt, makes the duality of tragic characters/choir an authentic operator of "transition," by which the citizens who "agree" experience the fact that their agreement must undergo the challenge (the worth of the seventh regime) of that which refuses to be addressed as being agreeable.

4. In Gilles Deleuze and Félix Guattari, *What Is Philosophy?*, trans. Hugh Tomlinson and Graham Burchell (New York: Columbia University Press, 1994), philosophy, science, and art do not express, other than through a genuinely modernist misunderstanding, the truth of the risks of thought, but are aspects stabilized by distinct traditions (see page 197). The fact that the question of what philosophy is today involves the question of art and science implies that it entails specific modes of translation-betrayal that attain their inherent truth only in the radical distinction of their respective "worth."

5. Daniel Bensaïd, *Walter Benjamin: Sentinelle messianique* (Paris: Plon, 1990).

6. See Deleuze and Guattari, *What Is Philosophy?*, 109, where the French is translated as "for, rather than before." The groups that have ancestors, and to the extent that the ancestral voice is expressed in terms of conditions for the "measure" being constructed, are certainly present via their diplomats, but the Parliament itself has no ancestors. It is the place where the kind of politics is invented that Philippe Pignarre (*Les Deux Médecines* [Paris: La Découverte, 1995], 179) has described as having preserved, through its scientific and technical avatars, the relevance of the proverb "μη ξιλοκρινειν"—"one must not seek one's ancestors"—by which Pierre Vidal-Naquet has characterized the invention of Greek political space.

7. No doubt, this singularity is consistent with the critical obstacle to the transmission of mathematics: the forgetting of the fact that the definition of these beings allows them to exist only for someone who has already understood the requirements they satisfy and the obligations they engender. We could say that mathematical practice is one in which ontogenesis repeats phylogenesis, which is to say, where the risk and passion that have marked the coming into existence of a new mathematical being must, in

one way or another, be repeated for every new "learner." It is in this sense, and not because the definition of these beings would contain "everything" that is needed by the mathematician, that the etymology that enables "mathematics" and "learning" (μανθανειν) to communicate, is true.

8. Alfred North Whitehead, *Process and Reality* (New York: Free Press, 1969), 11.

9. Ibid., 4.

10. To complete the indirect parallel with *What Is Philosophy?* I would add that artistic practices need diplomacy but cannot delegate experts. Art puts at risk the sedentary components of experience in themselves and brings them into existence for themselves. And from this anything can follow, except an "artistic" knowledge claiming to promote its constraints and negotiate its scope and meaning for others.

11. It is significant that only mathematicians, who know how to conceptualize the beings they fabricate, recognize as their own those who work in the history of mathematics as well as those who develop notions of mathematical risk, mathematical objects, the mode of truth specific to mathematics. Once again, the singularity (regardless of the arrogance with which that singularity is sometimes presented) of a practice that never had to claim to be modern, that is, never had to disqualify other practices, in order to promote its requirements and ensure its obligations were recognized, stands out.

12. Gilles Deleuze, "Mediators," in *Negotiations 1972–1990*, trans. Martin Joughin (New York: Columbia University Press, 1997), 126–28.

28. THE FINAL CHALLENGE

1. Gilles Deleuze and Félix Guattari, *Anti-Oedipus: Capitalism and Schizophrenia,* trans. Robert Hurley, Mark Seem, and Helen R. Lane (Minneapolis: University of Minnesota Press, 1983), 247.

2. Ibid., 245–46. A relative limit, where "schizophrenia" would be an absolute limit, an exterior limit of capitalism but produced by capitalism itself, the culmination of its deepest tendency, which it must, through factitious reterritorialization, inhibit if it is to be able to function.

3. Ibid., 239–40.

4. Ibid., 381.

5. Ibid., 12.

6. Ibid., 37. "In his study *The Empty Fortress,* Bruno Bettelheim paints the portrait of this young child who can live, eat, defecate, and sleep only if he is plugged into machines provided with motors, wires, lights, carburetors, propellers, and steering wheels."

7. Residual concentration, we should say, which relates to the question of the "survival" of practices. Questions such as what a "Yoruba physicist"

or a feminist engineer might be belong to a possible but not probable future, not in the sense of an individual history, no matter how rich, but in the sense of the new risks to which they may be able to expose their practice.

8. Gottfried Wilelm Leibniz, *Confessio Philosophi: Papers concerning the Problem of Evil, 1671–1678,* trans., ed., and with an introduction by Robert C. Sleigh Jr., and with additional contributions from Brandon Look and James Stam (Hartford: Yale University Press, 2005), 99.

9. This was Whitehead's thesis in *Science and the Modern World* (New York: Free Press, 1977), 8–9: "The Reformation and the scientific movement were two aspects of the historical revolt which was the dominant intellectual movement of the later Renaissance. . . . It is a great mistake to conceive this historical revolt as an appeal to reason. On the contrary, it was through and through an anti-intellectualist movement. It was the return to the contemplation of brute fact; and it was based on a recoil from the inflexible rationality of medieval thought. In making this statement I am merely summarising what at the time the adherents of the old régime themselves asserted. For example, in the fourth book of Father Paul Sarpi's *History of the Council of Trent,* you will find that in the year 1551 the Papal Legates who presided over the Council ordered: 'That the Divines ought to confirm their opinions with the holy Scripture, Traditions of the Apostles, sacred and approved Councils, and by the Constitutions and Authorities of the Holy Fathers; that they ought to use brevity, and avoid superfluous and unprofitable questions, and perverse contentions. . . . This order did not please the Italian Divines; who said it was a novity, and a condemning of School-Divinity, which in all difficulties, *useth reason.* . . .'" "Poor belated medievalists," Whitehead concludes, who used reason but were no longer understood by the dominant powers of their time. These powers were mobilized by war, and Whitehead helps us remember the connection between the mobilization of those who would now be referred to as "scientists" around "facts" said to be capable of overcoming the "reasons" of theologians, and the religious mobilization of the church in its struggle against the Reformation.

INDEX

robbery, 437–38n6; means and,
225; philosophy of, 401, 442n17,
456n9; on science/instruments,
401; whole, 219, 220, 221
Whose Body? (Sayers), 296
Wiame, Jean, 240
Wiener, Norbert, 274, 438n7
Wigner, Eugene, 159
Wimsey, Lord Peter, 296
Wittgenstein, Ludwig, 259
*Wonderful Life: The Burgess Shale and
the Nature of History* (Gould), 231

Yoruba, 327, 336, 338, 339, 362,
445–46n18, 451n9

Zermelo, Ernst, 124

Trained as a chemist and philosopher, **Isabelle Stengers** has authored or coauthored more than twenty-five books and two hundred articles on the philosophy of science. In the 1970s and 1980s, she worked with Nobel Prize recipient Ilya Prigogine, with whom she wrote *Order out of Chaos: Man's New Dialogue with Nature.* Her interests include chaos theory, the history of science, the popularization of the sciences, and the contested status of hypnosis as a legitimate form of psychotherapy. She is a professor of philosophy at the Université Libre de Bruxelles. Her books *Power and Invention: Situating Science* (1997), *The Invention of Modern Science* (2000), and *Cosmopolitics I* (2010) have been translated into English and published by the University of Minnesota Press.

Robert Bononno has translated more than a dozen books, including *Psychoanalysis and the Challenge of Islam* by Fethi Benslama (Minnesota, 2009) and *Decolonization and the Decolonized* by Albert Memmi (Minnesota, 2006).